T0406845

Key Concepts in Theme Park Studies

Florian Freitag · Filippo Carlà-Uhink ·
Salvador Anton Clavé
Coordinators

Böger · Clément · Lukas · Mittermeier · Molter ·
Paine · Schwarz · Staszak · Steinkrüger · Widmann

Key Concepts in Theme Park Studies

Understanding Tourism and Leisure Spaces

Coordinators
Florian Freitag
Institut für Anglophone Studien
Universität Duisburg-Essen
Essen, Germany

Filippo Carlà-Uhink
Historisches Institut
Universität Potsdam
Potsdam, Germany

Salvador Anton Clavé
Departament de Geografia
Universitat Rovira i Virgili
Vila-seca, Spain

ISBN 978-3-031-11134-1 ISBN 978-3-031-11132-7 (eBook)
https://doi.org/10.1007/978-3-031-11132-7

This Springer imprint is published by the registered company Springer Nature Switzerland AG
The registered company address is: Gewerbestrasse 11, 6330 Cham, Switzerland

Preface: An Experiment

Prefaces usually are where authors praise others for their contributions to the book and blame themselves for its faults. In the present case, however, we take both the praise and the blame. *Key Concepts in Theme Park Studies* is the result of an experiment—a transnational, transdisciplinary collaboration between 13 scholars from different disciplines and different university systems, speaking different languages, and at different stages in their careers. In various constellations and over the course of three years and four in-person meetings (all sponsored by the German Research Foundation DFG), we discussed and decided upon the scope and the structure of the entire book as well as its individual chapters and drafted, re-wrote, and edited texts. One person may have drafted one chapter and added a few thoughts and lines to others. Another person may have contributed different subsections to various chapters, which were then partly re-written and, in some cases, ultimately moved to different chapters by others. As a result of this repeated drafting, re-writing, and editing, it is, in the case of some passages, genuinely difficult for us now to tell exactly whose words these are—at least in this sense, our experiment has been a tremendous success.

It is precisely this experimental method, however, that also makes it hard for us now to fairly distribute the praises and the blames as well as the conventional roles of academic publishing. If the cover and the opening pages of *Key Concepts in Theme Park Studies* identify coordinators and contributors, then these are merely convenient fictions that we—once again, in a specific constellation—have come up with in order to comply with the requirements of our publisher. In truth, we should all be listed as authors of all the chapters and of the entire book: the pens were guided by many hands. In addition to the authorship and, hence, the blames and praises, what we also share, however, is the hope that this book will be of use to you as you enter or pay yet another visit to the world of theme park studies. Here is another convenient fiction:

you may pretend that there is a group of friends already inside, just waiting for you to show you around a bit.

Vila-seca, Spain Salvador Anton Clavé
Hamburg, Germany Astrid Böger
Potsdam, Germany Filippo Carlà-Uhink
Paris, France Thibaut Clément
Essen, Germany Florian Freitag
South Lake Tahoe, USA Scott A. Lukas
Kassel, Germany Sabrina Mittermeier
Mainz, Germany Céline Molter
Milton Keynes, UK Crispin Paine
Hildesheim, Germany Ariane Schwarz
Geneve, Switzerland Jean-François Staszak
Bonn, Germany Jan-Erik Steinkrüger
Ravensburg, Germany Torsten Widmann

Contents

List of Contributors

Salvador Anton Clavé Departament de Geografia, Universitat Rovira i Virgili, Vila-seca, Spain

Astrid Böger Institut für Anglistik und Amerikanistik, Universität Hamburg, Hamburg, Germany

Filippo Carlà-Uhink Historisches Institut, Universität Potsdam, Potsdam, Germany

Thibaut Clément Faculté des Lettres, Sorbonne Université, Paris, France

Florian Freitag Institut für Anglophone Studien, Universität Duisburg-Essen, Essen, Germany

Scott A. Lukas Department of Sociology and Anthropology, Lake Tahoe Community College, South Lake Tahoe, USA

Sabrina Mittermeier Fachgruppe Geschichte, Universität Kassel, Kassel, Germany

Céline Molter Institut für Ethnologie und Afrikastudien, Johannes Gutenberg-Universität Mainz, Mainz, Germany

Crispin Paine Department of Religious Studies, Open University, Milton Keynes, UK

Ariane Schwarz Institut für Medien, Theater und Populäre Kultur, Universität Hildesheim, Hildesheim, Germany

Jean-François Staszak Département de Géographie et Environnement, Université de Genève, Genève, Switzerland

Jan-Erik Steinkrüger Geographisches Institut, Universität Bonn, Bonn, Germany

Torsten Widmann Fakultät Wirtschaft, DHBW Ravensburg, Ravensburg, Germany

List of Figures

Immersion: Immersivity, Narrativity, and Bodily Affect in Theme Parks

Inclusion and Exclusion: Marginalization in Theme Parks

Labor: Working Conditions, Employment Trends, and the Job Market in the Theme Park Industry

Media and Mediality in Theme Parks

Methods: Facing the Challenges of Studying Theme Parks

Paratexts and Reception: Images of Theme Parks in Art, Popular Culture, and Discourse

List of Tables

Introduction: Defining Theme Parks and Assessing Theme Park Studies

Abstract The introduction offers a thorough discussion of various definitions of theme parks, a detailed reflection on the development of theme park studies, and a systematic presentation of the state of the art in theme park studies across various disciplines. It will further provide readers with a rationale for the organization of the book and the ways in which it can be used.

1 Introduction

"You are Disneyland" German rap artist Patrick Losensky told German-Nigerian journalist Malcolm Ohanwe in early 2021, during a debate about the use of controversial language in rap music (Hyped! 2021, 00:01:57; our translation). What Losensky sought to imply was that Ohanwe had grown up in a sheltered, middle-class environment and therefore had no idea of the hardships of street life that supposedly constitute the inspiration for much hip-hop music. "I'm going to Disneyland/Disney World!" is what NFL football players have exclaimed after winning the Super Bowl since the beginning of an ad campaign in 1987, thus defining a visit to the park as "the ultimate reward for a work well done" (Orvell 2012, 37). Disney has not only stood for the middle class and for a rewarding holiday, but also for Americanization, popularization, and trivialization (in 1992, a French stage director referred to Disneyland Paris as a "cultural Chernobyl"; Mnouchkine 2003). More generally, the theme park has served as a metaphor for touristification, commodification, and standardization ("Venice is the first urban theme park," a journalist noted in 2008; Kay 2008)—even in the academic domain (Barthel-Bouchier 2001). At the same time, it has also been used to creatively refer to extraordinary experiences, with astronaut Sally Ride describing her 1983 space flight as an "E-ticket ride" (see Begley 1983).

This work is contributed by Salvador Anton Clavé, Filippo Carlà-Uhink, Florian Freitag, Astrid Böger, Thibaut Clément, Scott A. Lukas, Sabrina Mittermeier, Céline Molter, Crispin Paine, Ariane Schwarz, Jean-François Staszak, Jan-Erik Steinkrüger, Torsten Widmann. The corresponding authors are Salvador Anton Clavé, Departament de Geografia, Universitat Rovira i Virgili, Vila-seca, Spain; Filippo Carlà-Uhink, Historisches Institut, Universität Potsdam, Potsdam, Germany; Florian Freitag, Institut für Anglophone Studien, Universität Duisburg-Essen, Essen, Germany.

F. Freitag et al., *Key Concepts in Theme Park Studies*,
https://doi.org/10.1007/978-3-031-11132-7_1

The theme park has also served as the setting and the subject of fiction, movies, art, video games, and songs—from Julian Barnes's *England, England* (1996) and Michael Crichton's *Westworld* (1973) to Banksy's *Dismaland* (2015), Frontier Developments' *Planet Coaster* (2016), and, somewhat ironically in the light of Losensky's remark, American rap artist Travis Scott's 2018 album *AstroWorld*, named after the defunct Six Flags park. And, of course, the theme park also remains a place: a workplace, a tourist destination, and a place of attachment (e.g. for fans), a representational space (i.e. a medium), a site of investment (i.e. a business and an industry), and a designed space that has impacted upon our lives in such powerful and manifold ways that these processes have received their own names in the literature (e.g. "Disneyfication," "Disneyization," and "theme-park-ification"; see Schickel 1968, 220; Bryman 2004; and Dear and Flusty 2002, 417, respectively).

Key Concepts in Theme Park Studies seeks to consider all of these various aspects of and approaches to theme parks. Indeed, the contributors to this volume are convinced that, for example, to explain the theming strategies of theme parks we need to pay attention to medial, ideological, and design aspects; to understand labor in theme parks we need insights from sociology, economics, and performance studies; and to grasp the temporalities of theme parks we cannot ignore either the historical, the cultural, or the management point of view—in short, we believe that it is only by transgressing disciplinary boundaries and by combining critical perspectives and analytical methods that we can arrive at a deeper understanding of the unique and multivalent cultural phenomenon referred to as the "theme park." In order to provide a sound basis for such a transdisciplinary, multi-faceted, and complex approach to theme parks, this introduction will (1) examine and compare various definitions of theme parks and related concepts such as immersive environments or amusement parks; (2) sketch the development and the institutionalization of theme park studies as a scholarly field; and (3) explain the organization of this book and suggest the various ways in which it can be used.

2 About Theme Parks

Although we may "know it when we see it"—that is, although we may have an intuitive idea of what a theme park is and what elements characterize such a place—theme park scholars have, as Tourism Research and Marketing already pointed out in 1995, often avoided offering precise definitions. While there are many similar formats, there are also enormous differences between individual theme parks, thus any definition may risk either excluding too many theme parks or including too many themed spaces or amusement parks, and may end up being inaccurate or unproductive. The problem has been exacerbated by the continuous spread of theming and immersivity and their application to an ever broader and more diverse range of both private and commercial, as well as public and non-commercial spaces and environments, from shopping malls, hotels, and restaurants to airports, museums, and private apartments. Hence, early definitions such as the one by Coltier, who identifies a theme park as "a closed universe whose purpose is to succeed in the encounter between the dreamy atmosphere it creates and the visitor's desire for *dépaysement*" (1985, 24), have become too inclusive: today, theme parks arguably constitute a subgroup within the broader category of themed or immersive environments (see Steinecke 2009; Lukas 2013; Anton Clavé 2022).

Nevertheless, in order to analyze the similarities and differences and to identify developments and trends, theme park scholars need a working definition of their object of study that serves as a heuristic tool. Therefore, several critics have resorted to compiling lists of characteristics that, taken together and in combination, aim to differentiate theme parks not just from themed environments, but from other related places as well. Thus, H. Jürgen Kagelmann, for example, defines the theme park as

> an enclosed, spacious, artificially constructed, stationary collection of various attractions, games, and other forms of entertainment that is predominantly located on the outskirts of large cities/metropolitan areas, that is open year-round, and that is commercially structured. [...] The constitutive characteristic of theme parks is their thematic unity, that is, either the park as a whole or individual self-contained parts of it are shaped by certain recognizable motifs, themes, characters etc. (1993, 407–408; our translation)

Kagelmann's definition may be considered rather restrictive, as seasonally operating parks such as Terra Mítica (Benidorm, Spain), publicly owned, non-commercial parks such as Taman Mini Indonesia Indah (Jakarta, Indonesia), as well as parks that have a primarily educational function such as living history museums (see Dreschke 2010) are all excluded.

Salvador Anton Clavé, in turn, has proposed a total of twelve characteristics, each of which distinguishes the theme park from other, related places: "(1) They have a thematic identity that determines recreational alternatives; (2) They contain one or more themed areas; (3) They are organized as closed spaces or with controlled access; (4) They have a great capacity to attract families; (5) They contain enough rides, shows and systems of movement to create a visit that lasts on average some 5–7 h; (6) They present atmospheric forms of entertainment (musicians, characters or actors who perform in the street "free of charge"); (7) They have an important commercial vocation (fundamentally food and beverages and shops); (8) They have high levels of investment per unit of ride or show capacity; (9) They have high-quality products, service, maintenance and standards of cleanliness; (10) They manage their productive and consumer processes centrally; (11) They incorporate technology as much in the production processes as in those of consumption; (12) Generally, though exceptions do exist, they have a single ("pay-one-price") admission system" (2007, 28–29).

Like Kagelmann, Anton Clavé puts particular emphasis on theming (his first characteristic), which allows him to distinguish the theme park from the amusement park. Anton Clavé's fifth point, in turn, separates theme parks from other commercial themed environments such as shopping malls or restaurants, as does his third point: entering a theme park means crossing a threshold that is not only often architecturally marked by some sort of tunnel or portal, but also ritually marked by the purchase and validation of a ticket (see Freitag and Schwarz 2015/2016). By crossing the threshold, visitors enter an area that is supposedly different from the "outside" world, a difference that is sometimes highlighted by the adoption of particular symbols such as a theme park-specific currency or tickets representing passports (see Harwood 2002, 52; Ingram 2003, 55–57). At the same time, the third point also allows us to differentiate between theme parks and national parks, natural reserves, and life spaces (see Lukas 2010), which do not have a clearly marked architectural boundary. Importantly, Anton Clavé's set of characteristics does not require rides as a necessary element of a theme park, thus allowing us to apply this definition to parks that focus on other kinds of entertainment as well. A case in point is Puy du Fou in Les Epesses, the fourth most visited theme park in France after the two parks at Disneyland Paris and Parc Astérix, which features shows and spectacles rather than rides and roller coasters.

Another useful strategy can be to adopt a clearly structured and yet not too narrow definition and then identify relevant subcategories that further specify the object of research. In the case of theme parks, many "subcategories" have been proposed, based, for example, on the represented themes or economic factors. In the former case, this has led to labels such as "cultural theme park" (Schlehe and Uike-Bormann 2010), "ethnic theme parks" (Hen 2013), "historical theme parks" (Carlà 2016), or "religious theme-park" (Lippy 1994, 228). Of course, these labels are somewhat misleading—beyond the difficulty intrinsic in using concepts such as culture that are extremely hard to define—in that what theme parks actually reproduce are rather the popular and most easily recognizable images of particular cultures, peoples, and historical eras (see Carlà-Uhink 2020, 6–7). Nonetheless, in individual research projects such definitions can be very useful to identify the precise scope of the study and to narrow down the number of parks to be considered, limiting it to a precise set of parks sharing distinct characteristics.

When parks are categorized on the basis of economic criteria, in turn, classifications frequently take into account the parks' size (e.g. their number of visitors and/or employees, surface, and amount of investment) or their specific target market(s) (Anton Clavé 2007, 28). The latter approach has been particularly successful, leading to a threefold classification of theme parks that is very widespread in the literature:

1. Local parks, visited almost exclusively by people living in the immediately adjacent region;
2. Regional parks, attracting visitors from a greater distance and yet still mostly within a one-day trip;
3. Destination parks, attracting visitors from much further away—and often visitors staying for more than one day; in this case, it is possible that people travel a long way only to visit the theme parks.

This classification allows for many further considerations and approaches, as the distance the visitors are willing to travel is often connected to the investment, the management, and the design choices of the parks. Thus, local parks are often less heavily themed, do not offer nighttime entertainment or accommodation, and focus more on rides, whereas destination parks develop a much more complex and varied attractions mix, feature extensive opening hours and hotels, and require much higher yearly investments. In any case, what is important is that scholars explicitly state which definition they are going to adopt at the beginning of their work.

3 About Theme Park Studies

Problems of definition have not prevented critics from discussing theme parks, of course. In fact, some of the earliest scholarly writings were published even before the term "theme park" (first attested in 1960, according to the *Oxford English Dictionary*) had broadly established itself: the earliest critical studies of Disneyland thus still referred to the place as an "atmospheric park" (Schickel 1968, 22), an "amusement park" (Finch 1973, 396), or an "entertainment park" (Landau 1973, 591). From the start, however, contributions to the scholarly debate about theme parks came from all over the (Western) world and from a wide array of scholarly disciplines and intellectual traditions, as the following overview illustrates: Richard Schickel's biographical *The Disney Version: The Life, Times, Art and Commerce of Walt Disney* (1968; revised editions in 1985 and 1997); Christopher Finch's medial analysis, Louis Marin's semiotic reading, and Roy Landau's architectural discussion of Disneyland (and Walt Disney World) in Finch's *The Art of Walt Disney from Mickey Mouse to the Magic Kingdoms*, Marin's *Utopiques: Jeux d'espaces*, and Landau's "Mickey Mouse the Great Dictator: The Disney Game as a Control System" (all published in 1973); semiotician Umberto Eco's postmodernist-inflected "Viaggio nell'iperrealtà" through California and Florida ("Travels in Hyperreality"; originally printed in 1975 in the Italian magazine *L'Espresso*); Millicent Hall's "Theme Parks: Around the World in 80 Minutes" (1976) and Richard V. Francaviglia's "Main Street, U.S.A.: The Creation of a Popular Image" (1977), both published in the American scholarly journal *Landscape: A Magazine of Human Geography*; Michael R. Real's Marxist study of Disneyland visitors in *Mass-Mediated Culture* (1977); Jean Baudrillard's analysis of Disneyland (and American culture in general) as postmodern in "La précession des simulacres" (1978); and the various articles collected in *The Journal of Popular Culture*'s "special issue" on theme and amusement parks, published in 1981 (King 1981). As a glance at the tables of contents of such recent edited collections on theme parks as *A Reader in Themed and Immersive Spaces* (ed. Scott A. Lukas; 2016a) or *Time and Temporality in Theme Parks* (ed. Filippo Carlà-Uhink et al.; 2017) confirms, the disciplinary and intellectual variety of theme park research has, if anything, become even more pronounced over the years.

At the same time, early studies of theme parks also established research paradigms that are only beginning to be questioned and overcome. For instance, in the past, theme park criticism has often focused on Western (and specifically the U.S.) as well as large destination (and specifically Disney and, to a lesser extent, Universal) parks. To be sure, the fact that especially very early writings on theme parks mostly focused on Disneyland (Anaheim, California) is no coincidence: up until the 1970s, there simply weren't too many other parks to write about, either in the U.S. or elsewhere. Over the years, however, the Disney parks' enormous impact on the industry and their easy accessibility (in the geographic, linguistic, and cultural sense of the term) to Western scholars has led to a serious critical imbalance, with smaller and non-Western parks having been comparatively neglected. In contrast, recent studies such as David Cardell's *Family Theme Parks, Happiness and Children's Consumption* (2015), Maribeth Erb and Chin-Ee Ong's *Theming Asia* (2017), Maria F. Piazzoni's *The Real Fake* (2018), Crispin Paine's *Gods and Rollercoasters* (2019), Filippo Carlà-Uhink's *Representations of Classical Greece in Theme Parks* (2020), and Florian Freitag and Chang Liu's *History in Chinese Theme Parks* (2022) explicitly go beyond the established canon of theme parks, either by focusing on a specific theme and how the latter is portrayed in smaller, local parks (Cardell, Paine, and Carlà-Uhink), or by concentrating on theme parks or themed environments in one specific non-Western location (Erb and Ong, Piazzoni, as well as Freitag and Liu).

Somewhat related to this new focus on smaller, local, and non-Western parks is the growing number of theme park studies published in languages other than English. Early analyses written in, for example, French or Italian (see the works by Marin, Eco, and Baudrillard listed above) were all subsequently translated into English; this option may be precluded in the future, however, due to the sheer number of publications in Chinese, French, German, and Spanish, among others. Consider, for example, Jigang Bao's *Zhu Ti Gong Yuan Yan Jiu* (2015); Estelle Sohier, Alexandre Gillet, and Jean-François Staszak's *Simulations du monde: Panoramas, parcs à theme et autres dispositifs immersifs* (2019); H. Jürgen Kagelmann's "Themenparks" (2004); or Salvador Anton Clavé's *Parques temáticos: Más allá del ocio* (2005). As a result, the status of English as the lingua franca of theme park studies is increasingly challenged and the field is slowly becoming more and more multilingual.

Sabrina Mittermeier's *A Cultural History of the Disneyland Theme Parks* (2021) and Florian Freitag's *Popular New Orleans* (2021a, 132–231), in turn, respond to Alan Bryman's implicit call for historical theme park studies ("the parks are not inert texts"; 1995, 83) by tracing the development of Disney's New Orleans-themed spaces in the U.S. and Japan (Freitag) and the evolution of Disney's "castle park" form from Anaheim to Orlando, Tokyo, Paris, Hong Kong, and Shanghai (Mittermeier). *The Journal of Popular Culture*'s 1981 "special issue" on theme and amusement parks (King 1981) had, in fact, featured a diachronic study of Cedar Park (Sandusky, Ohio; see Hildebrandt 1981), but subsequently historiographical approaches to individual attractions, themed areas, parks, or park chains would remain the exception (see, however, Wiener 1994; Francaviglia 1999; Rahn 2000; Davis 1997; Foglesong 1999; Koenig 2007). Adams (1991), Kagelmann (1993), Young and Riley (2002), Mitrasinovic (2006), and Philips (2012), in turn, laid the groundwork for a history of the theme park form as such—i.e. an analysis of the antecedents of the contemporary theme park and its general development after the mid-1950s. Early studies, particularly those by European scholars (Marin 1973; Eco 1975; Baudrillard 1978) and their American followers (for example, Fjellman 1992; Bukatman 1991), had viewed theme parks as a distinctly postmodern phenomenon. More recent work has either further specified this categorization—e.g. by discussing Disneyland as a product of the "age of white flight" (Avila 2004) or the "jet age" (Schwartz 2020)—or stressed the continuities between the theme park and earlier themed and immersive spaces, with, for instance, Norman Klein (2004) and Angela Ndalianis (2004) analyzing theme parks as "electronic baroque" (Klein) or "neo-baroque" (Ndalianis).

Both Freitag and Mittermeier are also interested in the reception of theme parks, as are Thibaut Clément (*Plus vrais que nature: Les parcs Disney ou l'usage de la fiction dans l'espace et le paysage*, 2016), Janet Wasko (*Understanding* Disney, revised edition 2020, 243–258), and especially Rebecca Williams (*Theme Park Fandom*, 2020) and Abby S. Waysdorf (*Fan Sites*, 2021)—all of these critics thus seek to heed the repeated calls for "triangulated" or "integrated" theme park studies that simultaneously take into account the parks themselves as well as their production (by employees) and reception (by visitors, fans, and artists; see Raz 1999, 6; Wasko 2001, 152; Lukas 2016b, 168). As in the case of theme park historiography, one pioneering study—Real's *Mass-Mediated Culture* (1977)—had long remained largely without any emulators beyond a handful of scholarly accounts looking at theme park audiences (in addition to the list of relevant studies for the Disney parks offered in Wasko 2020, 216–222, see, for example, the works by Hjemdahl 2002, 2003) and employees (see, for example, Smith and Eisenberg 1987; Van Maanen 1991; Kuenz 1995; Lukas 2007). Apart from the rising interest in theme park fandom (next to Williams and Waysdorf, see also Baker 2016; Waters 2016; Kiriakou 2017; Godwin 2017; Waysdorf and Reijnders 2018; Baker 2018), new approaches such as data-driven modeling (Yuan and Zheng 2018) and management (Singh et al. 2004), service experience analysis (Dong and Siu 2013), visitor satisfaction (Ryan et al. 2010), motivational factors exploration (Bakir and Baxter 2011), measuring visitors' spatial mobility (Birenboim et al. 2013; Huang et al. 2020), and measuring emotion (Orellana et al. 2016; Mitas et al. 2022) have recently contributed to further diversifying research on theme park audiences.

Performance and the Disney Theme Park Experience (2019), edited by Jennifer A. Kokai and Tom Robson, is not only interested in the performances of both visitors and employees, but also constitutes the latest in a list of publications that examine the contribution of specific media to the theme park's plurimediality. In fact, in *The Disney Version* (1968) Schickel had already described Disneyland both as "a new and unique medium" (17) and a "mixed-media show[...]" (325), but it wasn't until the 1990s that scholars started to investigate the roles of individual media within this media mix, e.g. architecture (e.g. Marling 1991), music (e.g. Carson 2004; Camp 2017), film (e.g. Freitag 2017; Gottwald and George 2020), and language (e.g. Freitag 2021b). The 2000s, in turn, saw the first detailed investigations of specific theme park paramedia, with Jay P. Telotte's *Disney TV* (2004) being followed by David L. Pike's "The Walt Disney World Underground" (2005) and Mathew J. Bartkowiak's "Behind the Behind the Scenes of Disney World" (2012; both on guided tours), Stephen Yandell's "Mapping the Happiest Place on Earth" (2012; on theme park souvenir maps), and Arthur Soto-Vásquez's "Mediating the Magic Kingdom" (2021; on Instagram)—but, as of yet, no comprehensive book-length study on the topic. Likewise, while the reception of, for example, the turn-of-the-century parks on Coney Island in media, from literature and movies to photography and painting, has already been documented and analyzed in anthologies such as Louis J. and John Parascandola's *A Coney Island Reader* (2014) or Robin Jaffee Frank's *Coney Island: Visions of an American Dreamland* (2015) and in studies such as Lauren Rabinovitz's *Electric Dreamland* (2012), the medial reception of contemporary theme parks has so far remained the subject of journal articles and book chapters that focus on their portrayal in one specific medium (see, for example, Schweizer 2016; Gottwald and Turner-Rahman 2019; Makai 2019; Graf-Janz 2022 on the depiction of theme parks in video games).

Along with their growing economic importance for regions and tourism destinations, theme parks have also increasingly become the subject of economic research. Economic geographical approaches to qualify and quantify the impacts of theme parks on regional economic structures and tourism markets were at first dominant (e.g. Fichtner and Michna 1987; Thomas-Morus-Akademie 1998). Over time, the understanding of the theme park industry as a discrete global branch of industry became widely accepted (Anton Clavé 2007). Even though in managerial economics theme parks are mostly studied along with amusement parks and other visitor attractions (Pechlaner et al. 2006; Swarbrooke 1995), Pieter Cornelis succeeded with his works in integrating theme park-related issues within managerial economic research (Cornelis 2011a, 2017).

Additionally, an array of empirical, applied, and theoretical studies on theme park economics and operations topics has populated the most recent academic inquiries about the theme park industry. Among them are financial analysis (Liu 2008); pricing and revenue management (Yoonjoung Heo and Lee 2009); computerized management and decision support systems (Tsai and Chung 2012); investment decision analysis (Li et al. 2018); amusement ride control (Woodcock 2014); corporate social responsibility (Holcomb et al. 2010); employment and human resources analysis (Milman and Dickson 2014); marketing and branding impact studies (Cornelis 2010); public–private development partnerships (Shen et al. 2006); the impact of social media (Fotiadis and Stylos 2017); and, more recently, issues related to the impact of COVID-19 (Gabe 2021).

Further key issues in this growing design, operational, and managerial-oriented field of theme park research are service quality (Tsang et al. 2012; Fotiadis and Vassiliadis 2016), safety (Woodcock 2008), digital transformation (Brown et al. 2013), and planning. With respect to the latter, topics such as carrying capacity (Zhang et al. 2017), location of services (Min et al. 2017), investment priorities (Dzeng and Lee 2007), the impact of new attractions (Cornelis 2011b), and accessibility (Carl 2020) are among the most significant. Economic research on theme parks has also responded to the broader developments and challenges of a global society by investigating, in both theoretical and operational studies, the digitization of immersive consumer experiences (Jung et al. 2015; Waysdorf and Reijnders 2018; Wei et al. 2019), sustainability dimensions and impacts of theme park developments (Wang et al. 2017), or ethical issues such as animal rights in the theme park industry (Shani and Pizam 2010). New sources of information about managerial aspects of theme parks have appeared that may be of interest not only to customers and operators but also to researchers. For example, the online database https://queue-times.com/, created in 2014, provides waiting times and crowd data for the largest theme parks in the world.

Finally, due to their wide and ambitious scope, monographs such as Salvador Anton Clavé's *The Global Theme Park Industry* (2007), Scott A. Lukas's *Theme Park* (2008), and David Younger's *Theme Park Design & the Art of Themed Entertainment* (2016) all offer more or less comprehensive introductions to theme parks and may thus signal the coming-of-age of theme park studies as a scholarly field. Indeed, what distinguishes Anton Clavé's, Lukas's, and Younger's from similarly broad studies such as Mark Gottdiener's *The Theming of America* (1997), Brian Lonsway's *Making Leisure Work* (2009), Albrecht Steinecke's *Themenwelten im Tourismus* (2009), Jan-Erik Steinkrüger's *Thematisierte Welten* (2013), and Scott A. Lukas's *Immersive Worlds Handbook* (2013) is that they do not discuss theme parks alongside other themed and immersive spaces (a tradition that goes back as far as Eco's "Travels in Hyperreality"), but rather identify the theme park as a unique, distinct form worthy of its own academic field. Likewise, the foundation of the *Journal of Themed Experience and Attractions Studies* in 2018 as well as the establishment of specialized degree programs and tracks of study—e.g. at the University of Central Florida, the University of Indianapolis, and the California Institute of Arts (all in the United States); Staffordshire University and Falmouth University (UK); the Baden-Wuerttemberg Cooperative State University Ravensburg (Germany); or Breda University of Applied Sciences (Netherlands)—have marked significant steps toward an institutionalization of theme park studies in academia.

4 About This Book

Of course, the monographs by Anton Clavé, Lukas, and Younger as well as the study programs listed above all remain somewhat committed to their authors' and departments' respective "home" disciplines, from tourism, geography, and hospitality management to anthropology and cultural studies as well as architecture and design. Indeed, it seems rather doubtful that there will ever exist a single, all-encompassing theory of theme parks along with a corresponding, commonly accepted methodology for their study—or that such a unified, streamlined approach to theme parks even should exist. What should exist, however, is a more, and more intensive, inter- and transdisciplinary collaboration between scholars in this field. This is the spirit with which the editors and authors of this book undertook the endeavor of writing the first transdisciplinary introduction to theme parks. Three main criteria were adopted in developing the project that eventually led to the volume you are now reading.

Firstly, after what has been stated in the previous sections, it was clear to us from the beginning that it would take a broad and diverse group of scholars from different disciplines, university systems, speaking different languages, and at different stages in their careers to discuss in a successful and rich way the diversity, complexity, and the interrelations of the various approaches that need to be adopted to analyze, discuss, and hopefully understand theme parks. With respect to their disciplinary background and training, for example, the editors and contributors to this volume represent the fields of American Studies, Anthropology, Drama and Performance,

Table 1 The contributors involved in *Key Concepts in Theme Park Studies* (in alphabetical order)

Name	Affiliation	Discipline
Salvador Anton Clavé	U Rovira i Virgili (Vila-seca, Spain)	Geography and Tourism Studies
Astrid Böger	U Hamburg (Hamburg, Germany)	American Studies
Filippo Carlà-Uhink	U Potsdam (Potsdam, Germany)	Ancient History
Thibaut Clément	Sorbonne U (Paris, France)	American Studies
Florian Freitag	U Duisburg-Essen (Essen, Germany)	American Studies
Scott A. Lukas	Lake Tahoe CC (South Lake Tahoe, USA)	Anthropology
Sabrina Mittermeier	U Kassel (Kassel, Germany)	American Studies
Céline Molter	JGU Mainz (Mainz, Germany)	Anthropology
Crispin Paine	Open U (Milton Keynes, UK)	Museum Studies
Ariane Schwarz	U Hildesheim (Hildesheim, Germany)	Performance Studies
Jean-François Staszak	U Genève (Geneva, Switzerland)	Geography
Jan-Erik Steinkrüger	U Bonn (Bonn, Germany)	Geography
Torsten Widmann	Baden-Wuerttemberg Cooperative State U (Ravensburg, Germany)	Tourism Studies

Economics, Geography, History, Museum Studies, and Tourism Studies (see Table 1).

Funding from the German Research Foundation (DFG) allowed us to regularly meet up for writing workshops and to invite additional guests. This was instrumental in fulfilling our second requirement: given the necessity of a truly transdisciplinary approach, one which moves beyond the mere gathering and juxtaposition of concepts, points of view, and approaches from different fields, it was necessary for the authors to not simply contribute individual chapters, but that every single chapter, and ultimately the entire volume, would be the product of genuine collaboration, of collective thinking, structuring, and drafting. It is only via the hardships of this kind of group work, which does not allow taking individual disciplinary "rules" or "postulates" for granted, that interdisciplinary communication can be transformed into transdisciplinary cooperation.

This is the point that also determined the third and last—but certainly not the least—criterion. Cooperative writing means continuously discussing, negotiating, and explaining concepts, models, and assumptions that may be commonly accepted within one's own discipline, but not in others. During our meetings, we very quickly realized that such concepts, models, and assumptions can become real battlefields as soon as we take them out of the comfort zone of their disciplinary culture; and some of our earliest discussions, e.g. on terms such as "space," "place," and "culture," revealed how much time and effort would be needed to find, if not a common denominator, then at least conceptual bridges between vastly different scholarly traditions. As soon as this became clear, we decided that the book should have the structure that it now has: a volume that employs a transdisciplinary approach to a series of "key concepts" in theme park studies.

As a result of the intense discussions during our meetings, the list of twelve "key concepts" initially suggested by the editors eventually grew to 16. Some of these are directly inspired by and reflect the research paradigms mentioned in the previous section of this introduction: for example, responding to the call for historical theme park studies, the chapter on ANTECEDENTS seeks to reconstruct the history of themed, immersive environments from Antiquity and the Middle Ages to the early twentieth century and to survey the economic, aesthetic, and cultural development of theme parks since the mid-twentieth century. Taking their cue from the growing interest in the production of theme parks by employees and their reception by visitors and fans, the chapters on LABOR and VISITORS discuss labor practices prevalent in theme parks across the globe and the various roles of visitors (customer, pilgrim, performer, fan, and experiencing body) in the theme park industry, respectively. The chapters on MEDIA and PARATEXTS AND RECEPTION, in turn, seek to provide common frameworks for the various articles and book chapters on the contribution of individual media and art forms to theme parks' plurimediality and the role of the theme park in transmedial networks, as well as for the various studies on individual theme park paratexts and the parks' reception in other media.

A second set of "key concepts" is dedicated to terms and concepts that have been regularly used in theme park studies, yet frequently without much awareness of how complicated their use really is beyond individual academic fields. Thus, the chapter on ATTRACTIONS, for instance, identifies the different types of attractions in the theme park and discusses how they come together in a park-specific "attractions mix" to (ideally) form one coherent, immersive narrative space. In the chapter on IMMERSION, we use the concept of immersivity to describe theme parks' capacity to engender or induce an immersed state of mind in visitors and focus on the roles played by narrativity and bodily affect in the parks' offers of immersivity. Drawing on the long debate about the term, especially in tourism and museum studies, the chapter on AUTHENTICITY examines the sometimes-parallel use of vastly different conceptualizations of authenticity in theme park marketing, design and performance, and reception. The chapter on THEMING, finally, constitutes a central intersection within the book, not only to define theming, discuss general characteristics and popular subcategories of themes, but also to reflect on the limits and future of theming.

A final group of "key concepts" uses fundamental characteristics of theme parks—from their spatiality and temporality to their commerciality and their (cultural and economic) embeddedness as well as disconnection—and subjects these to an examination through a radically transdisciplinary prism. For example, the chapter on SPACE is dedicated to the spatial dimension of theme parks and investigates their being within and structuring space, including the ways in which the park's space is both materially produced by the planners and engineers involved in its design and performatively constructed by visitors according to how and where they act in the park. Likewise, the chapter on TIME discusses the complex temporality of theme parks, in which several distinct temporal layers—"represented time," "experienced time," "managed time," and "external time"—continuously interact and manifest themselves in specific material objects. Both the spatiality and the temporality of theme parks are implicated in their planning and design as well as in their operational strategies, which are examined as part of their commerciality in the chapter on ECONOMIC STRATEGY. The chapters on WORLDVIEWS and INDUSTRY, in turn, reflect on how the theme park is culturally and economically connected to the larger world. The former analyzes the values, ideologies, and worldviews reflected in the parks' design and implemented through rules of behavior inside the park. The latter disentangles the system of the global theme park industry by outlining the different types of theme parks, the importance of the main theme park companies and brands in the evolution of the industry, and the role of the many diverse players that participate in the delivery of the theme park experience. By contrast, the chapter on INCLUSION AND EXCLUSION emphasizes the disconnect of theme parks, discussing how exclusions based on class, race, gender, sexuality, or able-bodiedness manifest themselves at both physical or spatial as well as symbolic levels in the theme park.

To the German Research Foundation DFG and our publisher, we described this volume as an introduction to theme park studies addressed to a broad audience ranging from undergraduate students to established scholars, and from newcomers to the field to theme park experts. To us, however, *Key Concepts in Theme Park Studies* rather constitutes an invitation—an invitation to all of those mentioned above, as well as to anyone else who is interested in the topic, to join us in the scholarly discussion about theme parks. Incidentally, our invitation comes with directions: the chapter on METHODS, identifying various challenges of studying theme parks, contains a few practical hints we wish someone had given us before we embarked on our first theme park-related research project. We hope that they, as well as the other chapters in this volume, will be of assistance to you as you accept our invitation. If this book were a theme park, a sign at its entrance would tell you that "Here You Leave Established Scholarly Paths and Enter the World of Transdisciplinary, Global Academia." If you choose to come in, we promise you it will be a bumpy, but fun ride.

References

Adams, Judith A. 1991. *The American Amusement Park Industry: A History of Technology and Thrills*. Boston: Twayne.

Anton Clavé, Salvador. 2005. *Parques temáticos: Más allá del ocio*. Barcelona: Ariel.

Anton Clavé, Salvador. 2007. *The Global Theme Park Industry*. Wallingford: CABI.

Anton Clavé, Salvador. 2022. Themed Visitor Attractions. In *Encyclopedia of Tourism Management and Marketing*, ed. Dimitrios Buhalis. Cheltenham: Edward Elgar. https://www.elgaronline.com/view/book/9781800377486/b-9781800377486.themed.visitor.attractionss.xml. Accessed 28 Aug 2022.

Avila, Eric. 2004. *Popular Culture in the Age of White Flight: Fear and Fantasy in Suburban Los Angeles*. Berkeley: University of California Press.

Baker, Carissa Ann. 2016. Creative Choices and Fan Practices in the Transformation of Theme Park Space. *Transformative Works and Cultures* 22. https://journal.transformativeworks.org/index.php/twc/article/view/974/693. Accessed 28 Aug 2022.
Baker, Carissa Ann. 2018. *Exploring a Three-Dimensional Narrative Medium: The Theme Park De Sprookjessprokkelaar, the Gatherer and Teller of Stories.* Diss., University of Central Florida.
Bakir, Ali, and Suzanne Baxter. 2011. "Touristic Fun": Motivational Factors for Visiting Legoland Windsor Theme Park. *Journal of Hospitality Marketing & Management* 20: 407–424.
Bao, Jigang. 2015. *Zhu Ti Gong Yuan Yan Jiu.* Beijing: Ke Xue Chu Ban She.
Barthel-Bouchier, Diane. 2001. Authenticity and Identity: Theme-Parking the Amanas. *International Sociology* 16 (2): 221–239.
Bartkowiak, Mathew J. 2012. Behind the Behind the Scenes of Disney World: Meeting the Need for Insider Knowledge. *The Journal of Popular Culture* 45 (5): 943–959.
Baudrillard, Jean. 1994 [1978]. The Precession of Simulacra. In *Simulacra and Simulation*, trans. Sheila Faria Glaser, 1–42. Ann Arbor: The University of Michigan Press.
Begley, Sharon. 1983. Sally Ride Dies: 1983 Newsweek Profile of Space Pioneer. *Newsweek*, June 27. http://www.newsweek.com/sally-ride-dies-1983-newsweek-profile-spacepioneer-207036. Accessed 28 Aug 2022.
Birenboim, Amit, Salvador Anton Clavé, Antonio Paolo Russo, and Noam Shoval. 2013. Temporal Activity Patterns of Theme Park Visitors. *Tourism Geographies* 15 (4): 601–619.
Bryman, Alan. 1995. *Disney and His Worlds.* London: Routledge.
Bryman, Alan. 2004. *The Disneyization of Society.* London: Sage.
Brown, Amber, Jacqueline Kappes, and Joe Marks. 2013. Mitigating Theme Park Crowding with Incentives and Information on Mobile Devices. *Journal of Travel Research* 52 (4): 426–436.
Bukatman, Scott. 1991. There's Always Tomorrowland: Disney and the Hypercinematic Experience. *October* 57: 55–78.
Camp, Gregory. 2017. Mickey Mouse Muzak: Shaping Experience Musically at Walt Disney World. *Journal of the Society for American Music* 11 (1): 53–69.
Cardell, David. 2015. *Family Theme Parks, Happiness and Children's Consumption: From Roller-Coasters to Pippi Longstocking.* Diss., Linköping University.
Carl, Morgan. 2020. *"It's an Accessible World After All": Evaluation of Amusement Park Accessibility and Accommodations for Guests with Disabilities.* Diss., Eastern Kentucky University.
Carlà, Filippo. 2016. The Uses of History in Themed Environments. In *A Reader in Themed and Immersive Spaces*, ed. Scott A. Lukas, 19–29. Pittsburgh: ETC.
Carlà-Uhink, Filippo. 2020. *Representations of Classical Greece in Theme Parks.* London: Bloomsbury.
Carlà-Uhink, Filippo, Florian Freitag, Sabrina Mittermeier, and Ariane Schwarz, eds. 2017. *Time and Temporality in Theme Parks.* Hanover: Wehrhahn.
Carson, Charles. 2004. "Whole New Worlds": Music and the Disney Theme Park Experience. *Ethnomusicology Forum* 13 (2): 228–235.
Clément, Thibaut. 2016. *Plus vrais que nature: Les parcs Disney, ou l'usage de la fiction de l'espace et le paysage.* Paris: Presses de la Sorbonne Nouvelle.
Coltier, Thierry. 1985. Les parcs à thèmes. *Espaces* 74: 24–26.
Cornelis, Pieter C.M. 2010. Effects of Co-Branding in the Theme Park Industry: A Preliminary Study. *International Journal of Contemporary Hospitality Management* 22 (6): 775–796.
Cornelis, Pieter C.M. 2011a. *Attraction Accountability: Predicting the (Un)Predictable Effects of Theme Park Investments?!* Breda: NRIT Media.
Cornelis, Pieter C.M.. 2011b. A Management Perspective on the Impact of New Attractions. *Journal of Vacation Marketing* 17 (2): 151–162.
Cornelis, Pieter C.M. 2017. *Investment Thrills: Managing Risk and Return for the Amusement Parks and Attractions Industry.* Nieuwegein: NRIT Media.

Davis, Susan G. 1997. *Spectacular Nature: Corporate Culture and the Sea World Experience.* Berkeley: University of California Press.
Dear, Michael J., and Steven Flusty. 2002. The Spaces of Representation. In *The Spaces of Postmodernity: Readings in Human Geography*, ed. Michael J. Dear and Steven Flusty, 415–418. Oxford: Blackwell.
Dong, Ping, and Noel Yee-Man Siu. 2013. Servicescape Elements, Customer Predispositions and Service Experience: The Case of Theme Park Visitors. *Tourism Management* 36: 541–551. https://doi.org/10.1016/j.tourman.2012.09.004. Accessed 28 Aug 2022.
Dreschke, Anja. 2010. Playing Ethnology. In *Staging the Past: Themed Environments in Transcultural Perspectives*, ed. Judith Schlehe, Michiko Uike-Bormann, Carolyn Oesterle, and Wolfgang Hochbruck, 253–267. Bielefeld: Transcript.
Dzeng, Ren-Jye, and Hsin-Yun Lee. 2007. Activity and Value Orientated Decision Support for the Development Planning of a Theme Park. *Expert Systems with Applications* 33: 923–935.
Eco, Umberto. 1986 [1975]. Travels in Hyperreality. In *Travels in Hyperreality: Essays*, trans. William Weaver, 1–58. San Diego: Harcourt Brace Janovich.
Erb, Maribeth, and Chin-Ee Ong, eds. 2017. Theming Asia: Theme Park Experiences in East and Southeast Asia. *Tourism Geographies* 19 (2): 143–300.
Fichtner, Uwe, and Rudolf Michna. 1987. *Freizeitparks: Allgemeine Züge eines modernen Freizeitangebotes, vertieft am Beispiel des Europapark in Rust/Baden.* Freiburg: N.P.
Finch, Christopher. 1973. *The Art of Walt Disney from Mickey Mouse to the Magic Kingdoms.* New York: Harry N. Abrams.
Fjellman, Stephen M. 1992. *Vinyl Leaves: Walt Disney World and America.* Boulder: Westview.
Foglesong, Richard E. 1999. *Married to the Mouse: Walt Disney World and Orlando.* New Haven: Yale University Press.
Fotiadis, Anestis K., and Nikolaos Stylos. 2017. The Effects of Online Social Networking on Retail Consumer Dynamics in the Attractions Industry: The Case of "E-da" Theme Park, Taiwan. *Technological Forecasting and Social Change* 124: 283–294.
Fotiadis, Anestis K., and Chris A. Vassiliadis. 2016. Service Quality at Theme Parks. *Journal of Quality Assurance in Hospitality & Tourism* 17 (2): 178–190.
Francaviglia, Richard V. 1977. Main Street, U.S.A.: The Creation of a Popular Image. *Landscape: A Magazine of Human Geography* 21 (3): 18–22.
Francaviglia, Richard V. 1999. Walt Disney's Frontierland as an Allegorical Map of the American West. *The Western Historical Quarterly* 30 (2): 155–182.
Frank, Robin Jaffee. 2015. *Coney Island: Visions of an American Dreamland, 1861–2008.* New Haven: Yale University Press.
Freitag, Florian. 2017. "Like Walking into a Movie": Intermedial Relations between Disney Theme Parks and Movies. *The Journal of Popular Culture* 50 (4): 704–722.
Freitag, Florian. 2021a. *Popular New Orleans: The Crescent City in Periodicals, Theme Parks, and Opera, 1875–2015.* New York: Routledge.
Freitag, Florian. 2021b. "This Way or That? Par ici ou par là?": Language in the Theme Park. *Visions in Leisure and Business* 23 (1). https://doi.org/10.25035/visions.23.01.07. Accessed 28 Aug 2022.
Freitag, Florian, and Chang Liu, eds. 2022. *History in Chinese Theme Parks. Cultural History* 11 (2): 121–218.
Freitag, Florian, and Ariane Schwarz. 2015/16. Thresholds of Fun and Fear: Borders and Liminal Experiences in Theme Parks. *OAA Perspectives* 23 (4): 22–23.
Gabe, Todd. 2021. Impacts of COVID-Related Capacity Constraints on Theme Park Attendance: Evidence from Magic Kingdom Wait Times. *Applied Economics Letters* 28 (14): 1222–1225.
Godwin, Victoria L. 2017. Theme Park as Interface to the Wizarding (Story) World of Harry Potter. *Transformative Works and Cultures* 25. https://journal.transformativeworks.org/index.php/twc/article/view/1078/871. Accessed 28 Aug 2022.

Gottdiener, Mark. 1997. *The Theming of America: Dreams, Visions and Commercial Spaces.* Boulder: Westview.

Gottwald, Dave, and Benjamin George. 2020. Cinematography in the Landscape: Transitional Zones in Themed Environments. *Landscape Research Record* 9: 49–64.

Gottwald, Dave, and Greg Turner-Rahman. 2019. The End of Architecture: Theme Parks, Video Games, and the Built Environment in Cinematic Mode. *International Journal of the Constructed Environment* 10 (2): 41–60.

Graf-Janz, Viktoria. 2022. Verzauberung, Vergnügen—Verblendung? Die konsumkritische Rezeption von Freizeitparks in Fiktion. In *Konsumvergnügen: Die Populäre Kultur und der Konsum*, ed. Dirk Hohnsträter and Stefan Krankenhagen, 123–140. Berlin: Kadmos.

Hall, Millicent. 1976. Theme Parks: Around the World in 80 Minutes. *Landscape: A Magazine of Human Geography* 21 (1): 3–8.

Harwood, Edward. 2002. Rhetoric, Authenticity, and Reception: The Eighteenth-Century Landscape Garden, the Modern Theme Park, and Their Audiences. In *Theme Park Landscapes: Antecedents and Variations*, ed. Terence G. Young and Robert Riley, 49–68. Washington, D.C.: Dumbarton Oaks Research Library and Collection.

Hen, Rai. 2013. *The Middle Class in Neoliberal China: Governing Risk, Life-Building, and Themed Spaces.* New York: Routledge.

Hildebrandt, Hugo John. 1981. Cedar Point: A Park in Progress. *The Journal of Popular Culture* 15 (1): 87–107.

Hjemdahl, Kirsti Mathiesen. 2002. History as Cultural Playground. *Ethnologia Europaea* 32 (2): 105–124.

Hjemdahl, Kirsti Mathiesen. 2003. When Theme Parks Happen. In *Being There: New Perspectives on Phenomenology and the Analysis of Culture*, ed. Jonas Frykman and Nils Gilje, 149–168. Lund: Nordic Academic Press.

Holcomb, Judy, Fevzi Okumus, and Anil Bilgihan. 2010. Corporate Social Responsibility: What are the Top Three Orlando Theme Parks Reporting? *Worldwide Hospitality and Tourism Themes* 2: 316–337.

Huang, Xiaoting, Minxuan Li, Jingru Zhang, Linlin Zhang, Haiping Zhang, and Shen Yan. 2020. Tourists' Spatial-Temporal Behavior Patterns in Theme Parks: A Case Study of Ocean Park Hong Kong. *Journal of Destination Marketing & Management* 15. https://doi.org/10.1016/j.jdmm.2020.100411. Accessed 28 Aug 2022.

Hyped! 2021. Clubhouse Talk: Fler vs. Malcolm! Teil 7. *YouTube*, February 16. https://www.youtube.com/watch?v=ZoJ4p4igpjU. Accessed 8 Nov 2021.

Ingram, Susan. 2003. Public Entertainment in Nineteenth-Century London. In *Placing History: Themed Environments, Urban Consumption and the Public Entertainment Sphere*, ed. Susan Ingram and Markus Reisenleitner, 33–64. Vienna: Turia + Kant.

Jung, Timothy, Namho Chung, and M. Claudia Leue. 2015. The Determinants of Recommendations to Use Augmented Reality Technologies: The Case of a Korean Theme Park. *Tourism Management* 49: 75–86.

Kagelmann, H. Jürgen. 1993. Themenparks. In *Tourismuspsychologie und Tourismussoziologie*, ed. Heinz Hahn and H. Jürgen Kagelmann, 407–415. Munich: Quintessenz.

Kagelmann, H. Jürgen. 2004. Themenparks. In *ErlebnisWelten: Zum Erlebnisboom in der Postmoderne*, ed. H. Jürgen Kagelmann, Reinhard Bachleitner, and Max Rieder, 160–180. Munich: Profil.

Kay, John. 2008. Welcome to Venice, the Theme Park. *The Times Online*, March 1, https://www.thetimes.co.uk/article/welcome-to-venice-the-theme-park-nj8hzntbdpf. Accessed 28 Aug 2022.

King, Margaret J., ed. 1981. *Amusement/Theme Parks. The Journal of Popular Culture* 15 (1): 56–179.

Kiriakou, Olympia. 2017. "Ricky, This Is Amazing!": Disney Nostalgia, New Media Users, and the Extreme Fans of the WDW Kingdomcast. *Journal of Fandom Studies* 5 (1): 99–112.

Klein, Norman. 2004. *The Vatican to Vegas: A History of Special Effects*. New York: New Press.
Koenig, David. 2007. *Realityland: True-Life Adventures at Walt Disney World*. Irvine: Bonaventure.
Kokai, Jennifer A., and Tom Robson, eds. 2019. *Performance and the Disney Theme Park Experience: The Tourist as Actor*. Cham: Palgrave Macmillan.
Kuenz, Jane. 1995. Working at the Rat. In *Inside the Mouse: Work and Play at Disney World*, ed. Jane Kuenz, Susan Willis, Shelton Waldrep, and Stanley Fish, 111–162. Durham: Duke University Press.
Landau, Royston. 1973. Mickey Mouse the Great Dictator: The Disney Game as a Control System. *Architectural Digest* 43 (9): 591–595.
Li, Tao, Jiaming Liu, and He Zhu. 2018. The International Investment in Theme Parks: Spatial Distribution and Decision-Making Mechanism. An Empirical Study for China. *Tourism Management* 67: 342–350.
Lippy, Charles H. 1994. *Being Religious, American Style: A History of Popular Religiosity in the United States*. Westport: Praeger.
Liu, Yi-De. 2008. Profitability Measurement of UK Theme Parks: An Aggregate Approach. *International Journal of Tourism Research* 10: 283–288.
Lonsway, Brian. 2009. *Making Leisure Work: Architecture and the Experience Economy*. New York: Routledge.
Lukas, Scott A. 2007. How the Theme Park Gets Its Power: Lived Theming, Social Control, and the Themed Worker Self. In *The Themed Space: Locating Culture, Nation, and Self*, ed. Scott A. Lukas, 183–206. Lanham: Lexington.
Lukas, Scott A. 2008. *Theme Park*. London: Reaktion.
Lukas, Scott A. 2010. From Themed Space to Lifespace. In *Staging the Past: Themed Environments in Transcultural Perspectives*, ed. Judith Schlehe, Michiko Uike-Bormann, Carolyn Oesterle, and Wolfgang Hochbruck, 135–153. Bielefeld: Transcript.
Lukas, Scott A. 2013. *The Immersive Worlds Handbook: Designing Theme Parks and Consumer Spaces*. New York: Focal.
Lukas, Scott A., ed. 2016a. *A Reader in Themed and Immersive Spaces*. Pittsburgh: ETC.
Lukas, Scott A. 2016b. Research in Themed and Immersive Spaces: At the Threshold of Identity. In *A Reader in Themed and Immersive Spaces*, ed. Scott A. Lukas, 159–169. Pittsburgh: ETC.
Makai, Péter Kristóf. 2019. Three Ways of Transmediating a Theme Park: Spatializing Storyworlds in Epic Mickey, the Monkey Island Series and Theme Park Management Simulators. In *Transmediations: Communication across Media Borders*, ed. Niklas Salmose and Lars Elleström, 164–185. London: Routledge.
Marin, Louis. 1984 [1973]. *Utopics: The Semiological Play of Textual Spaces*, trans. Robert A. Vollrath. New York: Humanity.
Marling, Karal Ann. 1991. Disneyland, 1995: Just Take the Santa Ana Freeway to the American Dream. *American Art* 5 (1–2): 168–207.
Milman, Ady, and Duncan Dickson. 2014. Employment Characteristics and Retention Predictors among Hourly Employees in Large US Theme Parks and Attractions. *International Journal of Contemporary Hospitality Management* 26 (3): 447–469.
Min, Deedee Aram, Kyung Hoon Hyun, Sun-Joong Kim, and Ji-Hyun Lee. 2017. A Rule-Based Servicescape Design Support System from the Design Patterns of Theme Parks. *Advanced Engineering Informatics* 32: 77–91.
Mitas, Ondrej, Helena Mitasova, Garrett Millar, Wilco Boode, Vincent Neveu, Moniek Hover, Frank van den Eijnden, and Marcel Bastiaansen. 2022. More is Not Better: The Emotional Dynamics of an Excellent Experience. *Journal of Hospitality & Tourism Research* 46 (1): 78–99.
Mitrasinovic, Miodrag. 2006. *Total Landscape, Theme Parks, Public Space*. Burlington: Ashgate.
Mittermeier, Sabrina. 2021. *A Cultural History of the Disneyland Theme Parks: Middle Class Kingdoms*. Chicago: Intellect.
Mnouchkine, Ariane. 2003. Disneyland Resort Paris, France: 1992. *Time*, August 10. http://content.time.com/time/specials/packages/article/0,28804,2024035_2024499_2024904,00.html. Accessed 28 Aug 2022.

Ndalianis, Angela. 2004. *Neo-Baroque Aesthetics and Contemporary Entertainment*. Cambridge: MIT Press.

Orellana, Alicia, Joan Borràs, and Salvador Anton Clavé. 2016. Understanding Emotional Behaviour in a Theme Park: A Methodological Approach. In *Consumer Behavior Tourism Symposium Book of Abstracts: Experiences, Emotions and Memories. New Directions in Tourism Research*, ed. Serena Volo and Oswin Maurer, n.p. Bruneck: Competence Centre in Tourism Management and Tourism Economics (TOMTE).

Orvell, Miles. 2012. *The Death and Life of Main Street: Small Towns in American Memory, Space, and Community*. Chapel Hill: The University of North Carolina Press.

Paine, Crispin. 2019. *Gods and Rollercoasters: Religion in Theme Parks Worldwide*. London: Bloomsbury.

Parascandola, Louis J., and John Parascandola, eds. 2014. *A Coney Island Reader: Through Dizzy Gates of Illusion*. New York: Columbia University Press.

Pechlaner, Harald, Thomas Bieger, and Klaus Weirmaier, eds. 2006. *Attraktions-Management: Führung und Steuerung von Attraktionspunkten*. Wien: Linde.

Philips, Deborah. 2012. *Fairground Attractions: A Genealogy of the Pleasure Ground*. London: Bloomsbury.

Piazzoni, Maria Francesca. 2018. *The Real Fake: Authenticity and the Production of Space*. New York: Fordham University Press.

Pike, David L. 2005. The Walt Disney World Underground. *Space and Culture* 8 (1): 47–65.

Rabinovitz, Lauren. 2012. *Electric Dreamland: Amusement Parks, Movies, and American Modernity*. New York: Columbia University Press.

Rahn, Suzanne. 2000. Snow White's Dark Ride: Narrative Strategies at Disneyland. *Bookbird: A Journal of International Children's Literature* 38 (1): 19–24.

Raz, Aviad E. 1999. *Riding the Black Ship: Japan and Tokyo Disneyland*. Cambridge: Harvard University Press.

Real, Michael R. 1977. *Mass-Mediated Culture*. Englewood Cliffs: Prentice-Hall.

Ryan, Chris, Yeh (Sam) Shih Shuo, and Tzung-Cheng Huan. 2010. Theme Parks and a Structural Equation Model of Determinants of Visitor Satisfaction: Janfusan Fancyworld, Taiwan. *Journal of Vacation Marketing* 16 (3): 185–199.

Schickel, Richard. 1968. *The Disney Version: The Life, Times, Art and Commerce of Walt Disney*. New York: Simon and Schuster.

Schlehe, Judith, and Michiko Uike-Bormann. 2010. Staging the Past in Cultural Theme Parks: Representations of Self and Other in Asia and Europe. In *Staging the Past: Themed Environments in Transcultural Perspectives*, ed. Judith Schlehe, Michiko Uike-Bormann, Carolyn Oesterle, and Wolfgang Hochbruck, 57–91. Bielefeld: Transcript.

Schwartz, Vanessa. 2020. *Jet Age Aesthetic: The Glamour of Media in Motion*. London: Yale University Press.

Schweizer, Bobby. 2016. Visiting the Videogame Theme Park. *Wide Screen* 6 (1). https://widescreenjournal.files.wordpress.com/2021/06/visiting-the-videogame-theme-park.pdf. Accessed 28 Aug 2022.

Shani, Amir, and Abraham Pizam. 2010. The Role of Animal-Based Attractions in Ecological Sustainability: Current Issues and Controversies. *Worldwide Hospitality and Tourism Themes* 2: 281–298.

Shen, Li-Yin, Andrew Platten, and X.P. Deng. 2006. Role of Public Private Partnerships to Manage Risks in Public Sector Projects in Hong Kong. *International Journal of Project Management* 24 (7): 587–594.

Singh, Varun, Tarun Singh, D. Langan, and Praveen Kumar. 2004. A Framework for Internet GIS Based Computerized Visitor Information System for Theme Parks. In *Proceedings: The 7th International IEEE Conference on Intelligent Transportation Systems*, 679–683. Piscataway: IEEE.

Smith, Ruth C., and Eric M. Eisenberg. 1987. Conflict at Disneyland: A Root-Metaphor Analysis. *Communication Monographs* 54 (4): 367–380.

Sohier, Estelle, Alexandre Gillet, and Jean-François Staszak, eds. 2019. *Simulations du monde: Panoramas, parcs à theme et autres dispositifs immersifs*. Geneva: MétisPresses.

Soto-Vásquez, Arthur. 2021. Mediating the Magic Kingdom: Instagram, Fantasy, and Identity. *Western Journal of Communication* 85 (5): 588–608.

Steinecke, Albrecht. 2009. *Themenwelten im Tourismus: Marktstrukturen—Marketing—Management—Trends*. Munich: Oldenbourg.

Steinkrüger, Jan-Erik. 2013. *Thematisierte Welten: Über Darstellungspraxen in Zoologischen Gärten und Vergnügungsparks*. Bielefeld: Transcript.

Swarbrooke, John. 1995. *The Development and Management of Visitor Attractions*. Oxford: Elsevier.

Telotte, J.P. 2004. *Disney TV*. Detroit: Wayne State University Press.

Thomas-Morus-Akademie, ed. 1998. *Kathedralen der Freizeitgesellschaft: Kurzurlaub in Erlebniswelten. Trends, Hintergründe, Auswirkungen*. Bergisch-Gladbach: Thomas-Morus-Akademie Bensberg.

Tourism Research and Marketing. 1995. *Theme Parks: UK and International Markets*. London: Tourism Research and Marketing.

Tsai, Chieh-Yuan, and Shang-Hsuan Chung. 2012. A Personalized Route Recommendation Service for Theme Parks Using RFID Information and Tourist Behavior. *Decision Support Systems* 52 (2): 514–527.

Tsang, Nelson K.F., Louisa Y.S. Lee, Alan Wong, and Rita Chong. 2012. THEMEQUAL: Adapting the SERVQUAL Scale to Theme Park Services. A Case of Hong Kong Disneyland. *Journal of Travel & Tourism Marketing* 29 (5): 416–429.

Van Maanen, John. 1991. The Smile Factory: Working at Disneyland. In *Reframing Organizational Culture*, ed. Peter J. Frost, Larry F. Moore, Meryl Reis Louis, Craig C. Lundberg, and Joanne Martin, 58–77. Newbury Park: Sage.

Wang, Jen Chun, Yi-Chieh Wang, Li Ko, and Jen Hsing Wang. 2017. Greenhouse Gas Emissions of Amusement Parks in Taiwan. *Renewable and Sustainable Energy Reviews* 74: 581–589.

Wasko, Janet. 2001. *Understanding Disney: The Manufacture of Fantasy*. Cambridge: Polity.

Wasko, Janet. 2020. *Understanding Disney: The Manufacture of Fantasy*, 2nd ed. Cambridge: Polity.

Waters, Richard D. 2016. Facilitating the "Charged Public" through Social Media: A Conversation with Disney Cruise Line's Castaway Club Members. In *Public Relations and Participatory Culture: Fandom, Social Media and Community Engagement*, ed. Amber L. Hutchins and Natalie T.J. Tindall, 181–192. New York: Routledge.

Waysdorf, Abby S. 2021. *Fan Sites: Film Tourism and Contemporary Fandom*. Iowa City: University of Iowa Press.

Waysdorf, Abby, and Stijn Reijnders. 2018. Immersion, Authenticity and the Theme Park as Social Space: Experiencing the Wizarding World of Harry Potter. *International Journal of Cultural Studies* 21 (2): 173–188.

Wei, Wei, Ruoxi Qi, and Lu Zhang. 2019. Effects of Virtual Reality on Theme Park Visitors' Experience and Behaviors: A Presence Perspective. *Tourism Management* 71: 282–293.

Wiener, Jon. 1994. Tall Tales and True. *The Nation*, January 31, 133–135.

Williams, Rebecca. 2020. *Theme Park Fandom: Spatial Transmedia, Materiality and Participatory Cultures*. Amsterdam: Amsterdam University Press.

Woodcock, Kathryn. 2008. Content Analysis of 100 Consecutive Media Reports of Amusement Ride Accidents. *Accident Analysis & Prevention* 40: 89–96.

Woodcock, Kathryn. 2014. Human Factors and Use of Amusement Ride Control Interfaces. *International Journal of Industrial Ergonomics* 44: 99–106.

Yandell, Stephen. 2012. Mapping the Happiest Place on Earth: Disney's Medieval Cartography. In *The Disney Middle Ages: A Fairy-Tale and Fantasy Past*, ed. Tison Pugh and Susan Aronstein, 21–38. New York: Palgrave Macmillan.

Yoonjoung Heo, Cindy, and Seoki Lee. 2009. Application of Revenue Management Practices to the Theme Park Industry. *International Journal of Hospitality Management* 28 (3): 446–453.

Young, Terence, and Robert Riley, eds. 2002. *Theme Park Landscapes: Antecedents and Variations*. Washington, D.C.: Dumbarton Oaks Research Library and Collection.

Younger, David. 2016. *Theme Park Design & the Art of Themed Entertainment*. N.P.: Inklingwood.
Yuan, Yuguo, and Weimin Zheng. 2018. How to Mitigate Theme Park Crowding? A Prospective Coordination Approach. *Mathematical Problems in Engineering*. https://doi.org/10.1155/2018/3138696. Accessed 28 Aug 2022.
Zhang, Yingsha, Xiang (Robert) Li, and Qin Su. 2017. Does Spatial Layout Matter to Theme Park Tourism Carrying Capacity? *Tourism Management* 61: 82–95.

Antecedents, Origins, and Developments: A History of Theme Parks from Antiquity to the Twenty-First Century

Abstract The history of theme parks is usually said to have begun in 1955 with the opening of Disneyland in Anaheim, California. While it is true that Disneyland represents a first in some respects, in the past few years, scholarship has also highlighted that Disney's creation was not an entirely new idea, and that antecedents of contemporary theme parks—like follies, pleasure gardens, and world's fairs—date back centuries. This chapter reconstructs the history of themed, immersive environments created for entertainment and leisure, from Antiquity and the Middle Ages to the early twentieth century, highlighting the elements that have resurfaced in the modern theme park. Additionally, the global success of theme parks (and thus their spread outside of North America and Europe) has brought about, in each region, a confluence of local forms of antecedents—such as Chinese gardens and palaces—which this chapter will also investigate within their specific local contexts. The second part of the chapter will then offer a broad perspective on the economic, aesthetic, and cultural development of theme parks since the mid-twentieth century, discussing the main trends of theme park history in the different regions of the world until today.

1 The Question of Origins and Development

The questions of when, where, and how theme parks originated are both complex and contested, especially because the answers to these questions are of commercial, as well as popular and academic, interest. In the field of location-based entertainment, where novelty, innovation, and uniqueness are paramount, the claim to offer something unprecedented—or at least something that had been unprecedented at some point, and thus made history—constitutes a powerful marketing tool that many theme parks have sought to employ. Visitors to Knott's Berry Farm in Buena Park, California, for instance, have long been greeted by a sign above the entrance that proclaims Knott's to be "America's 1st Theme Park," although on its website the

This work is contributed by Filippo Carlà-Uhink, Jean-François Staszak, Astrid Böger, Salvador Anton Clavé, Florian Freitag, Jan-Erik Steinkrüger, Thibaut Clément, Scott A. Lukas, Sabrina Mittermeier, Céline Molter, Crispin Paine, Ariane Schwarz, Torsten Widmann. The corresponding author is Filippo Carlà-Uhink, Historisches Institut, Universität Potsdam, Potsdam, Germany.

F. Freitag et al., *Key Concepts in Theme Park Studies*,
https://doi.org/10.1007/978-3-031-11132-7_2

park advertises itself somewhat more modestly as "California's 1st Theme Park." The website of Santa's Village Amusement Park in East Dundee, Illinois, meanwhile, makes a point of mentioning that the very first Santa's Village created by its founder Glenn Holland in Skyforest, California, was opened "six weeks before Disneyland," conveniently leaving out the fact that a Santa Claus Land had opened in Santa Claus, Indiana, nine years prior (see Samuelson and Yegoiants 2001, 146). Most famously, Disneyland itself claimed to be unique and unprecedented: in the premiere episode of Disney's weekly TV show *Disneyland*, broadcast on October 27, 1954, and entitled "The Disneyland Story" (see PARATEXTS AND RECEPTION), host Walt Disney declared that he hoped the park would "be unlike anything else on this earth. A fair, an amusement park, an exhibition, a city from *The Arabian Nights*, a metropolis from the future—in fact, a place of hopes and dreams, facts and fancy all in one" (Florey and Jackson 1954, 04:20–04:36).

Indeed, there is reason to argue that the theme park as it has appeared since the mid-twentieth century displays singular characteristics that define it as something new and different from other entertainment forms that had appeared before (Anton Clavé 2007, 21–27); it is important to stress that technological progress in the twentieth century allowed for the development of new and more complex forms to ensure excitement and entertainment—and in particular make theming much more immersive than in the previous medial and technological frame (see MEDIA). To contemporaries, the first theme parks seemed so different from what had preceded them that, much like Disney during "The Disneyland Story," they struggled to find a proper name for the places: although the *Oxford English Dictionary* dates the first use of the term "theme park" to the year 1960 (in the *American Peoples Encyclopedia Year Book*), the first critical studies of Disneyland still referred to the place as an "atmospheric park" (Schickel 1968, 22) or an "amusement park" (Finch 1973, 396). By the mid-1970s, however, the term "theme park" had established itself in critical discourse (see Hall 1976).

And yet the claims of newness, unprecedentedness, and distinctiveness that arrived with the theme park during the mid-twentieth century were hardly new. Almost exactly half a century earlier, the creators of Coney Island's famous Steeplechase Park (1897–1964), Dreamland (1904–1911), and particularly Luna Park (1903–1946) had sought to promote their ventures as decent, respectable places "for your mother, your sister, your sweetheart" and thus to establish a dichotomy between the "new and improved" Coney Island of the amusement park era and the primarily male-oriented and morally questionable concessions, sideshows, and entertainments that preceded and—all protests to the contrary notwithstanding—would continue to surround and eventually outlive those parks (see Sterngass 2001, 256). Ironically, much like Disneyland would later attempt to distance itself from supposedly seedy and poorly organized playgrounds and amusement parks such as those of Coney Island (see Weinstein 1992), Coney Island's parks used similar rhetoric of newness to differentiate themselves from the supposedly dirty and chaotic "Sodom by the Sea."

In the case of Coney Island, much of the press bought the ballyhoo. As Sterngass reports, between 1903 and 1908 "at least fifteen major articles appeared in national magazines serenading Coney Island with title words such as 'apotheosis,' 'awakening,' and 'renaissance'" (2001, 256). Fifty years later, Disneyland, with its intricate ties to the media and particularly the television industry (see Telotte 2004), made its own news: during the televised opening of Disneyland, host Art Linkletter not only described the park as nothing short of the "eighth wonder of the world" (Jackson et al. 1955, 02:45); he and co-host Bob Cummings also introduced their wives and children to underline the fact that, unlike earlier amusement parks, Disney's new park was intended for the entire family.

Perhaps even more importantly, Disneyland's promotional narrative of newness was readily adopted into scholarly criticism. For example, the early biographies of Walt Disney by Christopher Finch (1973) and Bob Thomas (1976) both emphasized Disney's personal dislike of conventional amusement parks in general and particularly Coney Island, thus implying Disneyland's difference from them. And in one of the earliest discussions of theme parks in American academic publishing, Millicent Hall's 1976 article "Theme Parks: Around the World in 80 Minutes," the author both asserts the uniqueness of theme parks against "other types of American entertainment" such as the amusement park and confirms the status of Disneyland as the very first theme park (3). Early European theme park critics such as Louis Marin (*Utopiques: Jeux d'espace*, 1973), Umberto Eco ("Viaggio nell'iperrealtà," 1975), and Jean Baudrillard ("La précession des simulacres," 1978) in turn stressed not so much the newness or uniqueness of theme parks (and particularly Disneyland), but rather their inherent Americanness: Baudrillard, for instance, describes the Anaheim park as a "digest of American life" and a "panegyric of American values" (12). American critics such as Margaret J. King ("The New American Muse," 1981a and "Disneyland and Walt Disney World: Traditional Values in Futuristic Form," 1981b) happily agreed. And yet this was nothing new, either. International visitors to Coney Island, from José Martí ("Coney Island," 1881) to Maxim Gorky ("Boredom," 1907) had written about Coney's Americanness with the very same mixture of awe and disdain that Europeans would later use for Disneyland ("Those people eat quantity; we, class," Martí wrote; 2014, 65).

Well-established and commercially useful as it was, the rhetoric surrounding the newness and Americanness of theme parks would not be critically contested until the early 1990s, when scholars such as Karal Ann Marling or Raymond M. Weinstein started to investigate the roots of Disneyland in railroad fairs and amusement parks, respectively (Marling 1991; Weinstein 1992). A decade later, Terence Young and Robert Riley's *Theme Park Landscapes: Antecedents and Variations* (2002) further complicated the discussion by adding not only European pleasure gardens and American agricultural fairs, but also traditional Chinese painting and landscape gardening to the list of antecedents that preceded and inspired modern theme parks. Miodrag Mitrasinovic's *Total Landscape, Theme Parks, Public Space* (2006) then cast an even wider net, going as far back as imperial Rome to establish the origins of theme parks. All these studies have substantially enriched our knowledge of the antecedents and origins of the theme parks; at the same time, however, they risk promoting what might be called a "teleological perspective," i.e. the idea that all these different places and forms of entertainment found their ultimate accomplishment and "destiny" when they generated the complete and perfected form of the modern theme park in the mid-twentieth century. Such an approach is, of course, unacceptable from a scholarly perspective and must be avoided.

Research into the development of the theme park *since* its supposed inception in 1950s southern California also started in the early 1990s (see Kagelmann 1993; Davis 1996), but has remained embryonic (see, however, Anton Clavé 2007; Younger 2017). And yet, the mere fact that contemporary autothemed theme park spaces (see PARATEXTS AND RECEPTION) evoke both Coney Island-style amusement parks (e.g. Toyville Trolley Park, located in the New York City-themed "American Waterfront" section of Tokyo DisneySea, Japan) as well as 1950s Disneyland (e.g. "Tomorrowland" at Disneyland, Anaheim; see Freitag forthcoming) suggests that the contemporary theme park may be, as a Disney spokesperson stated as early as 1982 with reference to Epcot (Orlando, Florida), "as different from Disneyland as (Disneyland) was from Coney Island" (qtd. in Weinstein 1992, 131). The parks' promotional narratives may thus have offered scholars an initial idea of where to begin the history of the theme park, but critics have yet to agree on the major turning points and peripeties of the later story. To be sure, numerous studies have traced the histories of individual parks (see, for example, Hildebrandt 1981 on Cedar Point, Sandusky, Ohio, or Davis 1997 on SeaWorld, San Diego, California) or even of specific themed lands (see, for example, Wiener 1994 on Disneyland's "Frontierland" or Freitag 2021 on Disneyland's "New Orleans Square"). The big picture, however, is still somewhat blurry. It may prove to be as complex and contested as that of the origins of theme parks.

As it would be impossible to present a complete story of all enclosed leisure spaces in the different regions of the world in this chapter, we will mainly concentrate on the European and North American traditions—certainly the most frequently studied in critical literature—while, at the same time, also glancing at the Far Eastern tradition, with a special focus on China and Japan. Although we do not claim to reach any form of completeness on the subject, we hope to show that it is impossible to speak of one "history of theme parks" and their antecedents, and that the genealogical relationships of leisure spaces throughout the centuries are as diverse as they are complex and interconnected.

2 Western Antecedents from Antiquity to the Eighteenth Century

2.1 Antiquity and the Middle Ages

While the theme park per se is a twentieth century phenomenon, all of its characteristics and aspects had already appeared in previous built spaces, starting from Antiquity, both in Europe and around the Mediterranean basin, as well as in the Far East. The tradition of the garden as an enclosed space designed to be pleasant, relaxing, and contemplative, which has been identified as a significant antecedent for theme parks both in the Western and the Eastern world, was first developed in the Ancient Near East and pharaonic Egypt (Keßler 2012, 124–125). Persian gardens and hunting parks, organized along strict aesthetic and philosophical principles, the *pairi-daeza*, are famously the origin of the word Paradise. In ancient Greece, this model of the garden was continued, and in democratic Athens it was made accessible to all citizens: in the fifth century BCE, Kimon organized gardens around the Lykeion and the Academy, which were used by philosophical schools, in particular those of Plato and Aristotle. Before that, these parks had been used for sports, theater, and other events of the city's civic and religious life. Even more suggestive of a theme park is the "Agora des Italiens" on the Greek island of Delos, a sort of club for the Italians and Romans living and working on the island in the second century BCE. It is a closed structure, and even if positive evidence is missing, one may assume that access to it was regulated, as in a modern theme park, perhaps restricted to Italians and Romans. Within the "Agora," libraries, baths, gardens, and statues allowed visitors to relax in a controlled environment that celebrated their identity (Trümper 2008; Carlà-Uhink 2017, 298–300). Additionally, technical discoveries were already being used to astonish and entertain visitors. In his *Automata* from the first century CE, Heron of Alexandria describes several devices created for this very purpose: temple doors opening automatically when a fire was lit upon an altar; an organ operated by a windmill; a mechanical theater that was completely automatized and performed a show lasting around 10 min (thus not dissimilar from animatronic shows in modern theme parks), and the first vending machine of history, which offered holy water in exchange for a coin.

The most famous antecedent of theme parks in the ancient world, however, is the Villa Adriana in Tivoli. Built in the second century CE by the Roman emperor Hadrian on a surface of over 77 ha, the Villa consists of many different buildings and structures, evoking the Roman Empire (the "theme") in its entirety. One example is the so-called Canopus, a replica and evocation of a water channel that connected the Egyptian cities of Alexandria and Canopus, where one could find copies of famous sculptures, displayed in a way which—as in modern theme parks—had little to do with their original sense and location, and much more with the creation of a new aesthetic landscape that included well-known elements from the past: "general demonstrations of education and actualized references had a higher priority than a direct understanding of the piece of art itself" (Knell 2008, 110).

Since living history museums are also, as we will see below, crucial antecedents of the theme park, it is worth mentioning that "living history," in the form of reenactment of past episodes, whether real or invented, was also practiced in the Roman world: spectacles organized in the amphitheaters could take the form of reenactments of ancient battles (including naval battles) or invented battles between ancient peoples who in reality never clashed with each other. These were immersive reconstructions of past times and civilizations, and in this sense they can also be considered an antecedent of the theme park and its shows. Similarly, in the medieval world, the "sacra rappresentazione" ("sacred representation"), which generally took place during the Easter week, provided a form of immersive reenactment of the Scriptures.

The religious component at the origin of immersive environments is indeed highly significant: it has been shown that the development of forms of ersatz Holy Lands in the U.S. for Protestant American "pilgrims" has strongly contributed to the development of "religious theme parks" (see Rowan 2004). Yet this idea of constructing immersive religious worlds to allow pilgrims to visit the Holy Land without traveling to the Near East was already present in medieval Catholicism. While replicas of the Sepulcher of Christ were known throughout Europe in the High Middle Ages, the most important example are the "Sacri Monti" built in northwestern Italy from the fifteenth century onwards. The oldest and most significant among these, located in Varallo, displays scenes from the life and passion of Christ and culminates in a representation of Paradise. Special points of view and openings allowed visitors, as with later dark rides at theme parks (see MEDIA), to experience the episodes from a first-person perspective: in the Nativity scene, for instance, visitors saw the scene from the perspective of the baby Jesus, as if the Three Kings were kneeling in front of them.

It has also been argued that the late medieval banquet or feast was a multimedia event that could be compared to contemporary immersive worlds.

> The mixed media of feasts breaks down the increasingly porous modern boundaries between high and decorative art, theatre, and music. [...] The immersion of medieval audiences in the interactive world of banqueting led both theorists and planners to grapple with the problem of how precisely such experiences might shape audiences' characters, a concern that arises explicitly in the discourses of ethics and courtesy and implicitly in the actual staging of feasts. (Normore 2015, 4)

This also applies, as scholarship has highlighted, to medieval jousts and tournaments, where performances of dance and bravura, displays of exotic animals, and exhibitions of representatives of other peoples and cultures played an important role in the performance of wealth and power (Szabo 2006, 28–29).

2.2 The Evolution of Garden Architecture

During the Renaissance, it was once again garden architecture in particular that was employed to create enclosed spaces (with limited access) that sought not only to entertain visitors, but also to communicate specific messages, often of a moral nature. The most famous example is probably the Sacro Bosco of Bomarzo, Italy. Commissioned by the prince Vicino Orsini and completed in 1547, the park covers hectares of woods and features buildings and sculptures inspired by Classical Antiquity. Their selection, however, is unusual and has generated huge discussion among scholars; while some have suggested that the entire park should be read as a symbolic path of initiation, most scholars now agree that its main purpose must have been to entertain and astonish guests and visitors: "the narrative allusions or topoi of historical gardens may be less important than their status as unstable, multisensual environments in which the visitor is enjoined and empowered to make his or her own decisions" (Morgan 2016, 139). This particularly applies to the "casa pendente," a leaning building whose floor, walls, and ceiling do not meet perpendicularly, and which thus creates a strong sense of loss of balance in those who enter it. Perhaps equally famous is the "Orco," a huge stone face with an open mouth, through which visitors can reach an inner room with a table and benches. Within the room, the specific acoustic environment distorts any sound, including the voices of visitors, thus creating an uncanny and scary environment.

Such gardens became common throughout Europe over the following two centuries and featured ever more of such "follies"—extravagant buildings realized for decoration, but which suggest another function, generally connected with their external appearance. The neoclassical style in particular contributed greatly to the creation of landscapes of pleasure in which imitations of ancient buildings, reproductions of ruins, or simulacra in the Italian style could be found. These were especially frequent in Great Britain. Chiswick House and Gardens from the 1720s, for example, sought to recreate a Roman landscape: its owner, Lord Burlington, bought ancient Roman statues (allegedly also from Villa Adriana) that were placed next to copies of ancient and modern Roman-Italian buildings (e.g. the Ionic Temple, a reproduction of the Pantheon).

Antiquity was a popular, but not the only theme: Painshill Park in Surrey, England (1738–1773), for instance, comprises not only ancient and Italian, but also Chinese elements (very popular at that time), medieval themes (the Gothic tower, the Gothic temple, and the Abbey), as well as a Turkish tent. Ancient Egypt, represented by pyramids and obelisks, was also popular. It has been argued that many of these motifs had specific symbolic meanings, either in connection with Freemasonry or as metaphorical passages from darkness to light. While this may be true in some cases, there is no doubt that all these landscapes were also meant to be immersive, themed environments with historical and "cultural" themes, aimed at entertaining and educating carefully selected visitors, as access was not open to everybody (Harwood 2002, 55–59).

> To move through Kew or Painshill, Belton or Badminton in the eighteenth century, and to see the rotundas and inscribed obelisks, the pagodas and Turkish tents, the ruins and hermitages that populated such sites was to be carried back and forth across space, time, cultures, and ideas. To respond to them most effectively was to employ the dynamic mental mechanics of memory and association and to bring to bear upon them the learned, interpretative lineaments of taste. And it is certainly the case that memories and related ideas of association continue to lie at the core of our experiences in theme parks. (Harwood 2002, 51–52)

The biggest and most elaborate parks of this kind were, obviously, those commissioned by the European royal families. Kew Gardens in London, realized in 1757–1772 in precisely the form in which it can still be admired today, is structured as a journey through the world and through time; William Chambers, the architect, included the famous Chinese pagoda, the Temples of Aeolus, Arethusa, and Bellona, as well as the Ruined Arch. Once a week, Kew Gardens was opened to the public. Since the entry was free but transportation to the park was quite expensive, access to the gardens remained, as in the case of modern theme parks, restricted to the middle class or those living nearby (Quaintance 2002; see INCLUSION AND EXCLUSION). Around the same time, Park Sanssouci in Potsdam, Germany, was constructed for

Frederick the Great, and worked in roughly the same way. Here, too, ruins on the Ruinenberg, inspired by the Forum Romanum, allowed a panoramic view of the castle within the park, and the classicizing temples and obelisks coexisted with two Chinese-themed structures, the Chinese House and the Dragon House. Frederick William IV added more buildings, among them the Roman Baths, built in 1829–1840 in Pompeian style, although never equipped with water. In Versailles, between 1783 and 1788, Marie Antoinette commissioned the Hameau de la Reine, a private area of leisure for the queen in the shape of a rustic village with a farmhouse, a mill, a barn, etc.

Another form of a park was the pleasure garden, a primarily commercial institution which developed in the seventeenth century, often in connection with country taverns, and often also as a place of prostitution. The pleasure garden provided entertainment to everyone who could afford to go there and also often displayed classicizing sculptures, follies, etc. Vauxhall Gardens, perhaps the most famous example, likely opened sometime before 1660 (the first mention is from 1662). Originally the gardens, then called "Spring Gardens," were free, their maintenance being financed by the sale of food and drinks within the venue. Yet they could be reached from London by boat only and, as in the case of Kew, this ensured that only those who could afford transportation visited the site. In 1785 an admission fee was introduced, which was quickly raised so as to control visitors' social standing ever more strictly. Meanwhile, new entertainments had been introduced (balloon rides, fireworks, shows, etc.), and season tickets were also available. Closed for the first time in 1840, the gardens reopened one year later before closing forever in 1859 (Coke and Borg 2011).

The Prater in Vienna, Austria, developed in a rather similar manner. Originally a hunting area, it was opened to the people in 1766 by the emperor. Drinking and eating facilities appeared almost immediately, and by the end of the century the Prater was not only open 24 h, but also hosted fireworks shows, balloon rides, etc. Commercial pleasure gardens were also successful in the U.S. and could already be found in New York in the eighteenth century. Their owners made their money from selling food and drinks rather than from charging an entrance fee. Due to competition, they constantly had to offer new entertainments, from circus acts to music, balloon rides to fireworks, and mechanical rides (Schenker 2002). Among the latter, the carrousel developed from a device used for training knights and hunters, as well as for the entertainment of the courts, into a machine devoted entirely to entertainment that was mainly aimed at children. The first one was opened in 1620 in Plowdiw, in the Ottoman empire. Starting from the eighteenth century, carrousels appeared more and more frequently in such popular contexts as commercial parks or at yearly markets (Szabo 2006, 131–132; Szabo 2009, 278–279).

Also crucial to the evolution of amusement parks, albeit in a different way, were the developments in France, exemplified by the Jardins de Tivoli. Starting in 1766, Simon Gabriel Boutin built a park with several houses, plants of different origins, and follies of various kinds. The complex was named "Tivoli" after the Italian town where both the Villa Adriana and the Villa D'Este were located, the latter a sixteenth century villa with magnificent gardens and many sculptures taken from the Villa Adriana. During the French Revolution, Boutin was guillotined and his gardens confiscated. They were opened to the public in 1795, with panoramas, marionette shows, magic lanterns, and even a bath; in this form, the park remained open to the public, even after Boutin's heirs recovered their property, until 1842. Tivoli was not the only park to become public due to the Revolution; it serves as an example of how this political development made such environments instantly accessible to a much broader public. As a result of the French Revolution, Germany saw the development of the Volkspark ("people's park"), public gardens which were designed to bring people into contact with nature and enjoy themselves. Yet they also functioned as safety valves, places of controlled compensation and education (with areas for sport, strolling, and sculptures aimed at historical education), lest what happened in France should repeat itself in Germany. The first Volkspark was opened in Munich in 1789; in the following decades, they increasingly targeted workers, allowing them to get some rest, practice sports, and spend their free time (Keßler 2012, 136).

2.3 The Redoute Chinoise in Paris

While individual buildings, follies, and attractions in these parks were themed, the gardens and parks as a whole were generally not. One early exception to this was the Redoute Chinoise, which opened in 1781 as part of the Saint Laurent fair, on the Parisian boulevards, much smaller than Tivoli or Vauxhall and themed to China through its architecture, decorations, and even exotic plants (Heulhard 1818; Alayrac-Fielding 2017). At the entrance stood a pagoda, similar to those constructed in aristocratic gardens; next to it, a cave reproduced a supposedly natural Chinese setting and housed a café. A "Japanese road," decorated with large Chinese porcelains, led to the dance hall, celebrated as "the largest piece of Chinese architecture that has yet been executed in France" (N.N. 1784, 253); the ceiling was decorated with Chinese lanterns and paintings by French artist François Boucher (famous for his Chinoiseries). Visitors could enjoy the "Chinese swing," pushed by two people dressed in a Chinese style. There was also a Chinese merry-go-round, decorated with parasols and pagodas, dragons, and Chinese characters—similar to the one installed by Marie Antoinette in 1776 at the Petit Trianon—a Chinese acrobatic show, a Chinese shadow-puppet show, and Chinese fireworks. The Redoute offered several types of supposedly Chinese experiences to the public, as well as an immersion into what was not a Chinese world, but rather a world of Chinoiseries. The Redoute did not claim to duplicate an existing Chinese location, but instead used Chinese motifs as they appeared in Western paintings and Chinese porcelains to provide a chic, fashionable, exotic, and picturesque setting for pleasure and entertainment. As such, it was more about fun than authenticity (see AUTHENTICITY).

3 Eastern Antecedents

Also rather different is the evolution of pleasure gardens and "themed environments" in Asia, which for a much longer time than their Western equivalents were accessible to only a limited group of visitors. In Japanese gardens, for example, the technique of "mitate" was "used to associate objects of ordinary life with mythological or classical images familiar to all literate people" (Yamaguchi 1991, 58). Such gardens have apparently been common in Japanese culture from the first century BCE to the present day, and intended to represent sometimes a distant abode of the gods, at times (such as a 1629 garden in Tokyo) the much-missed homelands of lords required to live in the capital. Joy Hendry points out (2000a, 195) that the same approach underlies those *tema paku* that display foreign or past places today (see below).

In China, the Yuanming Yuan ("Gardens of Perfect Brightness"), also known as the Old Summer Palace, were built during the eighteenth and nineteenth centuries some 8 km north of Beijing as the main residence of the emperors of the Qing Dynasty. Its name is misleading: the complex was used all year round and was not a palace, but rather a large walled area, five times bigger than the Forbidden City in Beijing, which included gardens, lakes, canals, and numerous buildings (palaces, pavilions, temples, galleries, etc.). It served not only as the residence of the emperor but also as a ceremonial place for interactions with the Chinese higher administration and the reception of tributaries and embassies, including European ones, such as the Dutch mission in 1795. The Summer Palace also contained a library and a museum that hosted a collection of precious antiques, artworks, and books from China and abroad.

At least some parts of the Summer Palace could be considered a theme park *avant la lettre*. Of the more than 150 scenic units of the Yuanming Yuan, many replicated famous Chinese landscapes or buildings: for instance, craftsmen from Southern China copied five large southern-style gardens. The units were meant to be walked in and/or viewed from designated viewpoints and pavilions. Such themed environments were observed, visited, and experienced by the emperor, his family, and his guests for the purpose of education and entertainment: the very strict Chinese imperial protocol did not allow the emperor to meet his people and travel abroad—hence, these themed environments provided him with the only way to escape from his role and experience the world, at least virtually. Therefore, usage of the Summer Palace was restricted to the emperor, his family, and domestics (more than 500 eunuchs). It was never opened to the public and only a very few privileged guests were allowed to visit. A "real" farm with domestic animals, a rice field, a weaving mill, and a silkworm facility were operated by court eunuchs, not so much for production but to give the emperor the opportunity to observe farm work and the daily life of his subjects. In the same spirit, a mock village with a "'Disney' market" (Wong 2016, 139) was simulated, where eunuchs played the role of shopkeepers, customers, thieves, and policemen, quarreling and making business for the education and pleasure of the emperor and his family.

The most famous part of the Yuanming Yuan was a park called "The Western Mansions." This remote and isolated section included about 30 "Western attractions"—not just Western-style buildings, but also fountains, an aviary with exotic birds, a labyrinth, and French-style gardens, all designed for the emperor Qianlong by Italian Jesuit and painter Giuseppe Castiglione, and constructed mainly between 1747 and 1751, with additions until 1773. The so-called "Palaces" were mainly baroque facades to be looked at from the outside, and museums, or rather storage facilities, where the emperor collected his precious artifacts imported from Europe, including clocks, automata, and six Beauvais tapestries sent by Louis XVI. Baroque architecture was given a light oriental touch by decorative details such as colored glazed tiles, but was based on European materials (stone), designs (columns), and techniques (glass windows). Borrowing from European gardens and theater, some of these scenic spots used not only facades but also painted panels representing urban landscapes and forced perspectives to create depth and give the emperor the illusion of watching a real European city. Just one building served as a residence, namely that of the Fragrant Lady, one of the Emperor's concubines, who came from Turkestan. A contemporary engraving shows her seated with the Emperor in front of a building known as the Belvedere, wearing a European outfit; another painting shows Chinese ladies in the same costume playing checkers in a Europeanized interior: the Western Mansions were thus also a place to experience the European way of life. There are striking similarities between the Chinese farm and village in Yuanming Yuan and Marie Antoinette's Hameau de la Reine, as well as between the Western mansions in Yuanming Yuan and the so-called Chinese pagodas or tea houses in the European aristocratic gardens of the eighteenth century, yet theming in Yuanming Yuan operated on a scale which had no equivalent in Europe at that time.

The Old Summer Palace was looted and burned by French and British troops during the Second Opium War in October 1860 as a retaliation for the torture and execution of 20 Western prisoners. As they were made of stone rather than wood, Yuanming Yuan's Western Mansions were not entirely destroyed by the sacking and the fire: some of their ruins are still visible today. Several wooden buildings have been reconstructed, and the canals and lakes have been restored. A New Yuanming Palace was opened in 1997 in Zhuhai (Guangdong), which reproduced 18 scenic spots from the original Summer Palace at the same scale. In fact, the Old Summer Palace has now become a real theme park (Campanella 2008, 258–262).

4 Western Antecedents in the Nineteenth and Twentieth Centuries

4.1 The Sublime and the Picturesque

Starting with the mid-eighteenth century, the development of new aesthetic standards for appreciating natural beauty, such as the sublime or the picturesque, impacted the structure and appearance of public pleasure grounds. Early articulations of the idea of the sublime include Burke's *Philosophical Enquiry into the Origin or Our Ideas of the Sublime and Beautiful* (1757) and Kant's *Observations on the Feelings of the Beautiful and the Sublime* (1764). The picturesque, meanwhile, found its most vigorous expression in such works as William Gilpin's *Remarks on Forest Scenery and Other Woodland Views* (1794) and Sir Uvedale Price's *Essay on the Picturesque, as Compared with the Sublime and the Beautiful* (1794). A joint product of Enlightenment ideas (such as primitivism) and Romanticism, this reevaluation of nature as wild and free would lead to new landscaping traditions, away from the formal style previously favored. Somewhat reflecting the inherent ambivalence of the picturesque, which Gilpin defines as "that kind of beauty that would look well in a picture" (1798, 238), landscape architects would now strive for artful and illusionistic representations of seemingly natural landscapes. Those principles came to be most clearly articulated by English landscape gardeners—most notably Lancelot "Capability" Brown, whose work provided British landscape architecture with its basic tenets. His serene, natural-looking parks intended to make the best possible use of the site's "capabilities" (hence Brown's moniker), complete with unobstructed views of the countryside and meandering pathways meant to convey an element of surprise and discovery. Brown's successors would eventually place more emphasis on flowers and exotic plants—as in Humphry Repton's gardens—as well as more visually arresting and overtly picturesque elements such as ruins or bridges, as Price in particular recommended (Figueiredo 2014).

It was not long before John C. Loudon became the instigator of a British parks movement, bringing private pleasure gardens into the public realm as well as promoting parks as central to urban planning efforts. Other nations were quick to take note and public parks soon came to be assigned clear practical functions, at once sanitary and social: in their joint effort to promote the good health (and productivity) of working-class laborers and pacify the urban poor, these parks not only testified to the dominant class's values but also to a perceived and growing need to accommodate the masses' leisure time to the point that their construction soon became a matter of civic duty. Unsurprisingly, public parks were first systematically developed in England: while originally financed privately, they were soon funded directly by municipal governments—as pioneered by Sir Joseph Paxton's (1803–1865) Birkenhead Park (1847), near Liverpool. Birkenhead Park was modeled after Regent's Park in London, which had initially been developed in the 1810s as a residential development for a planned 56 villas and was opened to the public for 2 days a week in 1835. France and the United States picked up on the effort almost simultaneously. While the French Revolution had already forced the opening of private pleasure gardens like the Jardins du Luxembourg or Parc Monceau in Paris, Baron Georges Haussman's ambitious plans for the city included a number of parks that were designed as authentically public realms and clearly drew on Napoleon III's earlier exile in London: originally outside the city limits (such as the Bois de Boulogne, 1857; the Jardin d'Acclimatation, 1860; or the Bois de Vincennes, 1860), these parks would later find their way into the city (Parc des Buttes-Chaumont, 1867; Parc Montsouris, 1869) and were all erected under the supervision of engineer Adolphe Alphand. In the United States, the indefatigable Frederick Law Olmsted spawned an entire park movement, whose starting point and best-known example is New York's Central Park (1857). The legacy of the U.S. park movement is immense: public parks not only contributed to the development and democratization of spaces for leisure, but they also point to the increasing place of recreation in American society that theme parks later came to symbolize and capitalize upon.

In his Preliminary Report for the preservation of Yosemite Valley (1864), Olmsted also provided the earliest rationale for the creation of national parks—though Yellowstone was the first piece of federal land to earn the status in 1872. While original calls for preservation made national parks ill-suited to accommodate anyone but a few elite tourists, the Sierra Club (founded in 1892 by naturalist John Muir, widely held as the "father of natural parks") did much for the construction of roads, camping grounds, and other accommodations to democratize their access during the 1920s. Complementing the larger, more central urban parks were a myriad of neighborhood parks that were created in the late nineteenth century, and whose primary purposes were now physical exercise as well as close proximity to residential neighborhoods, especially in the more deprived parts of towns. These public efforts were soon to compete with private initiatives: in approximately the same period, operators of the electric trolleys started building picnic grounds and amusement parks at the end of their lines to attract riders on weekends. As a result, most major Northern (and a handful of Southern) cities boasted such trolley parks by 1910 (Wolcott 2012, 10). Not long after, the Great Depression forced the government back into action, with the Work Projects Administration investing a billion dollars in the creation of 40,000 sports and recreation facilities (e.g. skating rinks, swimming pools, and school gymnasiums) within half a dozen years.

Equally as significant, Olmsted's urban parks not only inaugurated America's new age of mass leisure, but they also laid out the foundational principles later found in theme parks: Olmsted's creations were meant to foster a sense of "gregariousness" (Olmsted 2010, 224) or togetherness that early critics found central to Disneyland's success (Moore 2004). They also strove for the same brand of immersive escapism that was especially conducive to the visitor's "suspension of disbelief" (Figueiredo 2014, 150). The pastoral style favored by Olmsted (in essence, the illusion of the countryside) was meant to visually keep the neighboring city as far away as possible, in order to allow for the contemplative reverie and "unconscious recreation" that he deemed the best antidote to the demoralizing influence of urban living.

4.2 European Amusement Parks

Public parks in Paris had picked up directly from the Jardins de Tivoli. While the latter's novelty effect had worn off and the public had lost interest by the 1840s, Tivoli represents the first combination of entertainment and landscaped settings and as such, probably the first Western articulation of our modern understanding of an amusement park. And while entry was not free, they opened up forms of entertainment that were previously available only for nobility to vast portions of the public, making them the harbinger of a new age of mass leisure. Once a private garden, Tivoli opened to the public in 1795 and soon came to accommodate live entertainment (e.g. folk dances, parades, fantoccini, shadow theater, singers, and fortune tellers), games (not only hide and seek, blind man's bluff, but also see-saws, greasy poles, and in a nod to the amusements of the nobility that inspired the park, ring-tilts), as well as spectaculars (such as the especially lavish "Descente d'Orphée aux Enfers," created in 1798), fireworks, and nighttime illuminations. Opportunities for intimate encounters were plentiful in moderately lit parts of the park. The attractions on offer gradually expanded to include modern wonders of science and engineering such as balloon flights or increasingly complex mechanical rides, notably early versions of the Ferris wheels (known as "katchelis") or roller coasters (which first appeared in 1817 as "Promenades aériennes" in a competing park).

Not long after the closure of Paris's second Tivoli in 1842 (political instability, high operation costs, and the public's fickle tastes made them precarious affairs), the concept found a permanent home in Copenhagen's own version of the place, whose original name—Tivoli and Vauxhall—evoked its French and English inspirations. Founded in 1846 in Copenhagen by Georg Carstensen, the son of Denmark's consul general in Algiers, today's Tivoli still incorporates the same blend of live entertainments, amusements, and rides in a fanciful, landscaped setting (for a complete history of Tivoli Gardens, see Langlois 1991). Incidentally, Copenhagen's Tivoli also traces the clearest line from English pleasure gardens and French "jardins spectacles" to the modern theme park industry, as Walt Disney himself paid the place a visit in 1951 in preparation for his own effort at Disneyland (Marling 1997, 47).

4.3 Pageants and Spectaculars

No account of theme parks' history would be complete without a reference to pageants and spectaculars—especially Buffalo Bill's Wild West Show and the countless competitors that it inspired. Preceding the Western film genre, William Cody's show soon established itself as one of the major purveyors of Western imagery in the U.S.—one that would become central to theme parks, from Knott's Berry Farm to Disneyland's "Frontierland," to name but a few. Likewise, the show's fictionalized rendition of American history helped turn the latter into a subject of popular entertainment. The show's capacity to transport the audience to a different time and place owed much to its monumental staging (Cody's was an epic rendition of the West) and its recurring claims to authenticity, presenting reenactments of historical events as fine substitutes to the real thing (see AUTHENTICITY). By the early twentieth century, Cody's version of the West had so effectively captured popular imaginations that Wild West pageantry was somewhat elevated to the status of patriotic entertainment. Theodore Roosevelt's 1901 inauguration ceremony thus staged a "pageant made up of traditional military and civilian elements," including such representatives of Western life as General Custer's old Seventh Cavalry or six "wild" Indians in full tribal dress—among whom was Geronimo, received "as a star and not as a murderer" (McVeigh 2007, 27–28).

Many other spectaculars followed—most prominently not only the Real Wild West of Miller Brothers 101 Ranch (1907–1927), but also the Campbell-Bailey-Hutchinson Circus and Wild West Combined (1920–1922), or the Wheeler Brothers Greater Shows and Famous Stampede Wild West (1921). Buffalo Bill's show, however, provided the prototype upon which all of these would be modeled. A former gold prospector, army scout, and buffalo hunter for the Kansas Pacific Railroad (hence his alias), Cody's life had become the stuff of legend by the late 1860s, loosely providing the material for Ned Bluntine's 1869 dime novel *Buffalo Bill, the King of the Border Men*—the first of 1,500 Buffalo Bill dime novels published during Cody's lifetime. Building on the tendency of Western folk heroes to self-mythologize, Cody would go on to play himself in Ned Bluntine's 1872 play *The Scouts of the Prairie* (McVeigh 2007). The show's triumphant reception inspired Cody to organize his "Buffalo Bill Combination" in 1873—a troupe of cowboy and Indian actors who dramatized adventures from the dime novels and reenacted events in Western history.

By the time Cody introduced his first proper "Wild West Show" in 1883, the show's formula had been standardized to include three acts, including an Indian horse race (with as many as ten Indian competitors), "historical" scenes (such as reenactments of the Pony Express or attacks on a stage coach), and demonstrations of "real" or "natural skills" such as bronco riding, rope demonstrations, and shooting acts, notably by legendary sharpshooter Annie Oakley (Warren 2005, 222). Another characteristic of the show was its claim to realism, as indicated by its cast of illustrious figures from Western history (such as Sitting Bull or Calamity Jane, among many other such legendary names) or lavishly illustrated programs. The programs contained articles on Western history, Native spirituality, and portraits and biographies of great Civil War scouts and frontiersmen. In a statement evocative of the show's documentary ambitions, the booklet for the 1893 season stated that

> The exhibitions given by 'Buffalo Bill's Wild West' have nothing in common with the usual professional exhibitions. Our aim is to make the public acquainted with the manners and customs of the daily life of the dwellers in the far West of the United States, through the means of actual and realistic scenes from life. (qtd. in Kasson 2001, 115)

As original in tone and content as they might have been, Cody's shows truly expanded on Phileas Taylor Barnum's pioneering efforts in the circus business—only for Cody to eventually "out-Barnum Barnum" (Ashby 2012, 80). A consummate showman, Barnum first made a name for himself as a museum operator and impresario—one famously prone to exaggerations and half-truths. True to his taste for sensationalism, his New York museum's collection was meant as much to edify as to entertain popular audiences, prefiguring a blend later found at Disney theme parks (Culhane 1991, 36) and contributing enormously to the development of American popular culture. Puppet shows and models of great cities rubbed shoulders with waxworks and, perhaps most famously, exotic creatures and human curiosities such as the "Feejee Mermaid" (in reality, the mummified upper body of a monkey sewn onto a fish tail) or General Tom Thumb ("the smallest person that ever walked alone").

Following the museum's second fire and permanent closure in 1868, Barnum's venture to Americanize the circus and modernize its mode of operation eventually gave rise to the first "industrial-entertainment complex": Barnum's business was the first to extend near-industrial production modes to the entertainment sector, with the adoption of technological, managerial, and logistical innovations that emphasized scale, efficiency, and profitability. By tailoring his circus to train transportation,

Barnum capitalized on the railroad's expanding network, enabling him to tap into the national market and deploy operations at a scale never seen before. Barnum associate W.C. Coup's game-changing flatcar and wagon system especially "enabled laborers to roll fully loaded circus wagons on and off the train" and allowed the circus to evolve into a fully optimized "mammoth railroad outfit" (Davis 2012, 33) whose dominant position was later furthered by a merger with rival James Anthony Bailey (1881). Much in keeping with changes typical of the U.S. industrial age, the unprecedented scale of Barnum's railroad circus required the development of complex managerial structures along new lines of labor division, with highly specialized teams operating in the fields of advertising, market research (to determine the circus's route and stops) and, crucially, logistics: such were the circus's supply and transportation feats that it repeatedly elicited the curiosity of the military, both at home and abroad (Davis 2012, 47). The sheer scale and impact force of Barnum's circus helped establish it as a prime agent of the nascent mass culture in the U.S.

Barnum's efforts were thus directed at increasing the circus's respectability and mass-appeal as a family-friendly entertainment: to eschew the designation of his show as a circus and then-common charges of vulgarity, he named his show P.T. Barnum's Great Traveling Museum, Menagerie, and World's Fair—though in reality, with its display of female performers' bodies, his spectacle still walked a fine line between respectability and titillation (Ashby 2012, 76–77). Just as characteristic of its mass-audience orientation was the show's awe-inspiring scale—as reflected in the gradual introduction of the two-, and later, three-ring circus, not to mention additional attractions such as the menagerie or its outer hippodrome track. The goal for the three-ring circus was to accommodate ever-larger audiences, with more expensive seats facing the center ring and others relegated to the outer rings' side shows. With two shows ranging from 20 to 24 acts, each of the circus's stops typically drew a cumulative audience of 10,000–15,000 spectators.

Alongside circus attractions, spectaculars also often developed controlled "catastrophe scenarios"—a theme that would also become recurrent in theme parks. An important example are the pyrodramas representing the eruption of Mount Vesuvius and the destruction of Pompeii. The most famous was produced by James Pain, a British fireworks manufacturer, and together with the special effects featured dances and athletic displays. It toured throughout the world, and was presented regularly at Coney Island between 1879 and 1914. A one-time publicity stunt was the Crash at Crush, organized in 1896, when two trains were made to collide in front of a crowd of 40,000 spectators. In spite of an unexpected explosion after the collision and two ensuing deaths, the stunt was a huge success and thus found many imitators over the following years. The controlled catastrophe or "riskless risk" (Wright 2007, 247) has thus become a staple of the modern entertainment industry and theme parks that has a clear and strong genealogy in nineteenth century spectacle culture.

4.4 Panoramas and Dioramas

Of course, Cody's appeal to audiences on the U.S. East coast was built on previous representations of the American West—such as touring exhibitions of monumental paintings like Albert Bierstadt's *The Rocky Mountains, Lander's Peak* (1863) and, even earlier, Catlin's Indian Gallery (1838), whose paintings were often accompanied by live Indians performing ritual dances or songs, in a "mixture of instruction and commercial entertainment" that some have called "possibly the first version of the Wild West Show" (Buscombe 1988, 61). Typical of Bierstadt (and others associated with the Rocky Mountain School of Painting), nature's every detail is captured with new clarity and accuracy by an "all-knowing eye," in a clear replication of the sort of "optical inventories" (Oettermann 1997, 132) already on display in the panorama.

The panorama (and its distant cousin, the diorama) indeed form part of a long string of immersive "spaces of illusion" (Grau 2003)—albeit the first to operate as mass media and adopt "the methods of industrial, profit-oriented production" (Grau 2003, 66). Pioneered by the English painter Robert Barker, who patented his invention in 1787 under the name of "la nature à coup d'œil," the panorama first opened as a permanent attraction in London inside a purpose-built rotunda in Leicester Square in 1793. Perched on a viewing platform, visitors were thus able to contemplate a gigantic, circular painting (on average 300 ft in length and 40–60 ft high), whose edges remained carefully hidden from view to enhance the scene's sense of reality. The canvas's top rim was concealed by an umbrella-like velum that allowed the rotunda's skylight to bounce off the canvas, as if the painting itself were the source of light. Initially hidden by the platform itself, the bottom edge later came to blend with "faux terrains" filled with three-dimensional objects, an innovation introduced in 1830 by Charles Langlois. Favorite themes included cityscapes (viewed from above, as if in a balloon or a spire), foreign landscapes, and historical scenes (most prominently, famous battles)—the latter category allowing panoramas to serve as imaginary journeys in both space and time.

Panoramas were quick to enthrall audiences internationally, and soon developed as attractions in not only Britain, but also Germany, Austria, the U.S., and France, where their success was such that they briefly became the object of speculative panorama stock companies. By then, canvas production had attained near-industrial efficiency, allowing operators to set up several exhibitions a year, and even international tours. Perhaps the most spectacular of all of these panoramas was London's Colosseum—at the time the biggest that had ever been attempted. After the bankruptcy of its original owner and near-certain demolition, the Colosseum reopened in 1845, this time complete with a whole array of amusements meant to make it a one-stop attraction for a Londoner's every whim. Reflecting Victorian tastes for curios, it accommodated a museum of sculptures (available for purchase), an exterior promenade along simulated ruins, greenhouses filled with exotic plants, a Swiss cottage flanked by a real waterfall, an artificial ice rink, and a reproduction of the grotto of Adelsberg (Oettermann 1997, 132–139).

Building on mass audiences' tastes for illusion, which was evident from the interest in panoramas, dioramas were invented in the 1820s by Louis Daguerre—then a stage designer—though most likely inspired by Franz Niklaus König's diaphanorama pictures. Standing on a rotating plate, the audience was moved from one scene to another (and occasionally a third) and treated to a 15-min light show, for example simulating the passage from day to night. Even more spectacular were Daguerre's "double-effect" dioramas (invented in the mid-1830s): scenes painted on both the front and the back of a translucent canvas simulated motion when hit by special light effects, thus allowing scenes to come to life. One such diorama, the Church of Saint-Etienne du Mont, for example, begins to fill with worshippers as darkness falls (Oettermann 1997).

Another way for operators to suggest motion was to unroll long, continuous landscape paintings past the audience, as if observed from a car window. While invented almost simultaneously in England, Germany, and the U.S., the emphasis of the "moving panorama" on travel made it especially well-suited to a New World consumed with territorial expansion—to the point where it has been called "an art form for American tastes" (Oettermann 1997, 323). John Banvard's famed mobile panorama of the Mississippi River (introduced in 1846 and advertised as being "three-mile long") paved the way for many imitators, firmly entrenching the medium's association with westward migration with three themes that dominated the industry: travel on the Mississippi, overland routes to the West, and booming San Francisco. The incomparably superior experience of the moving panorama out a train's window eventually dealt the moving panorama a fatal blow in the 1860s (Warren 2005, 258).

4.5 Ethnographic Exhibitions, Zoos, and Living History Museums

The panorama's spirit lived on, augmented by live animals in yet another inventory of the natural world, the zoo—whose modern incarnation owes much to German entrepreneur Carl Hagenbeck. Beginning in 1874, Hagenbeck expanded from the international trade in exotic animals (supplying zoos and circuses worldwide, including Barnum's) to the staging of touring ethnographic exhibitions (*Völkerschau*). Rather notably, some of these were in the style of Cody's Wild West Show, such as his Oglala-Sioux exhibition (1910)—Hagenbeck's most successful venture in the area, with 1.1 million visitors (Ames 2009, 65). Not content with the exhibitions' usual presentation, Hagenbeck added to an already successful genre by creating immersive, fully furnished "native villages"—artful, yet natural-looking presentations of the natives' daily lives (in reality, the humans and even animals were thoroughly trained to perform the drama of their everyday activities). Eventually branching out into the zoo sector, Hagenbeck designed his Tierpark in Hamburg (Germany, 1907) as a new benchmark for zoological gardens around the world: setting out to demonstrate the practicality of open-air acclimation pens (as opposed to the comparatively expensive, heated buildings of earlier menageries), Hagenbeck thus hoped to encourage the construction of private or public zoos around the world and boost his animal trade business in the process (Rothfels 2012, 182).

Breaking from the scientifically oriented zoos, two major innovations guided the Tierpark's design: they not only made Hagenbeck "a kind of storyteller and the creator of a fictional universe" but "aligned [his zoo] with the kind of 'creative geography' that was being pioneered in the cinema and would later inform the Disney theme parks, as well" (Ames 2009, 126, 173). One such design principle was the zoo's use of immersive landscaping to create the illusion of animals in the wild. Animals (and occasionally people) were staged as part of tableaus, "colossal open-air panorama[s]" (Ames 2009, 141) representing the animal's natural habitats and drawing from Hagenbeck's own experience with traveling animal panoramas—such as his wildly successful Norwegian sea panorama, filled with seals, sea lions, polar bears, and Arctic birds (1896). Just as important was the use of "open enclosures" which replaced iron fences and bars with faux terrain—ponds and hidden moats that kept animals and visitors safely separated yet sustained an illusion of proximity.

This was the key to the Tierpark's second guiding principle: its entertainment value. In a "controlled modulation of distance and proximity, risk and safety, destabilization and reassurance" (Ames 2009, 179), the park aimed for a sense of thrills and wonder, not just as a result of the visitors' seemingly close contact with animals (and especially dangerous ones, as with Predators' Ravine), but also of the captivating (yet unscientific) mixing of prey and predators in the same display space. As made clear here, Hagenbeck's animal park was infused with theatricality and drama, and much in the style of his earlier native villages, the presentation of animal life remained primarily subordinated to a narrative: in a tale of Edenic nature, "animals would live beside each other in harmony and the fight for survival would be eliminated," as Hagenbeck had it (Rothfels 2012, 163; Steinkrüger 2013, 186–202). The zoo's entertainment orientation was soon made even more evident with the opening of the Tierpark's adjacent twin attraction, H. H. Park (pronounced "ha-ha" Park) in 1913—an "American style" amusement park built in collaboration with an outside company and open to the zoo's visitors for the price of a single admission.

From their very beginnings, theme parks have also been "time parks" (see TIME) and have created immersive environments that are temporally located elsewhere; beginning with Disneyland, theme parks have also displayed strong ideological underpinnings, seeking to reinforce the identity of the visitors (see WORLDVIEWS). It is not surprising, then, that living history museums have also played an important role in the genesis of theme parks. The first open-air, living history museum was Skansen, opened in Stockholm, Sweden, in 1891. Its creator, Artur Hazelius, felt the need to document and preserve the folkloristic traditions of all the various regions of Sweden, which he thought were threatened by puritanism, touristic development, and urban migration during the process of industrialization. After opening an indoor folklore museum, Hazelius, inspired by the Swedish and Norwegian cabinets at the industrial exhibition in Paris of 1857, started buying buildings from all over Sweden and had them moved to Stockholm, together with the plants and animals from the same regions. The result was a reproduction of Sweden, its fauna, flora, and traditional architecture; the tableaux were even staffed so as to showcase and preserve folk music and dances, handicraft traditions, etc. (Anderson 1984, 17–23; Conan 2002).

Due to the great success of Skansen, similar national open-air museums were quickly built in other northern European countries; later, the genre expanded to the U.S. Indeed, when Hazelius created a tableaux for the Swedish section of the Centennial Exhibition in Philadelphia (1876), it generated a great deal of interest on the other side of the Atlantic. At the beginning of the twentieth century, for instance, George Francis Dow dedicated himself to reproducing and displaying the folklore of New England in Salem, Massachusetts, thus creating the first American open-air museum. Such projects quickly attracted the attention of wealthy Americans such as Henry Ford, who funded the restoration of the Wayside Inn, or more famously John D. Rockfeller, Jr., who started the monumental project of Colonial Williamsburg. Williamsburg meant recreating an entire city in a historical style, not just a single building. By offering a much larger immersive environment, it allowed visitors to "time travel" to one specific time and place. Rockefeller agreed in 1926 to invest in the project; since 1932, Colonial Williamsburg has also featured staff in historical costumes, thus providing a further and crucial aspect of immersion: historical interpretation (Anderson 1984, 25–33; Wallace 1996, 13–15; Chappell 2002, 123–142).

The model provided by Colonial Williamsburg quickly expanded to many other living history museums created in the first half of the twentieth century, for example the Old Sturbridge Village, opened in 1946 in Sturbridge, Massachusetts, and inspired by the Scandinavian living history museums. The most famous among them, however, is Plimoth Plantation, established in 1947 in Plymouth, Massachusetts, as a reconstruction of the village of the Pilgrim Fathers. When Plimoth Plantation was expanded in 1959, it was peopled by mannequins; only in the 1960s was historical interpretation added to the open-air museum (Anderson 1984, 45–52). It is therefore important to highlight that the years during which Walt Disney conceived, planned, built, and opened Disneyland were the years of the great success and expansion of open-air, living history museums. The discussion about (re)living history, employing live interpretation, and reconstructing historical environments was very lively between the 1940s and the 1960s and surely had a strong impact on Disney's musings about his own creation and representations of history.

4.6 Country Fairs, Agricultural Fairs, World Fairs

Starting in the nineteenth century and continuing into the present, country and agricultural fairs have become another form of popular attraction across many Western cultures. A largely local event lasting from a few days to three weeks, country fairs typically combine sports, live music performances, animal shows, cultural programs, and, very importantly, regional products and foods. Their overall theme is thus consumption and their chief attraction is wholesome family entertainment for a relatively modest entrance fee. Some of these fairs have become famous and internationally known, despite their local focus and limited theme. Thus, the Houston Livestock Show and Rodeo presents the world's largest livestock exhibitions and rodeo every year, and attracts huge audiences despite hefty entrance fees. A local fair such as the Higher Ground Fair in Wyoming, on the other hand, is not very well known beyond the small town of Laramie where it is staged each fall, even though its stated mission is to showcase all six Rocky Mountain states encompassing Colorado, Idaho, Montana, New Mexico, Utah, and Wyoming as well as the Native First Nations. Mainly emphasizing local agriculture and gardening in a high-altitude environment, the small fair only lasts for 2 days and offers a marketplace for traditional and folk art as well as a variety of foods from the region. For entertainment, there are live music performances, kids' programs, and animal competitions such as the popular obstacle course race, with mostly willing lamas and alpacas. Typical of similar, recently established local fairs, the Higher Ground Fair also offers a space for community activism, for instance by offering educational programs and public discussions on such vital topics as shelter, energy, and food programs for the poor, among others. Hence, agricultural and country fairs play an important role in offering locals and visitors alike a space for casual interaction, consumption, and entertainment. What is more, they continue to serve an important function aimed at community formation and development.

By contrast, world's fairs and exhibitions are obviously much larger affairs with rather different goals, although they are also clearly performative in nature. Moreover, it is certainly possible to consider world exhibitions a form of mass entertainment, albeit literally a kind of world-class entertainment. This ambition is often conveyed by the fairs' names themselves, starting with the very first, "The Great Exhibition of the Works of Industry of All Nations," staged in London from May through October 1851, thus lasting for half a year, which has remained a typical running time for such massive events until today. The Great Exhibition attracted more than six million visitors and displayed the industrial and agricultural products of an impressive 28 nations. Famous for its central structure, popularly known as the Crystal Palace and chiefly designed by architect Joseph Paxton, the London exhibition pioneered its large-scale approach to exhibitions serving as an arena of peaceful competition for the leading nations of the world (an invaluable account of attractions in London over the previous 350 years is Altick 1978).

In 1901, U.S. president McKinley quipped that "expositions are the timekeepers of progress." And indeed, each world exhibition has introduced the latest technological innovations to curious mass audiences, ranging from Colt revolvers and the telegraph (London, 1851) to many other seminal inventions, including the hydraulic elevator (Paris, 1867), the typewriter (Philadelphia, 1876), and electric lighting (Paris, 1878). From these select examples, it becomes clear that world's fairs have always attempted to gauge the progress of cultures cast in a race for technological and industrial advancement, if not domination over other cultures. In this regard, it is striking that even to this day, world's fairs inevitably present the achievements of the world's leading nations next to those seen as still "developing," even though the rhetoric nowadays is far less imperialist or even militant than in the nineteenth century.

Instead, the baseline is one of ameliorative intent vis-à-vis other, problem-stricken regions in the world. For more than a century and a half, world exhibitions have served the larger purpose of presenting a certain world order that follows a clearly ideological agenda, be it imperialist as in the nineteenth century, capitalist as in the twentieth, or environmentalist as reflected by most exhibitions staged in the twenty-first century so far.

Another constant of most world expositions is the idealized environment they create, thus removing visitors from their daily surroundings and transporting them to a magical space promising a brighter future for all. A brilliant case in point is the World's Columbian Exposition, staged in Chicago in 1893 to commemorate the quadricentennial of Columbus's arrival in the New World. Its central area, referred to as the White City and chiefly designed by architect Daniel Hudson Burnham, provided an enchanted, idealized environment that helped many visitors forget, if only for a day, the less than desirable realities of the overcrowded and chaotic, slum-infested city that proudly hosted the fair.

4.7 Coney Island and Film Studios

Even more popular than the White City, however, was the Exposition's Midway Plaisance, a mile-long corridor of independently operated sideshows featuring exotic performances and mechanical amusements such as George Washington Gale Ferris Jr.'s Ferris Wheel. George C. Tilyou, a visitor to the fair from New York City, was so impressed with Ferris's invention that he installed a replica at Coney Island, where his parents operated a hotel. Owing to its proximity to the populous New York area, Coney Island had been a popular seaside resort since the first half of the nineteenth century (see Sterngass 2001), but the final decades of the century saw Coney Island evolve into a sort of permanent world's fair with grand hotels, piers, public plazas, and, around the turn of the century, several amusement parks that offered restaurants, concessions, ballrooms, sideshows, as well as carousels, dark rides, scenic railways (roller coasters), and other mechanical rides. Starting with the 1876 Philadelphia Centennial Exposition—whose 300-ft observation tower with its steam-powered elevators, as well as its Japanese, Brazilian, and U.S. government pavilions, were all moved to Coney Island after the closure of the fair—rides, shows, and entire buildings were frequently imported from world's fairs and their amusement zones to Coney. For instance, Coney Island inherited not only Tilyou's replica of the Ferris Wheel, which would eventually form part of his Steeplechase amusement park (1897–1964), from the 1893 World's Columbian Exposition, but also the neoclassical architectural design of the White City, which would be adopted at the Dreamland amusement park (1904–1911). The main attraction of Coney's third major amusement park, Frederic Thompson and Elmer Dundy's Luna Park (1903–1964), was the dark ride A Trip to the Moon, which Thompson and Dundy had originally created for the 1901 Pan-American Exposition in Buffalo and which played for one

Fig. 1 Illustration of A Trip to the Moon at Luna Park (Coney Island, New York City). *Image* GRANGER—Historical Picture Archive/Alamy Stock Photo

season at Tilyou's Steeplechase before Thompson and Dundy decided to use it to found their own park (see Register 2001; see Fig. 1).

Coney Island's amusement parks pioneered several concepts and strategies that would later be considered to have been invented by theme parks: they were all fenced in and charged a general admission to filter out undesirable customers (see INCLUSION AND EXCLUSION), enjoyed a symbiotic intermedial relationship with the early film industry, and entertained and dazzled visitors with their walkaround characters and parades during the day and their electric illumination at night. What is more, their eclectic offerings were often unified by a common architectural style—neoclassical in the case of Dreamland, Oriental in the case of Luna Park—or even a common theme such as the aquatic motif of Sea Lion Park (1895–1902; see Kasson 1983; Immerso 2002; Rabinovitz 2012). Drawing enormous crowds of up to five million visitors per season during their early years, Steeplechase, Luna Park, and Dreamland became the models for electric or amusement parks across the country and beyond (see Nye 1997; Rabinovitz 2012).

In the early twentieth century, entertainment and thrills also began to be offered by Hollywood film studios, which were promptly opened to paying visitors. As early as March 1915, Carl Laemmle, director of Universal Studios, opened his brand-new studio in Fernando Valley, known as Universal City, to the public. This was a way to diversify the studio's revenue (the entrance ticket cost 25 Cents), but mostly to advertise the studio and its stars and build a closer connection between the studio as a brand and the audience, at a time when the morality of the movie industry was under discussion. The tour included stunts and special effect shows, a visit to the zoo (all sorts of wild animals were used in the movies), and to the backlot sets. Visitors were allowed to wander from one stage to the next and, in some cases, to observe the process of filmmaking from the "visitors' observation platform." They might even have the chance to bump into a star and get an autograph, or at least mingle with extras and technicians. The introduction of sound recording put an end to studio tours, as the industry now needed quiet sets, which was incompatible with the presence of an audience used to cheering and booing the shows (see Jacobson 2017; Latsis 2019). Universal Studios did not reopen to the public until 1964, when it provided essentially the same kind of attractions as in 1915, including the simulated flash flood, which is still a hit today. In the following decades, however, film studios would adopt the forms and structures of theme parks, such as the Universal Theme Parks in Hollywood and Orlando.

5 The Evolution of Theme Parks in North America and the West since 1955

5.1 Destination Parks and Regional Parks in the U.S.

In the U.S., Disneyland's opening in 1955 greatly contributed to a drastic reconfiguration of the amusement and theme park sector, namely its concentration and division along national and regional lines. Factors that have contributed to the sector's concentration include the closure of local amusement parks in the late 1960s and early 1970s, which was partly the result of racial tensions, urban decay, and land speculation by former operators (see Wolcott 2012), as well as a series of sales throughout the 1980s and 1990s: initially lured by Disney's success, conglomerates eventually sold off their theme park divisions, opening new opportunities for mergers and acquisitions and leaving the sector in the hands of a few companies, with Disney left as the undisputed leader. For instance, the Six Flags parks were once the property of Bally (a pinball and slot machine manufacturer) and Warner Brothers, Paramount Parks the property of CBS, and SeaWorld that of publisher Harcourt Brace Jovanovich (Anton Clavé 2007). The situation has since led companies to operate almost exclusively in one segment of the U.S. market or the other: destination parks and regional parks.

As the designation suggests, destination parks are multi-gate (i.e. comprising several individual parks), multi-day destinations whose patronage is national and international (see ECONOMIC STRATEGY). The exclusive province of the Walt Disney Company and Universal, destination parks include Disneyland Resort (1955), Walt

Disney World (1971), Universal Orlando Resort (1989), and, to a lesser extent, Universal Studios Hollywood (1964)—all located in the Los Angeles and Orlando regions and each generating over five million entries per year. It was Walt Disney's original vision for his Florida "vacation kingdom of the world"—whose draws included not just a theme park, but a projected Experimental Prototype Community of Tomorrow (EPCOT), hotels, restaurants, and nighttime entertainment—that helped establish theme parks as multi-day destinations and made Orlando the world capital of themed entertainment. Walt Disney World now includes four theme parks and two water parks, not to mention Disney Springs (a shopping and nighttime entertainment complex, crucial to convincing visitors to stay overnight) and 36 resorts (26 of which are Disney-owned), adding up to 30,000 hotel rooms.

While Walt Disney World's scale remains unparalleled, both Universal Resort Orlando and Disneyland Resort in California have since adopted a similar positioning and developed into two-gate resorts (with the addition of Universal's Islands of Adventure in 1994, and Disney's California Adventure in 2001), complete with nighttime entertainment complexes and ample overnight accommodation. With no second gate and only its entertainment district Universal City Walk to attract multiday visitors, Universal Studios Hollywood has yet to develop into a fully fledged destination park. Thanks to their multiple gates, destination parks not only accommodate a growing number of visitors but also drive per capita spending by increasing the length of visits from one to several days while keeping the audience on property. Disney's Hollywood Studios (1989) and Disney's Animal Kingdom (1997) were thus both constructed in an effort to discourage visitors from attending competing parks—namely Universal Studios Florida and Busch Gardens Tampa (Younger 2016, 18). Likewise, the Walt Disney Company was quick to pull its support from a projected Mag-Lev train line from the Orlando airport to its Florida property once it became apparent the line might stop *en route* at competing sites (Foglesong 1999).

Regional parks, meanwhile, are often seasonal operations averaging from one to four million visits per year. Established in major metropolitan areas, their patronage is essentially local, as 80% of their visitors reside within a two-hour driving distance (see Anton Clavé 2007, 60). For the most part, they comprise sites established long before Disneyland or corporate ventures developed from the 1970s onwards. Three groups in particular have now come to dominate the regional parks market. After a period of aggressive expansion and a series of acquisitions, Six Flags has expanded from only seven parks in the mid-1980s to 24 locations across the U.S. (including water parks)—developing a winning formula heavy on rides and lighter on themes, the bulk of which are based on the parks' Warner Bros. and DC Comics licenses. Following its acquisition of Paramount Parks from CBS Corporation in 2006, Cedar Fair Entertainment Company currently owns and operates 13 locations. Most prominent among these are Cedar Point, Ohio, whose origins as a coastal resort date as far back as the 1870s, and Knott's Berry Farm, originally opened around 1940 and acquired from the Knott family in 1997. A third major operator includes SeaWorld Entertainment Inc. (formerly a subsidiary of brewing empire Anheuser-Busch), whose portfolio numbers twelve locations, including three SeaWorld theme parks (in San Diego, Orlando, and San Antonio, opened in 1964, 1973, and 1988, respectively) and two Busch gardens (in Tampa and Williamsburg, opened in 1959 and 1975, respectively).

While not destination parks per se, SeaWorld Entertainment's Orlando and Tampa locations directly benefit from their proximity to Disney's and Universal's operations, making them the second and third non-Disney and non-Universal top attractions in the U.S. (Knott's Berry Farm, with four million visitors, is first; see Rubin 2018, 31). Together with Disney and Universal, these three regional theme park companies now own and operate 19 of the top 20 most visited theme parks in North America (the sole exception being independently owned Hershey Park, in Hershey, Pennsylvania), with the top ten positions held exclusively by Disney and Universal (Rubin 2018, 31).

5.2 *The Development in Europe*

Drawing their inspiration from the first permanent ventures to have survived to this day (namely Tivoli, Denmark, 1843; and Alton Towers, UK, 1924), the first generation of family-owned parks developed in Europe during the 1950s and 1960s in the form of landscaped amusement grounds that were originally light on rides. Notable examples include Efteling, the Netherlands, opened in 1952, and Phantasialand, Germany, opened in 1967. The second generation took over in the 1970s, as mid-sized parks developed that placed a heavier emphasis on mechanical rides but whose appeal remained overwhelmingly regional. A case in point is Europa-Park, opened in 1975 by the Mack family as an extension of their ride-manufacturing business; other examples include Italy's Gardaland (1975) and Germany's Heide Park (1978).

In the 1980s, the third generation arose when the announcement of Disney's arrival in Europe led to a flurry of projects, sometimes at the initiative of local governments (as with France's Futuroscope or Puy du Fou), while forcing existing parks to expand their offerings (in terms of amusements, food and beverages, and retail) and open hotels to reposition themselves as resorts—as notably illustrated by Efteling, Alton Towers, and Europa-Park. In France alone, the decade witnessed the opening of Mirapolis (1987–1991), Zigofolis (1987–1991), Planète Magique (1989–1991), Futuroscope (1987), Puy du Fou (1989), Big Bang Schtroumpf (1989, now Walygator Park), and Parc Astérix (1989)—many of which eventually closed their doors after only a few years of operation. The next generation was inaugurated with the opening of Disneyland Paris in 1992, and has more IP-centric, resort-style operations that often, though not exclusively, involve large American media groups such as Disney, Universal (PortAventura Park's major shareholder from 1999 to 2004), and Warner (Parque Warner Madrid, 2002).

Four players have come to dominate the European market, and their portfolios (save for the Walt Disney Company) usually include dozens of operations beyond the realm of theme parks, such as zoos and aquariums, wax museums, and water parks. Those operators include EuroDisneyland SCA, which owns and operates

Parc Disneyland and Walt Disney Studios; Merlin Entertainments, whose operations—following the purchase of Legoland and the Tussauds Group—now include Legoland's eight locations across the world as well as Britain's Alton Towers; Grévin et Cie (Park Astérix and Walabi amusement parks in France, the Netherlands, and Belgium); and Parques Reunidos (Parque Warner de Madrid). Regardless of these groups' size and influence, estimates suggest that 85% of European parks remain independent operations—with the Walt Disney Company and the Merlin group being the only European park operators with a truly global presence (Anton Clavé 2007, 129).

5.3 General Trends

A few industry-wide trends may be discerned, which have emerged over the years. Resort-style parks are now standard industry practice across the U.S. and Europe, most notably as a result of the Walt Disney Company and other American operators going global. Attractions have also become more self-consciously designed as storytelling devices, conforming to Hollywood standards of a well-formed plot—a shift from "experiential" and "implicit story" to "explicit story" (that is, from primarily experiential or episodic to tightly plotted attractions; see Younger 2016, 98–100), as illustrated by Pirates of the Caribbean's many successive iterations (see ATTRACTIONS and IMMERSION). The result of Disney C.E.O. Michael Eisner's early career as a Paramount Pictures executive, this growing emphasis on story-driven parks, lands, or attractions was soon to evolve into a complete design style whose influence expanded well beyond the confines of Disney parks. The "new traditional style" (essentially the same detail-rich, immersive design style pioneered at Disneyland, only with a stronger focus on story) was, for example, adopted at Universal's Islands of Adventures, though it was also recently taken up in Disney's California Adventure's "Cars Land" (2012).

Other design styles have emerged in the past decades, including the "presentational style" (which replicates the feel, design, and informational approach of world's fairs, as in Epcot) or the "postmodern style" originally deployed at studio parks (an aesthetic based on the tongue-in-cheek embrace of the park environment's artificiality)—both of which have since fallen out of favor. Another category, the "themed amusement design style," is chiefly associated with regional parks such as Six Flags, where rides are only minimally themed (Younger 2016, 152–156).

Theme parks' increased emphasis on story has likewise allowed for their growing reliance on their parent companies' intellectual property, as attractions, lands, or sometimes entire parks are now exclusively devoted to specific transmedia franchises—as is notably illustrated by Universal's "Wizarding World of Harry Potter" (2010), Disney's Animal Kingdom's "Pandora: The World of Avatar" (2017), "Star Wars: Galaxy's Edge" at Disneyland (2019), Magic Kingdom (2019), and eventually Disneyland Paris (see MEDIA and THEMING). Probably in an effort to offer as seamless a transition as possible from the movie to the park experience, many IP-heavy attractions now build on new ride/movie combinations (sometimes enhanced with three-dimensional glasses)—a technological feat only available to the sectors' heavy-weights (see Disney's Hollywood Studio's Ratatouille: l'Attraction, 2014; Pandora's Flight of Passage, 2017; or Universal's Islands of Adventure's Harry Potter Escape from the Gringotts, 2014; or Harry Potter and the Forbidden Journey, 2014).

As theme park attendance keeps increasing, another trend likely to expand is for more VIP, "boutique" experiences in the style of SeaWorld's Discovery Cove. Disney's new Star Wars: Galactic Starcruiser hotel thus allows visitors to fully immerse themselves into the Star Wars universe and engage in role play in an interactive setting, away from the crowd of ordinary park-goers (see INCLUSION AND EXCLUSION).

6 The Evolution of Theme Parks in Eastern Asia

6.1 Japanese Theme Parks

In Eastern Asia, the theme park in its modern form first appeared in Japan, where audiences had been familiarized with the concept through Walt Disney's weekly TV show, broadcast on Japanese TV in the 1950s (see Raz 1999, 160; see PARATEXTS AND RECEPTION). Following a trip to the then newly opened Disneyland in Anaheim, entrepreneur and impresario Kunizo Matsuo contacted Walt Disney and his design team in order to enlist their help bringing Disneyland to Japan. Featuring copies of Disneyland's "Main Street, U.S.A.," Sleeping Beauty Castle, and Matterhorn Bobsleds, among others, Matsuo's Nara Dreamland opened in Nara (southern Honshu) in 1961, and was followed in 1964 by a sister park in Yokohama (eastern Honshu). Due to disagreements about licensing fees, Nara Dreamland was never officially associated with Disney and eventually closed in 2006, but remained a popular destination for urban explorers until its demolition in 2016 (see, among others, N.N. 2011).

It was not until over two decades after the opening of Nara Dreamland that Japan received an "official" Disney park in the shape of Tokyo Disneyland, opened in Tokyo (eastern Honshu) in 1983. Originally conceived as the centerpiece of a land reclamation project and owned and operated by the Japanese Oriental Land Co. under license from Disney, Tokyo Disneyland was mostly based on existing designs for Disneyland and Walt Disney World's Magic Kingdom (1971), following the explicit request of Oriental Land managers, who insisted on an "authentic" American experience and asked designers to not "Japanese" the park (see Brannen 1992, 216; Mitrasinovic 2006, 83). Although "localization" did occur on the levels of design, operation, and reception (see, for example, Van Maanen 1992; Raz 1999; Laemmerhirt 2013; Mittermeier 2021; Freitag 2021), the idea of Tokyo Disneyland offering an imaginative trip to the U.S. would prove to be trend-setting.

In fact, Tokyo Disneyland could be described as one of Japan's first "foreign country villages" (see Hendry 2000a, 94). Marked not only by a significant increase in per capita disposable income, the spread of the five-day workweek, and "a new attitude toward relaxation and recreation" (Brannen 1992, 217), but also by a continuing preference for short, domestic travel over longer journeys abroad, Japan's "bubble economy" of the 1980s led to the creation of a large number of so-called *tema paku* ("theme parks") or *gaikoku mura* ("foreign country villages"), which all sought to virtually transport visitors to specific (mostly European) countries. Examples include Oranda Mura (themed to the Netherlands and opened in Sasebo in 1983), Glückskönigreich (Germany, Obihiro, 1989), Marine Park Nixe (Denmark, Noboribetsu, 1990), Canadian World (Canada, Ashibetsu, 1991), Huis Ten Bosch (the Netherlands, Sasebo, 1992), Niigata Russian Village (Russia, Agano, 1993), and Parque Espana (Spain, Shima, 1994). With its individual "ports" referring to countries and regions from Europe and Arabia to Latin and North America, Tokyo DisneySea, opened as a second gate next to Tokyo Disneyland in 2001, has also been considered in the tradition of the *tema paku* (see Laemmerhirt 2013, 95).

In contrast to the Japanese Disney parks, however, the *tema paku* have generally focused less on rides and more on food, retail, shows, and craft demonstrations (often by natives of the respective countries), as well as intermedial references to literary works (Glückskönigreich, Canadian World, and Marine Park Nixe, for instance, featured references to the Brothers Grimm, Lucy Maud Montgomery's *Anne of Green Gables*, and Hans Christian Andersen, respectively; see Hendry 2000b, 212; see MEDIA). Also in contrast to Tokyo Disneyland and Tokyo DisneySea, the popularity of most of the "foreign country villages" faded rather quickly, with Glückskönigreich, Canadian World, and Niigata Russian Village all closing due to lack of attendance in the late 1990s and early 2000s.

In the twenty-first century, the theme park market in Japan has thus been overwhelmingly dominated by a small number of parks associated with multinational brands: Tokyo Disneyland, Tokyo DisneySea, and Universal Studios Japan, owned and operated by Universal Theme Parks and opened on reclaimed land in Osaka (southern Honshu) in 2001. In terms of attendance, these parks not only consistently occupy the top three spots in Japan and the Asia–Pacific region in general, but also regularly make it into the top five theme parks worldwide (see Rubin 2019, 10, 47).

6.2 *The People's Republic of China*

In the People's Republic of China, meanwhile, it was the Dengist reform policy and especially the establishment of Special Economic Zones (SEZs) from 1979 onwards that created the economic, social, and infrastructural conditions necessary for the arrival of the modern theme park. These conditions comprised an expanding economy with a capital influx from abroad, a growing middle class with sufficient leisure time and discretionary income, and a "suburban revolution" (Campanella 2008, 240–244; Campanella 2009, 80) that also brought a dense network of highways, gated communities, and shopping malls to China—a situation not entirely unlike that of 1950s California, where the modern theme park first emerged (see Hannigan 1998, 177). In fact, it was in Shenzhen SEZ, not far from Hong Kong in the Pearl River Delta, where what is commonly cited as the first theme park in mainland China opened in October 1989: Splendid China, a miniature park modeled upon Madurodam (The Hague), co-owned and co-operated by the public Overseas Chinese Town Enterprises (OCT; see Ren 1998, 69–71; Campanella 2008, 254–256).

In terms of its location, theme, and ownership, Splendid China was a harbinger of things to come. From the SEZs and coastal cities opened to overseas investment (particularly Shenzhen, Guangzhou, Zhuhai, and Shanghai), a veritable "theme park mania" would sweep the country during the 1990s, with many of the over 2,000 theme parks constructed during the decade being publicly (co-)owned and featuring "cultural" themes (see Stanley 2002, 289; Ap 2003, 195; Campanella 2008, 248–250). In Shenzhen, for instance, Splendid China was rapidly followed by three more OCT parks—China Folk Culture Villages (1991), Window of the World (1994), and Happy Valley (1998)—two of which focus on ethnographic displays of ethnic minorities and miniature models of world-famous sites. Moreover, the early-to-mid-1990s also saw the first unsuccessful attempts of internationally established theme park brands, most notably Disney, seeking to enter the Chinese market (Groves 2011, 138).

However, the boom of the early 1990s was eventually followed by a bust during the second half of the decade. Somewhat ironically, this bust was partly due to the success of Splendid China and the very design strategy that had appeared to stand behind this success, namely imitation. As John Ap explains,

> The misconception was that all one had to do was to replicate a successful theme park and it too would make profit. Little or no consideration was given to incorporating new or innovative facilities to give the park some individuality and distinctiveness. This led to a rampant oversupply of parks and excessive competition for attractions seeking to lure patronage from the same target market. (2003, 205)

Suffering from under-maintenance, low (repeat) visits, and little reinvestment, many parks had extremely short life cycles and closed down within a few years of their opening. By 2000, Shenzhen's OCT parks were still doing well, yet all but one of Shanghai's many parks had disappeared (see Ap 2003, 203).

In 1999, amidst this period of bust, the Walt Disney Company and the government of the Hong Kong Special Administrative Region signed a contract to build a theme park resort in the former British colony, which had been transferred back to China only two years prior (see Groves 2011, 139; Choi 2012). The opening of Hong Kong Disneyland in 2005 heralded the beginning of a second boom phase in the development of theme parks in China, this time focusing on chains of parks with mixed themes not only in coastal SEZs (Zhuhai, Xiamen) and Development Areas (Tianjin, Nantong, Shanghai, Ningbo, and Guangzhou), but also in inland "national central" cities (Beijing, Chengdu, Wuhan, Zhengzhou, and Chongqing). This boom phase initially saw increased activity from Chinese groups, three of which emerged as the leading theme park operators in the country (and among the top six theme park operators in the world; see Rubin 2019, 9): OCT, Chimelong, and Fantawild.

Since 2006, OCT has steadily expanded its Happy Valley Park in Shenzhen (opened in 1998) into a chain of parks in six of China's nine "national central cities" (Beijing, 2006; Chengdu and Shanghai, both 2009; Wuhan, 2012; Tianjin, 2013; and Chongqing, 2017). Chimelong, by contrast, has focused on the coastal cities of Guangzhou (Chimelong Paradise with adjacent Safari, Water, and Bird Parks, 2006) and Zhuhai (Chimelong Ocean Kingdom, 2014). Fantawild, finally, has developed no fewer than four branded chains: Adventure (nine parks, 2007–2015), Dreamland (five parks, 2010–2015), Water Park (five parks, 2014–2017), and Oriental Heritage (six parks, 2015–2018), which are frequently combined into multi-gate resorts. In 2012, for example, the group opened an Adventure Park in Zhuzhou, to which a Dreamland Park and a Water Park were added in 2015.

In 2016, following the opening of Disney's second theme park in China, Shanghai Disneyland, other non-Chinese park groups have also expanded their branded chains into China. Merlin, for example, has operated a Dungeon in Shanghai since 2018, and Six Flags and Universal were planning to open Six Flags parks in Zhejiang in 2019 and Chongqing in 2020, and a Studio park in Beijing in 2021, respectively. At the same time, Chinese groups have begun to expand beyond the national market, with Fantawild opening a Dreamland park in Isfahan (Iran) in 2014. The Chinese theme park market had not become the largest in the world by 2020 (as some analysts had predicted—see, for example, Rubin 2018, 41), mostly because the per capita spending remains significantly lower than in the U.S. It was also foreseen that in 2020 the number of visitors to the Asia and Pacific theme parks would grow bigger than those in the U.S. market, but at the moment of writing, the consequences of the COVID-19 pandemic crisis of 2020–2022 and its impact on local and international travel are still difficult to foresee (see INDUSTRY).

References

Alayrac-Fielding, Vanessa, ed. 2017. *Rêver la Chine: Chinoiseries et regards croisés entre la Chine et l'Europe aux XVIIe et XVIIIe siècles.* Toucoing: Invenit.

Altick, Richard D. 1978. *The Shows of London.* Cambridge, MA: Belknap Press.

Ames, Eric. 2009. *Carl Hagenbeck's Empire of Entertainments.* Seattle: University of Washington Press.

Anderson, Jay. 1984. *Time Machines: The World of Living History.* Nashville: The American Association for State and Local History.

Anton Clavé, Salvador. 2007. *The Global Theme Park Industry.* Wallingford: CABI.

Ap, John. 2003. An Assessment of Theme Park Development in China. In *Tourism in China*, ed. Alan A. Lew, Lawrence Yu, John Ap, and Zhang Guangrui, 195–214. New York: Haworth Hospitality Press.

Ashby, LeRoy. 2012. *With Amusement for All: A History of American Popular Culture since 1830.* Lexington: The University Press of Kentucky.

Baudrillard, Jean. 1994 [1978]. The Precession of Simulacra. In *Simulacra and Simulation*, trans. Sheila Faria Glaser, 1–42. Ann Arbor: The University of Michigan Press.

Brannen, Mary Yoko. 1992. "Bwana Mickey": Constructing Cultural Consumption at Tokyo Disneyland. In *Re-Made in Japan: Everyday Life and Consumer Taste in a Changing Society*, ed. Joseph J. Tobin, 216–234. New Haven: Yale University Press.

Buscombe, Edward, ed. 1988. *The BFI Companion to the Western.* New York: Atheneum.

Campanella, Thomas J. 2008. *The Concrete Dragon: China's Urban Revolution and What It Means for the World.* New York: Princeton Architectural Press.

Campanella, Thomas J. 2009. Mimetic Utopias: Themeing and Consumerism on China's Suburban Frontier. *New Geographies* 1: 80–85.

Carlà-Uhink, Filippo. 2017. *The "Birth" of Italy. The Institutionalization of Italy as a Region, 3rd-1st Century BCE.* Berlin: De Gruyter.

Chappell, Edward A. 2002. The Museum and the Joy Ride: Williamsburg Landscapes and the Specter of Theme Parks. In *Theme Park Landscapes: Antecedents and Variations*, ed. Terence G. Young and Robert Riley, 119–156. Washington, D.C.: Dumbarton Oaks Research Library and Collection.

Choi, Kimburley. 2012. Disneyfication and Localisation: The Cultural Globalisation Process of Hong Kong Disneyland. *Urban Studies* 49 (2): 383–397.

Coke, David E., and Alan Borg. 2011. *Vauxhall Gardens: A History.* New Haven and London: Yale University Press.

Conan, Michel. 2002. The Fiddler's Indecorous Nostalgia. In *Theme Park Landscapes: Antecedents and Variations*, ed. Terence G. Young and Robert Riley, 91–117. Washington, D.C.: Dumbarton Oaks Research Library and Collection.

Culhane, John. 1991. *The American Circus: An Illustrated History.* New York: Holt.

Davis, Janet M. 2012. The Circus Americanized. In *The American Circus*, ed. Susan Weber, Kenneth L. Ames, and Matthew Wittman, 23–53. New Haven: Yale University Press.

Davis, Susan G. 1996. The Theme Park: Global Industry and Cultural Form. *Media, Culture and Society* 18: 399–422.

Davis, Susan G. 1997. *Spectacular Nature: Corporate Culture and the Sea World Experience.* Berkeley: University of California Press.

Eco, Umberto. 1986 [1975]. Travels in Hyperreality. In *Travels in Hyperreality: Essays*, trans. William Weaver, 1–58. San Diego: Harcourt Brace Janovich.

Figueiredo, Yves. 2014. *Frederick Law Olmsted et le park movement américain.* Neuilly: Atlande.

Finch, Christopher. 1973. *The Art of Walt Disney from Mickey Mouse to the Magic Kingdoms.* New York: Harry N. Abrams.

Florey, Robert, and Wilfred Jackson, dir. 1954. The Disneyland Story. *Disneyland* 1 (1), October 27. Burbank: Walt Disney Productions.

Foglesong, Richard E. 1999. *Married to the Mouse: Walt Disney World and Orlando.* New Haven: Yale University Press.

Freitag, Florian. 2021. *Popular New Orleans: The Crescent City in Periodicals, Theme Parks, and Opera, 1875–2015.* New York: Routledge.

Freitag, Florian. Forthcoming. *"The Future That Never Was Is Finally Here": Depicting the Future in Disney's Tomorrowlands.*

Gilpin, William. 1798. *Observations on the Western Parts of England, Relative Chiefly to Picturesque Beauty.* London: T. Cadell jun. and W. Davies.

Gorky, Maxim. 2014 [1907]. Boredom. In *A Coney Island Reader: Through Dizzy Gates of Illusion*, ed. Louis J. Parascandola and John Parascandola, 90–94. New York: Columbia University Press.

Grau, Oliver. 2003. *Virtual Art: From Illusion to Immersion*, trans. Gloria Custance. Cambridge: MIT Press.

Groves, Derham. 2011. Hong Kong Disneyland: Feng-Shui Inside the Magic Kingdom. In *Disneyland and Culture: Essays on the Parks and Their Influence*, ed. Kathy Merlock Jackson and Mark I. West, 138–149. Jefferson: McFarland & Company.

Hall, Millicent. 1976. Theme Parks: Around the World in 80 Minutes. *Landscape: A Magazine of Human Geography* 21 (1): 3–8.

Hannigan, John. 1998. *Fantasy City: Pleasure and Profit in the Postmodern Metropolis.* London: Routledge.

Harwood, Edward. 2002. Rhetoric, Authenticity, and Reception: The Eighteenth-Century Landscape Garden, the Modern Theme Park, and Their Audiences. In *Theme Park Landscapes:*

Antecedents and Variations, ed. Terence G. Young and Robert Riley, 49–68. Washington, D.C.: Dumbarton Oaks Research Library and Collection.
Hendry, Joy. 2000a. *The Orient Strikes Back: A Global View of Cultural Display*. Oxford: Berg.
Hendry, Joy. 2000b. Foreign Country Theme Parks: A New Theme or an Old Japanese Pattern? *Social Science Japan Journal* 3 (2): 207–220.
Heulhard, Arthur. 1818. *La Foire Saint-Laurent, son histoire et ses spectacles*. Paris: A. Lemerre.
Hildebrandt, Hugo John. 1981. Cedar Point: A Park in Progress. *The Journal of Popular Culture* 15 (1): 87–107.
Immerso, Michael. 2002. *Coney Island: The People's Playground*. New Brunswick: Rutgers University Press.
Jackson, Wilfred, Stu Phelps, and John Rich, dir. 1955. *Dateline: Disneyland*. Burbank: Walt Disney Productions.
Jacobson, Brian R. 2017. *Studios before the System: Architecture, Technology, and the Emergence of Cinematic Space*. New York: Columbia University Press.
Kagelmann, H. Jürgen. 1993. Themenparks. In *Tourismuspsychologie und Tourismussoziologie*, ed. Heinz Hahn and H. Jürgen Kagelmann, 407–415. Munich: Quintessenz.
Kasson, John F. 1983. *Amusing the Million: Coney Island at the Turn of the Century*. New York: Hill and Wang.
Kasson, Joy S. 2001. Life-like, Vivid and Thrilling Pictures: Buffalo Bill's Wild West and Early Cinema. In *Westerns: Films Through History*, ed. Janet Walker, 109–130. New York: Routledge.
Keßler, Ute. 2012. Geschichte des Vergnügungsparks. In *Erlebnislandschaft: Erlebnis Landschaft? Atmosphären im Architektonischen Entwurf*, ed. Achim Hahn, 123–145. Bielefeld: Transcript.
King, Margaret J. 1981a. The New American Muse: Notes on the Amusement/Theme Park. *The Journal of Popular Culture* 15 (1): 56–62.
King, Margaret J. 1981b. Disneyland and Walt Disney World: Traditional Values in Futuristic Form. *The Journal of Popular Culture* 15 (1): 116–140.
Knell, Heiner. 2008. *Des Kaisers neue Bauten: Hadrians Architektur in Rom, Athen und Tivoli*. Mainz: Von Zabern.
Laemmerhirt, Iris-Aya. 2013. *Embracing Differences: Transnational Cultural Flows between Japan and the United States*. Bielefeld: Transcript.
Langlois, Gilles-Antoine. 1991. *Folies, Tivolis et attractions: Les premiers parcs de loisirs parisiens*. Paris: Délégation à l'action artistique de la ville de Paris.
Latsis, Dimitrios. 2019. Riding, Shooting, Viewing: Railroads, Amusement Parks, and the Experience of Place in Early Hollywood. *The Moving Image* 18 (2): 48–71.
Marin, Louis. 1984 [1973]. *Utopics: The Semiological Play of Textual Spaces*, trans. Robert A. Vollrath. New York: Humanity.
Marling, Karal Ann. 1991. Disneyland, 1995: Just Take the Santa Ana Freeway to the American Dream. *American Art* 5 (1–2): 168–207.
Marling, Karal Ann. 1997. Imagineering the Disney Theme Parks. In *Designing Disney's Theme Parks: The Architecture of Reassurance*, ed. Karal Ann Marling, 29–177. Paris: Flammarion.
Martí, José. 2014 [1881]. Coney Island. Trans. Esther Allen. In *A Coney Island Reader: Through Dizzy Gates of Illusion*, ed. Louis J. Parascandola and John Parascandola, 61–66. New York: Columbia University Press.
McVeigh, Stephen. 2007. *The American Western*. Edinburgh: Edinburgh University Press.
Mitrasinovic, Miodrag. 2006. *Total Landscape, Theme Parks, Public Space*. Burlington: Ashgate.
Mittermeier, Sabrina. 2021. *A Cultural History of the Disneyland Theme Park: Middle Class Kingdoms*. Chicago: Intellect.
Moore, Charles W. 2004. You Have to Pay for the Public Life. In *You Have to Pay for the Public Life: Selected Essays of Charles W. Moore*, ed. Kevin Keim, 111–141. Cambridge, MA: MIT.
Morgan, Luke. 2016. *The Monster in the Garden: The Grotesque and the Gigantic in Renaissance Landscape Design*. Philadelphia: University of Pennsylvania Press.
N.N. 1784. *Mémoires secrets pour servir à l'histoire de la république des lettres en France, ou Journal d'un observateur: Vol. 17*. London: John Adamson.

N.N. 2011. Nara Dreamland: Frequently Asked Questions. *Abandoned Kansai*, October 21. https://abandonedkansai.com/2011/10/21/nara-dreamland-frequently-asked-questions-faq/. Accessed 28 Aug 2022.

Normore, Christina. 2015. *A Feast for the Eyes: Art, Performance, and the Late Medieval Banquet*. Chicago and London: University of Chicago Press.

Nye, David E. 1997. *Electrifying America: Social Meanings of a New Technology, 1880–1940*. Cambridge: MIT.

Oettermann, Stephan. 1997. *The Panorama: History of a Mass Medium*, trans. Deborah Lucas Schneider. New York: Zone Books.

Olmsted, Frederick Law. 2010 [1870]. Public Parks and the Enlargement of Towns. In *Essential Texts*, ed. Robert Twombly, 201–252. New York: W.W. Norton.

Quaintance, Richard. 2002. Toward Distinguishing Theme Park Publics: William Chambers's Landscape Theory vs. His Kew Practice. In *Theme Park Landscapes: Antecedents and Variations*, ed. Terence G. Young and Robert Riley, 25–47. Washington, D.C.: Dumbarton Oaks Research Library and Collection.

Rabinovitz, Lauren. 2012. *Electric Dreamland: Amusement Parks, Movies, and American Modernity*. New York: Columbia University Press.

Raz, Aviad E. 1999. *Riding the Black Ship: Japan and Tokyo Disneyland*. Cambridge: Harvard University Press.

Register, Woody. 2001. *The Kid of Coney Island: Fred Thompson and the Rise of American Amusements*. New York: Oxford University Press.

Ren, Hai. 1998. Economies of Culture: Theme Parks, Museums and Capital Accumulation in China, Hong Kong, and Taiwan. Diss., The University of Washington.

Rothfels, Nigel. 2012. *Savages and Beasts: The Birth of the Modern Zoo*. Baltimore: Johns Hopkins University Press.

Rowan, Yorke M. 2004. Repackaging the Pilgrimage: Visiting the Holy Land in Orlando. In *Marketing Heritage: Archaeology and the Consumption of the Past*, ed. Yorke M. Rowan and Uzi Baram, 249–266. Walnut Creek: AltaMira Press.

Rubin, Judith, ed. 2018. *2017 Theme Index and Museum Index: The Global Attractions Attendance Report*. Burbank: Themed Entertainment Association and AECOM.

Rubin, Judith, ed. 2019. *2018 Theme Index and Museum Index: The Global Attractions Attendance Report*. Burbank: Themed Entertainment Association and AECOM.

Samuelson, Dale, and Wendy Yegoiants. 2001. *The American Amusement Park*. St. Paul: MBI.

Schenker, Heath. 2002. Pleasure Gardens, Theme Parks, and the Picturesque. In *Theme Park Landscapes: Antecedents and Variations*, ed. Terence G. Young and Robert Riley, 69–89. Washington, D.C.: Dumbarton Oaks Research Library and Collection.

Schickel, Richard. 1968. *The Disney Version: The Life, Times, Art and Commerce of Walt Disney*. New York: Simon and Schuster.

John, A.p. 2003. Chinese Theme Parks and National Identity. In *Theme Park Landscapes: Antecedents and Variations*, ed. Terence G. Young and Robert Riley, 269–289. Washington, D.C.: Dumbarton Oaks Research Library and Collection.

Steinkrüger, Jan-Erik. 2013. *Thematisierte Welten: Über Darstellungspraxen in Zoologischen Gärten und Vergnügungsparks*. Bielefeld: Transcript.

Sterngass, Jon. 2001. *First Resorts: Pursuing Pleasure at Saratoga Springs, Newport & Coney Island*. Baltimore: Johns Hopkins University Press.

Szabo, Sacha. 2006. *Rausch und Rummel: Attraktionen auf Jahrmärkten und in Vergnügungsparks. Eine soziologische Kulturgeschichte*. Bielefeld: Transcript.

Szabo, Sacha. 2009. Chillrides. In *Kultur des Vergnügens: Kirmes und Freizeitparks, Schausteller und Fahrgeschäfte. Facetten nicht-alltäglicher Orte*, ed. Sacha Szabo, 275–287. Bielefeld: Transcript.

Telotte, J.P. 2004. *Disney TV*. Detroit: Wayne State University Press.

Thomas, Bob. 1994 [1976]. *Walt Disney: An American Original*. New York: Hyperion.

Trümper, Monika. 2008. *Die "Agora des Italiens" in Delos: Baugeschichte, Architektur, Ausstattung und Funktion einer späthellenistischen Porticus-Anlage*. Rahden: Marie Leidorf.

Van Maanen, John. 1992. Displacing Disney: Some Notes on the Flow of Culture. *Qualitative Sociology* 15 (1): 5–35.

Wallace, Mike. 1996. *Mickey Mouse History and Other Essays on American Memory*. Philadelphia: Temple University Press.

Warren, Louis S. 2005. *Buffalo Bill's America: William Cody and the Wild West Show*. New York: Alfred A. Knopf.

Weinstein, Raymond M. 1992. Disneyland and Coney Island: Reflections on the Evolution of the Modern Amusement Park. *The Journal of Popular Culture* 26 (1): 131–164.

Wiener, Jon. 1994. Tall Tales and True. *The Nation*, January 31, 133–135.

Wolcott, Victoria W. 2012. *Race, Riots, and Roller Coasters: The Struggle over Segregated Recreation in America*. Philadelphia: University of Pennsylvania Press.

Wong, Young-tsu. 2016. *A Paradise Lost: The Imperial Garden Yuanming Yuan*. Singapore: Springer.

Wright, Talmadge. 2007. Themed Environments and Virtual Spaces: Video Games, Violent Play, and Digital Enemies. In *The Themed Space: Locating Culture, Nation, and Self*, ed. Scott A. Lukas, 247–270. Lanham: Lexington.

Yamaguchi, Masao. 1991. The Poetics of Exhibiting Japanese Culture. In *Exhibiting Cultures: The Poetics and Politics of Museum Display*, ed. Ivan Karp and Steven D. Levine, 57–67. Washington: Smithsonian Institution Press.

Young, Terence, and Robert Riley, eds. 2002. *Theme Park Landscapes: Antecedents and Variations*. Washington, D.C.: Dumbarton Oaks Research Library and Collection.

Younger, David. 2016. *Theme Park Design & the Art of Themed Entertainment*. N.P.: Inklingwood.

Younger, David. 2017. Traditionally Postmodern: The Changing Styles of Theme Park Design. In *Time and Temporality in Theme Parks*, ed. Filippo Carlà-Uhink, Florian Freitag, Sabrina Mittermeier, and Ariane Schwarz, 63–82. Hanover: Wehrhahn.

Attractions: How Rides, Retail, Dining, and Entertainment Structure the Theme Park

Abstract Attractions in theme parks are not just rides, but also include entertainment, retail, and food. This chapter traces the etymology of the term and uses case studies such as the long history of the Pirates of the Caribbean ride(s) at Disney theme parks or the Ice College show at Phantasialand. It also addresses the different types of attractions in the theme park, their various uses, and how they come together as a whole to (ideally) form one coherent, immersive narrative space. Furthermore, the chapter discusses the concept of an "attractions mix" as well as the role of attractions in marketing.

1 Introduction

While the word "attraction" is attested since the 1600s, with the meanings of "action or property of drawing (diseased matter) to the surface" and "pulling" or "drawing" together, the specific meaning of attraction as "a thing which draws a crowd, interesting or amusing exhibition" dates from 1829 (Online Etymology Dictionary). Since then, the word has been used to describe a variety of spaces, devices, technologies, performances, etc. It was first recorded within the context of tourism in the 1830s, when it was used to highlight the appeal and drawing power of specific sites in John Murray's (1838) *Handbook for Travellers on the Continent* and, later, in the English translations of Karl Baedeker's travel guides (for example, Baedeker 1867). The term's use in the theme park industry is largely the result of Disney's insistence that it be used in place of the term "rides"—a designation meant to convey Disneyland's emphasis on themes and atmosphere, which stood in contrast to the physical thrills found at fairs and amusement parks. Former president and vice chairman of Walt Disney Imagineering Marty Sklar explains that he

> began to realize what people didn't understand about Disneyland, from the questions that they'd ask. [...] I quickly found out that people were saying, "Now I want to go on the Mark

This work is contributed by Sabrina Mittermeier, Scott A. Lukas, Ariane Schwarz, Florian Freitag, Salvador Anton Clavé, Astrid Böger, Filippo Carlà-Uhink, Thibaut Clément, Céline Molter, Crispin Paine, Jean-François Staszak, Jan-Erik Steinkrüger, Torsten Widmann. The corresponding author is Sabrina Mittermeier, Fachgruppe Geschichte, Universität Kassel, Kassel, Germany.

F. Freitag et al., *Key Concepts in Theme Park Studies*,
https://doi.org/10.1007/978-3-031-11132-7_3

> Twain, and I want to go on the Jungle Cruise, but I don't want to go on any rides." I learned that they related rides in those days to the whips and the shoot-the-chutes and all the stuff that was in the old amusement parks, and they knew that Disneyland was something else. [...] They didn't really understand the word "rides" in connection with our kinds of shows and experiences. Because of this, we invented a whole new language and started calling them "attractions," "adventures," "experiences" and we wrote the word "ride" out of our vocabulary for many years. The public brought to the Park their preconceived expectations about the Mark Twain and the Steam Trains, and didn't consider them rides in the traditional amusement park sense. (qtd. in Janzen and Janzen 1998, 5–6)

Attractions have multiple and often overlapping functions in a theme park and it is important to understand that almost every component of the theme park experience—from rides and entertainment to merchandise and retail and food and beverage—can function as an attraction, as different visitors have different reasons to visit the theme park (see VISITORS). Clearly, all of these types of attractions are significant in the evolution of the theme park, especially given that scholarship so far has focused on the ride as the theme park's primary attraction form (Cartmell 1986; Lukas 2008, 116). There has also been a significant focus on the evolution, alteration, and adaptation of the theme park ride into new realms of technology, inter- and transmedia, and intellectual property domains (Lukas 2013; see MEDIA), but less so for other types of attractions in the theme park space.

Yet much like pleasure gardens, amusement parks, and their other antecedents, theme parks have always offered different types of attractions and entertainment (Mangels 1952, 5; see also ANTECEDENTS). For instance, the practice of a theme park adding a new attraction every season is based on a tradition established in the early European pleasure gardens of the 1700s and 1800s (Mangels 1952, 6–8). The world's fair tradition, most notably the tradition of the "midway" established at the World's Columbian Exposition of 1893, also contributed to the eclecticism of attractions—or what will be referred to here as the "attractions mix"—in amusement and theme parks (Lukas 2008, 2013). This chapter thus concentrates on the variety of attractions found in the theme park, seeking to explain their function, to identify the types of attractions and how they are combined, and to explore the unique function or draw of rides, retail, food, and entertainment through the lens of the "mix."

2 Attractions Mix

Indeed, a park's appeal to audiences lies in part in its "attractions mix"—the variety of themes, experiences, and ride types found in any given theme park. The attractions mix is typically measured along two lines: (1) the nature of the experiences on offer and (2) the attractions' level of sophistication and public profile. The variety of experiences on offer is notably evident in the Disney designers' broad distinction between "dreamers" and "screamers," where the former are more quiet, thematic rides and the latter roller coasters or other thrill rides (Raz 2000, 96). Additional subcategories of experiences may be drawn depending on the attraction type (e.g. rides and their associated transportation technologies: roller coasters and flumes; shows and parades; shops and restaurants) as well as the expected sensory or emotional response (e.g. thrill, fun, and wonder). The attraction's level of sophistication (in its set design, engineering, or technology and as evidenced in its budget, length, or prestige) is in turn reflected in Disney's A to E ranking: originally a by-product of the ticket book system in the park before the adoption of all-inclusive admissions in 1982, it has since become an industry-wide nomenclature, with A being the simplest and least popular attractions, and E the most popular, often sophisticated headliner rides.

Incidentally, A or lower-profile attractions sometimes serve as temporary placeholders to later make way for flagship attractions: on opening day, Disneyland's "Tomorrowland" was nearly empty and famously filled with corporate exhibits in lieu of proper "attractions." As William Mangels notes,

> the development of the amusement park [created] a demand for fun-producing devices [...], and the introduction of these devices, in turn, attracted an ever-growing number of people at the parks. As a consequence, inventors were prompted to devise new thrillers to satisfy the amusement-seeking public's appetite for novelties. The need for strikingly new features offered a challenge to [...] inventive genius. It was soon realized that the most popular device was that which gave the patron the greatest thrill along with the utmost assurance of safety. (1952, 53)

The attractions mix will vary from park to park, as well as from themed area to themed area. Theming, in this regard, exerts a strong influence on the nature of the mix for each of a park's areas: some themes show greater potential for multiple action-packed attractions, while others seem geared almost exclusively toward younger audiences—as the almost universal association of fairy tales with child audiences and child-friendly attractions suggests, with examples ranging from "Fantasyland" at Disney's parks and "Grimm's Enchanted Forest" at Europa-Park (Rust, Germany) to the "Fairytale Forest" at Efteling (Kaatsheuvel, Netherlands; see THEMING). And while themes help tie all the attractions together within a specific area, they can conversely act as differentiating devices between otherwise largely similar attractions: Disney's Magic Kingdom in Orlando, Florida, for example, boasts two mine train attractions, Big Thunder Mountain Railroad at "Frontierland" (opened in 1980) and The Seven Dwarfs Mine Train at "Fantasyland" (opened in 2014). Identically, Universal Studios Florida currently has two ride simulators, *The Simpsons* Ride (opened in 2007) and Despicable Me Minion Mayhem (opened in 2012). While these are the same ride type, they still diversify the park's attractions mix simply because of their different theming and tone.

Attractions mixes also vary from park to park: this is true not only of rival companies targeting distinct market segments, but also of parks operated by the same company. As an example of the first scenario, Legoland and Six Flags are situated at opposite ends of the spectrum. At the time of writing (2020), Legoland Deutschland (Günzburg, Germany) features 18 attractions where no age restrictions apply for accompanied children, and one for accompanied children aged eight or older. By contrast, Six Flags Magic Mountain in Los Angeles advertises 20 of its attractions as "thrill rides," whereas only twelve rides are advertised as "family rides" and 14 as "kids rides" only. Family or kids rides generally dominate the attractions mix in family-oriented parks, for example at Parc Astérix in Plailly, France (seven thrill rides, 14 family rides, and 13 kids rides), PortAventura Park in Vila-seca and Salou, Spain (five thrill rides, 11 family rides, and 27 kids rides), or, in spite of its name, Ferrari World in Abu Dhabi, UAE (eight thrill rides, nine family rides, and six kids rides).

Operators with large portfolios will generally vary the attractions mix from park to park, especially in the case of multi-gate resorts. This allows them to expand the palette of experiences available to visitors and cater to specific market segments (see ECONOMIC STRATEGY). The portfolio of Spain-based theme park operator Parques Reunidos, for instance, ranges from kid-centric parks (such as Story Land in Glen, New Hampshire, and Dutch Wonderland in Lancaster, Pennsylvania) to family-friendly (as with Motiongate or Bollywood Parks, both in Dubai, UAE) or thrill-oriented parks (Movie Park Germany in Bottrop-Kirchhellen). The Walt Disney Company was likely the first operator to systematically expand its portfolio by adjusting the attractions mixes of its parks, as illustrated by EPCOT (opened in 1982 in Orlando, Florida), whose educational tone and movie-centric attractions were meant to burnish Disney's credentials with adults and appeal to their more discerning tastes. Originally devoid of Disney characters, EPCOT was also the first of all the Disney parks to sell alcoholic beverages—an adult connection that persists to this day with EPCOT's International Food & Wine Festival (since 1995).

Combined with its overall tone (a postmodern, behind-the-scenes peek at the movie industry and a tongue-in-cheek, self-parodical take on Los Angeles kitsch), the thrill-heavy attractions mix of Disney's Hollywood Studios (opened in 1989 as Disney-MGM Studios Theme Park in Orlando, Florida) was likewise meant to remove exclusive associations between Disney and childhood and lure back teen audiences, which at the time were increasingly drawn to Universal Studios or Six Flags. The latter company had, beginning with the 1992 roller coaster Batman The Ride at Six Flags Great America (Gurnee, Illinois), targeted a teen audience with attractions branded to the DC Comics universe. In fact, "teenification" or the "gradual shift in a theme park's target audience towards only teenagers and young adults" (Younger 2016, 240; see also VISITORS) is one of the main reasons for the increasing focus on new thrill rides, which are generally considered to be more popular among teen audiences. This phenomenon is possibly related to thrill rides' height restrictions and, therefore, their "rite of passage" implications. Designer Eddie Sotto explains: "The thinking for a teenager is, 'I am old enough now that I am not going to be scared. I can do that ride, that I can handle it'" (2011).

3 Changing the Attractions Mix

"Teenification" is one reason why the attractions mix of a given theme park may change over time. In fact, a park's existing mix must often be adjusted to keep pace with the audience's changing demographics and tastes. Such changes to the mix may follow two broad directions: the expansion of new ride types on offer in a given park, and the re-theming of previous attractions. The addition of thrill rides in parks that were originally devoid of them—for teenification or other reasons—is one especially significant trend in the first area. Starting in the late 1970s, Efteling took a big change by adding mechanical thrill rides to its then largely atmospheric, slow-paced environment. Originally billed as educational parks, EPCOT and Disney's Animal Kingdom (Orlando, Florida) have likewise in the past tweaked their mix to include such thrill-heavy attractions as Test Track and Mission: Space (opened at EPCOT in 1998 and 2003, respectively) or Expedition Everest and Avatar: Flight of Passage (opened at Animal Kingdom in 2006 and 2017). Reflecting on Expedition Everest, designer Joe Rohde remarks that, while necessary to attract wider audiences, the ride's addition was potentially antithetical to the park's original theme and overall feel:

> We had spent years crafting a park that expressed the intrinsic value of nature through the use of wild landscape, free-roaming animals, and quietly moldering picturesque buildings, and now we were proposing to add a thrill ride, which by implication meant a roller coaster. Hundreds of feet of steel, thundering cars, screaming people! Every park needs a full palette of experiences, from the lush and romantic, to the light and humorous, to the sweeping excitement of physical thrills. We had always known that the park would require a thrill ride to be a fully fleshed-out theme park, but how to do this without compromising the park was the big challenge. (2006)

Typically, new thrill rides are billed as high-profile E attractions—a draw whose addition to existing parks is usually a major selling point for repeat visitors, also making singular attractions a valuable marketing tool. For example, the addition of Space Mountain to Disneyland in 1977 (adapted from the existing attraction at Magic Kingdom, opened in 1975) helped propel annual attendance past 10 million visitors for the first time, in a park that at the time catered mostly to regional audiences. The roller coaster was meant to allow Disneyland to compete with rival Southern California theme parks, whose faster thrill rides kept luring teenagers away from the park (Strodder 2012, 388). In a telling anecdote, the addition of a thrill ride (Disney's first, save for Disneyland's Matterhorn Bobsleds, opened in 1959) generated surprise among the audience, and many visitors mistook it upon its debut at Magic Kingdom for a slow, panoramic ride—a problem compounded by the fact that the ride unfolds entirely indoors and therefore hidden from view. In the attraction's first few weeks, unsuspecting visitors were thus occasionally injured, prompting the company to add audio warnings in the queue as well as a sign that specified "roller coaster" under the attraction's entrance marquee (Koenig 2007, 136).

Headliner attractions such as Space Mountain are also most likely to occupy prime real estate and to serve as weenies, i.e. iconic, usually vertical centerpieces that draw visitors in, visually express the land's theme, and help tie all the attractions together. The four mountain-shaped attractions at Disneyland (Space Mountain, Big Thunder Mountain, Splash Mountain, and the Matterhorn) serve as visual magnets: often located at the far ends of lands or the park's outer edges, such attractions help draw people all the way into the park, past shops or lower-profile attractions and, combined with high theoretical hourly ride capacities, prove critical to visitor flow management (see SPACE). Alternatively, flagship attractions can serve as gates, as with EPCOT's

Spaceship Earth, or as giant billboards meant to advertise the park, as with Six Flags parks' soaring skylines, which are visible from neighboring highways. At Universal Orlando Resort, headliner attractions for Harry Potter's Wizarding World are strategically split across two parks, Islands of Adventure and Universal Studios Florida—they thus require a two-park ticket, which is in effect a way to collect a premium on the whole experience.

Re-theming can likewise provide added relevance to and renewed interest in familiar attractions, bringing changes to the attractions mix at a minimal cost. Some re-themings consist of temporary makeovers (as with Disneyland's and Tokyo Disneyland's Haunted Mansion Holiday, themed after Tim Burton's 1993 *Nightmare Before Christmas*), while others are permanent, sometimes to make better use of the operator's intellectual property: in 2017, Disneyland Paris's Space Mountain was re-themed as Star Wars Hyperspace Mountain (after Disney's Star Wars franchise, acquired in 2012), and Disney California Adventure's Tower of Terror as *Guardians of the Galaxy*—Mission: Breakout! (after Disney's Marvel franchise, acquired in 2009). Six Flags has similarly resorted to refurbishing and re-theming attractions after its DC Comics franchise, as with its 2011 Green Lantern coaster at Six Flags Great Adventure (Jackson, New Jersey; formerly known as Chang and inaugurated in 1997 at Six Flags Kentucky Kingdom) or the 2016 The Joker ride at Six Flags Discovery Kingdom (Vallejo, California; formerly known as Roar, a wooden coaster inaugurated in 1999 and since retrofitted with steel tracks). Sometimes re-theming involves the relocation of a ride to a different themed area, as in the case of Tizona at Terra Mítica (Benidorm, Spain), a suspended looping coaster that became Titanide when it was moved from the "Iberia" to the "Grécia" section of the park (Carlà-Uhink 2020, 91–92).

In the case of transnational park operators, foreign locations are seldom exact duplicates of original parks, as variations often reflect local weather conditions or, more crucially, the cultural context. This results in different attractions mixes in parks that may appear to be very similar at first sight. For example, inclement weather has been invoked to explain the absence of a Disneyland Paris counterpart to Disneyland's and Magic Kingdom's Splash Mountain, although a version of the flume ride does exist at Tokyo Disneyland, where the weather is much more similar to Paris than to California or Florida. Likewise, some Disney park attractions have been relocated to other lands in foreign versions of the park, often resulting in dramatic changes to the lands' respective attractions mixes. The themes for the Haunted Mansion's American locations (at Disneyland's "New Orleans Square" and Magic Kingdom's "Liberty Square") are embedded in specifically American settings: a Southern plantation house and a Dutch Gothic manor of Hudson River Valley (a nod to Washington Irving's *Sleepy Hollow*; see Mittermeier 2021, 72).

However, while keeping the Dutch Gothic exterior, the attraction was moved to "Fantasyland" in Tokyo Disneyland due to Japanese conceptualizations of ghosts as magical creatures (Mittermeier 2021, 88). In Disneyland Paris, in turn, the attraction was moved to "Frontierland" and its exterior was changed into a pastiche of references to horror movies, including *Psycho* (Freitag 2017, 713–714). In Hong Kong, it turned into Mystic Manor in the park's "Mystic Point" area: owing to traditional

Chinese culture, all references to the afterlife were removed from the attraction, and the plot and setting instead place it in British colonialist Sir Henry Mystic's house (Mittermeier 2021, 153). While these relocations and glocalizations do not change the overall attractions mix of the parks, they do allow for different attractions mixes within individual lands and provide variety, especially for those visitors, or rather fans, who visit all Disney theme parks (see VISITORS). It is not only Disney that operates under these practices: reflecting local tastes for "cute" (*kawaii*) characters, "Universal Wonderland" at Universal Studios Japan (Osaka) revolves around non-Universal franchises that are largely absent from the company's other parks, including Snoopy and Sesame Street (licensed in the United States to competitors Knott's Berry Farm or SeaWorld Entertainment), as well as Hello Kitty (also the star of its parent company's own theme park, Sanrio Puroland in Tokyo).

4 Rides

While the theme park is, in and of itself, a narrative space—designer Joe Rohde calls designing these spaces "narrative placemaking" (2006; see IMMERSION, SPACE, and THEMING)—rides also tell smaller, self-contained stories that fit into the overarching story of the space that surrounds them. As Margaret King has argued: "Rides expand the narrative experience with appropriate physical sensations, never for effect alone, but always to advance the storyline" (2002, 3). Yet, while she rightly contends that it is not rides, but rather "the architecture, public space design, landscaping, musical cueing, detailing, and the use of symbols, archetypes, and icons [...] that define the essence of theme parks" (3), it is still the rides that draw the majority of visitors to their premises. Originally, while in an amusement park the ride is "a thing of sensory and kinetic delight, it throws people together and reminds them of their mortality [...] in the theme park it is often a part of the story being told through theming, something that affects the body but also the mind" (Lukas 2008, 98), theme park rides, then, are often as much about the physical as the mentally immersive experience, and so the most immersive ones are often those that actively involve their riders in the narrative. Rohde has pointed out that in a well-designed theme park space, visitors are "given roles within the narrative" and this extends to rides (Rohde 2006; see also IMMERSION and VISITORS).

The dark ride deserves particular attention here, as it is a form that has been themed from its inception: "In dark rides, participants board a buggy, train, or boat and enter a dark, enclosed space. The space is themed—a ghost train, a haunted house, a trip to the moon—and the vehicle on track allows the designers some control over the ways the story unravels" (Ndalianis 2010, 14). As Angela Ndalianis highlights here, the dark ride was the first form of amusement park ride that allowed for storytelling and was thus crucial for the development of the immersive, narrative space of the theme park. However, the rider's point of view, the most crucial aspect of this specific type of ride, must be clearly communicated to the visitors, for instance through the theming of the waiting area. In a famous example discussed by Suzanne Rahn, visitors to Snow White's Adventures in Disneyland's "Fantasyland" did not grasp that they were supposed to experience the ride from the point of view of the titular heroine. The visual absence of the princess only confused riders who "just wondered where the hell Snow White was," according to designer Ken Anderson (qtd. in Rahn 2000, 22). This shows the narrative limits of the classic dark ride: much like movies (whose camera eye the twisting and turning ride vehicles of some dark rides have been compared to; see Lonsway 2009, 121), the traditional dark ride may favor an omniscient perspective. Technologically more advanced attractions that allow for a more individual immersion do not struggle with such confusions of point of view. As Bukatman has noted,

> in the 1980s and 1990s [...] theme park rides and attractions became more narrative than, say, roller coasters had been. They were also extended. Waiting on line for Star Tours was part of the ride, as elaborate sets and amusing droids entertained but also grounded the spectacle. (1998, 266)

In the case of adaptations or remediations (e.g. from movies; see Bolter and Grusin 2000), dark ride designers may also rely on visitors' knowledge of the source material. Since the inception of Disneyland, rides have been based on existing intellectual properties (IPs), a trend that has since come to dominate the theme park landscape (see MEDIA). While Disneyland's "Fantasyland" rides from the 1950s and 1960s were the first forays into synergy, today's multimedia conglomerates routinely use theme park attractions as part of their marketing machines (Ndalianis 2010, 15–16), firmly integrating them into strategies of transmedia storytelling (Williams 2020, 106–116)—a point perhaps best illustrated by the famous Pirates of the Caribbean ride.

Pirates of the Caribbean originally opened in Disneyland's "New Orleans Square" in 1967. One of the last attractions directly overseen by Walt Disney before his death in 1966, it remains a staple of Disneyland theme parks to this day. The ride tells the story of pirates ransacking a village and is a sophisticated mix of audio-animatronic technology within cinematic set design along waterways traversed by boats holding up to 15 passengers each. Like several other classic Disney rides, it also comes with its own theme song, "Yo Ho (A Pirate's Life for Me)," written by George Bruns (music) and X Atencio (lyrics), a fact that has significantly contributed to its cult status. This cult status was reached in record time after the attraction's debut: when the Magic Kingdom opened in 1971 without a version of the ride, visitor demand pushed the Walt Disney Company to rectify the omission (Surrell 2005, 54), and two years later the Florida counterpart, albeit in a somewhat shortened version, made its debut. Pirates was one of the opening attractions at Tokyo Disneyland in 1983 as well as Parc Disneyland in Paris in 1992. Its fandom notwithstanding (see also VISITORS), the attraction underwent several changes over the years, in part due to changing social and cultural climates (see INCLUSION AND EXCLUSION).

Other changes came to Pirates due to its remediation (see MEDIA), among others, into film: kicking off with *Pirates of the Caribbean: The Curse of the Black Pearl* (2003), which held its premiere in Disneyland, the ride was turned into a massively successful film franchise, with the fifth installment released in 2017 and the sixth originally announced for 2022. The popularity of the films in general, and their main character Jack Sparrow (portrayed by Johnny Depp) in particular, led to the addition of audio-animatronics of Jack Sparrow and Captain Barbossa in Disneyland and Magic Kingdom in 2006, followed by Tokyo Disneyland in 2007 and Disneyland Paris in 2018—a process Williams (following Schweizer and Pearce 2016) has described as "a form of 'auto-textual poaching'" (2020, 106). This remediation process has come full circle with the opening of Pirates of the Caribbean: Battle for the Sunken Treasure in Shanghai Disneyland in 2016. This new installment of the ride tells a completely original story based on the characters and visual iconography of the film franchise and features a sophisticated ride system that allows vehicles to spin, travel sideways, and even move backwards, as well as the newest version of the audio-animatronic technology, large-scale set pieces, and media projections. Moreover, the ride is housed in "Treasure Cove," a land entirely dedicated to the *Pirates* films and thus a testament to the ongoing popularity of this media product (Mittermeier 2021, 175).

Transmedia storytelling has also influenced the way rides are designed, even extending to waiting areas. Emily Nelson has traced the development of queue lines in the Disney parks all the way back to Disneyland's opening day attraction Rocket to the Moon. The waiting area of this ride featured a pre-show film, as the waiting room for the theater made a regular switch-back style queue impractical (2016, 51). Waiting lines have continued to become more and more elaborately themed—most notably in Disneyland's "Fantasyland" (51); since the 1960s, attractions such as the Haunted Mansion have also regularly begun to incorporate pre-shows (55). As noted above, during the 1980s and 1990s all the queues in the Disney parks became more immersive, and nowadays almost all rides, not only in the Disney parks, extend their narrative to the queue (Lonsway 2009, 123)—often seeking to fully immerse visitors from the moment they set foot into the waiting area (Williams 2020, 105; see also SPACE).

5 Merchandise and Retail

Rides are far from the only attractions in a theme park, however. Shops are also integral to a park's power to attract visitors: the appeal of Disney parks' shops has expanded well beyond the parks' confines and eventually justified the creation of Disney stores in 1987. According to one of the oldest theme park clichés—which even provided the title for Banksy's 2010 documentary *Exit through the Gift Shop* (see PARATEXTS AND RECEPTION)—shops often serve as an extension of an attraction, allowing visitors to purchase a piece of a ride or show and take it home with them. For example, located at the exit of the Pirates of the Caribbean ride at Parc Disneyland in Paris, Le Coffre du Capitaine ("The Captain's Treasure Chest") sells "non-diegetic" items (candy and other products that could also be sold in any other shop in the park) and "extra-diegetic" merchandise (character plush toys in pirate costumes and other items that are "associatively themed" to the ride; see Younger 2016, 351). However, it also offers "diegetic merchandise" such as swords, guns, eye patches, and skull and crossbones flags, which appear to have been taken directly from the ride, as well as photographs of visitors taken during the ride in appropriately themed frames. It is in this sense that scholars like Umberto Eco have criticized theme park shops for deliberately conflating consumption and immersion and thus turning the act of buying into a way of participating in the overall experience: "you buy obsessively, believing you are still playing" (Eco 1986, 43; see also Marin 1984, 247; Fjellman 1992, 161–167; Gottdiener 1997, 75; Bryman 2004, 34).

In addition to building on the experience of an existing ride or show, however, theme parks have also developed strategies to turn shops into attractions of their own. These strategies range from (1) using shops as storytelling devices and (2) operating shops that in one way or another involve potential customers in the production process of the items for sale to (3) fostering interactive experiences built around high-end or "cult" merchandise. Historically, the opening of "New Orleans Square" at Disneyland in 1966 marked a significant step in the development of all these three types of "retail attractions." In fact, before the debut of the Pirates of the Caribbean and Haunted Mansion rides in 1967 and 1969, respectively, this land exclusively featured restaurants and shops or, as the original press kit described them, "Adventures in Shopping" (N.N. 1966, 4). Among these is Mlle. Antoinette's Parfumerie with its fictional owner Mlle. Antoinette, who not only lent her name to the shop but also lives right above it, as the themed balcony with a chaise longue, an elaborate mirror, and two shelves loaded with decorative fragrance bottles on the second floor of the building suggests. Similarly, the exterior and interior designs of Le Brave Petit Tailleur ("Brave Little Tailor") and La Girafe Curieuse ("The Curious Giraffe"), both at Parc Disneyland in Paris, feature numerous elements that invite visitors to explore the fictional stories behind the places, whether they are based on existing narratives, as in the case of Le Brave Petit Tailleur, or have been specifically created for the shop, as in the case of La Girafe Curieuse. As illustrated by La Bottega di Geppetto ("Geppetto's Shop") at Parc Disneyland or Ollivander's Wand Shop at Universal Studios Florida, this works particularly well when the source material—here, *Pinocchio*, *Harry Potter*, and their respective filmic adaptations—already contains a shop. In any case, none of these shops require visitors to make purchases before exploring and enjoying their stories.

Along with the neighboring Le Forgeron ("The Blacksmith"), Laffite's Silver Shop, and Cristal d'Orleans, Mlle. Antoinette's Parfumerie also exemplifies the second type of "retail attraction," in which visitors are somehow involved in the production of the merchandise. Indeed, most of "New Orleans Square"'s "Adventures in Shopping" originally featured what Younger refers to as "retail entertainment," where items "are created in front of the customers" (2016, 353). At Mlle. Antoinette's Parfumerie, for instance, a perfumer would "blend special fragrances to compliment a guest's personality, complexion and type" (N.N. 1966, 7), while at Laffite's Silver Shop a "metal craftsman" would make "silver charms and other jewelry on order from customers" (N.N. 1966, 6). Cristal d'Orleans with its glassblowing showcase is significant insofar as the shop would later spawn branches in other Disney parks, including Crystal Arts at the Magic Kingdom, Glass Slipper at Tokyo Disneyland, and Crystal Treasures at Shanghai Disneyland. All operated and managed by Arribas

Brothers (see Fjellman 1992, 166), these shops also exemplify the tendency to bring in independent lessees for "retail entertainment" (see Younger 2016, 348).

Other examples of "retail entertainment" have included candy making, caricaturing and painting, embroidering, face painting, silhouette making, watchmaking, and even neon sign making (see Younger 2016, 353). With the exception of face painting or "makeovers," as with Disneyland's Bibbidi Bobbidi Boutique (see Gutierrez-Dennehy 2019, 75–78), most recent "retail entertainment" shops have not only replaced the "watch the craftsman" with a "build your own" approach but have also extended the casual interaction between craftsmen/performer and visitors into a fully scripted show with special effects. At Savi's Workshop at Disney's Hollywood Studios, for example, visitors take part in a ten-minute ceremony during which they build their own customized Star Wars lightsaber in order to "continue the Way of the Force" (N.N. 2020). At Ollivander's Wand Shop, by contrast, it is supposedly the merchandise (i.e. the wand) that chooses the customer; and while purchasing the wand is not required, visitors may pre-order an "interactive" wand online before their visit that will then activate special effects in the park environment (see VISITORS).

Ollivander's thus simultaneously exemplifies the third type of "retail attractions," which is built around extremely high-end or "cult" merchandise that serves not only as a souvenir but also to interact with the park landscape and with theme park actors such as employees or other visitors. Here, too, Disneyland's "New Orleans Square" offers an early example in the shape of the One-of-a-Kind shop, which from 1966 to 1996 offered rare antiques and unique items. Operator and antique buyer Otto Rabby has described the One-of-a-Kind and its sister shop, Olde World Antiques at Magic Kingdom (1972–1996), as "'essentially attractions for our guests. [...] We want people to feel free to visit and to browse, to ask questions and to share antique anecdotes with the shop hosts and hostesses. It isn't necessary to buy antiques to enjoy them'" (qtd. in Younger 2016, 348).

Due to a switch in merchandise operating systems at Disney parks—from "Contributory Per Capita Operation," according to which individual stores were allowed to incur losses as long as the overall division was profitable, to "Independent Per Capita Operation," according to which each individual shop had to make a profit (see Younger 2016, 347)—stores like One-of-a-Kind or Olde World Antiques were replaced with more conventional retail locations during the 1990s. Yet the concept of using merchandise to foster fannish activities (see VISITORS) and more generally interaction among theme park actors and between theme park actors and the theme park landscape has survived. Universal's interactive wands may constitute the most technically advanced example, but there are also much simpler models, such as collecting autographs from character performers (see LABOR) or stamps from individual locations in the park. Certainly, the most successful development in this area has been Disney Pin Trading, which since 1999 has encouraged visitors to purchase and collect pins and then trade them with employees and other visitors, often in designated areas or during special events (see Younger 2016, 352). Beyond simply building upon existing attractions, Pin Trading, like other cases of "retail attractions," has thus created an attraction around retail.

Somewhat similar to temporary re-themings of rides (see above), "cult merchandise" can also come in the shape of limited-edition merchandise or souvenirs (pins, popcorn buckets, books, etc.). Such items have proven particularly popular among regular visitors and fans, who alert each other about when and where to find individual products via social media (see Soto-Vásquez 2021, 600). Limited-time only food items from pastries to milkshakes, in turn, may "serve as bright, vibrant content for Instagram foodie guides" (599) besides encouraging repeat visits, although eating and drinking can function as an attraction in other ways as well.

6 Themed and Character Dining

Eating and drinking are basic human needs, and since a theme park visit normally lasts more than a few hours—and due to the fact that visitors are usually not allowed to bring their own food and drink for commercial reasons—theme parks need to offer dining locations that meet their audience's dietary needs, tastes, and preferences (see also INCLUSION AND EXCLUSION). To fulfill this requirement, some theme parks simply bring in external chains: Knott's Berry Farm (Buena Park, California), for instance, features a Panda Express restaurant that is virtually indistinguishable from the company's other locations, while the addition of Starbucks to the Disney parks has integrated the chain into previously existing themed venues (such as Main Street Bakery). Most parks will use the opportunity to not only appropriately theme their dining venues and the food offered therein to allow them to contribute to visitor immersion, but to also turn the dining experience into an attraction itself. This is mostly done by combining dining and entertainment.

Dining venues in theme parks are usually themed in line with the rest of the area, whether they are table service restaurants (e.g. the Jungle Cruise-themed Skipper Canteen at Magic Kingdom), quick service restaurants (e.g. Tokyo Disneyland's Queen of Hearts Banquet Hall), or just kiosks, and can thus become a vital part of the immersion and world-building of the space (Williams 2020, 154). In addition to traditional theme park favorites such as popcorn, turkey legs, and corn dogs, attraction- or land-specific foods enhance the visitor's immersion, as with the otherworldly, glow-in-the-dark treats at "Pandora: The World of Avatar" in Disney's Animal Kingdom or the items for sale in "The Wizarding World of Harry Potter" in Universal Studios. Here, sweets such as Chocolate Frogs, Butterbeer, and Bertie Bott's Every Flavour Beans are sold; notably, the packaging and look of these also follow the art design of the films, proving a draw for many fans. As Williams has argued (2020, 153), the food and beverage items served, as well as the merchandise items being sold, can thus be seen as paratexts in the sense of Gray (2010), and by consuming them visitors are quite literally buying into the immersion.

Dining and entertainment have, in turn, been combined in a variety of ways in theme parks. In some cases, restaurants are strategically placed to allow their patrons to enjoy a nearby show or attraction while they are eating. The terraced sitting areas of Café des Visionnaires (1992–1993) and Café Hyperion in Parc Disneyland in Paris

Fig. 1 View of the Tower of Power slide from the Thai Bar restaurant at Siam Park (Costa Adeje, Spain). *Photograph* Greg Balfour Evans/Alamy Stock Photo

have thus doubled as viewing spaces for the live shows on the Royal Castle Stage and in the Videopolis theatre, respectively. Similarly, the Taverna Mykonos and the Bamboe Baai restaurant at Europa-Park or the Thai Bar restaurant at Siam Park (Costa Adeje, Spain) offer diners perfect views of the visual and acoustic spectacles of the Poseidon water coaster, the Piraten in Batavia dark ride, and the Tower of Power slide (see Fig. 1). Here, then, specific rides and their riders are "staged" and turned into a show for the entertainment of restaurant patrons (see also below).

Mostly, however, theme park "dining attractions" feature their own exclusive entertainment, either in the form of audio-animatronic shows or of professional live performers, especially characters. The history of these concepts goes all the way back to the early years of Disneyland (California): at Aunt Jemima's Pancake House in "Frontierland," for instance, actress Alyene Lewis made daily appearances as former slave-cum-restaurant owner and Quaker Oats mascot Aunt Jemima to welcome visitors, take pictures with them, and generally provide "streetmosphere" (Younger 2016, 368). This lasted from 1955 until the 1960s, when the performances were discontinued in the wake of the Civil Rights movement (see Freitag 2021, 148–153)—an overdue decision, given the character is rooted in minstrelsy. The Enchanted Tiki Room attraction in "Adventureland" (opened in 1963), in turn, evolved out of the idea of "a restaurant followed with an after-dinner bird show, but when the plan proved to be unworkable, the entertainment was expanded into an entire attraction" (Strodder 2012, 157). The concept of a restaurant featuring audio-animatronic entertainment was eventually realized at Magic Kingdom's Cosmic Ray's Starlight Café, where alien lounge singer Sonny Eclipse has been performing his 25-min loop for diners since 1995.

Yet the combination of dining and live performances would prove to be much more successful, especially in the shape of so-called "character meet and greets," which can be found not only at several theme park restaurants, particularly the Disney parks, but also at Universal Studios, Parc Astérix, and Europa-Park. At the Tusker House Restaurant at Disney's Animal Kingdom, for example, guests can meet Mickey Mouse and other Disney characters dressed in safari gear for breakfast and dinner. Tusker House is part of the "Africa" section of the park, and the space is themed in line with the rest of the Harambe Village it is set in. The restaurant doubles as the Safari Orientation Center for the nearby Kilimanjaro Safaris attraction, and guests are greeted with details like a board with the (fictional) names of the drivers listed in the entrance area. The buffet section of the restaurant is set up like a marketplace in the building's courtyard, with buffet tables themed as food stalls in the middle of a square, and the ceiling of the room is covered with colorful drapes to make it appear like an outdoor venue. As in other character dining locations, the characters roam the restaurant and visit the tables to have their pictures taken and interact with visitors.

As Williams has argued, these character meets are an attraction in and of itself for many theme park visitors, particularly those loyal fans that gain cultural capital (in Bourdieu's sense) from these experiences and often spend a great amount of time and effort in meeting "rare" characters (Williams 2020, 148). This particularly applies to Disney's and Universal's theme parks, where the characters that can be met are well-known figures from both animated and live action films and television and where the meet and greet thus functions as a form of celebrity encounter (Williams 2020, 137–143). Remarkably, this is the case despite the complex authenticity of such encounters, as pointed out by Schwarz (2016); even adult fans find value in them, especially as photos taken in these themed settings make for great souvenirs.

7 Park Entertainment

Although character meets, shows, parades, street performers, and all other forms of park entertainment form a central part of the attraction of theme parks, they are only now coming into the focus of theme park research (see Kokai and Robson 2019; Williams 2020). Phantasialand in Brühl, Germany, serves as a good example as it offers a characteristic roster of the types of shows offered in theme parks. Among them is the ice-skating show Ice College (2017), which combines choreographies on ice with a story about a group of girls and boys getting ready for their high school prom. The cast of characters includes such American stereotypes as the outsider girl who wants to join a group of cheerleaders, the popular boy and girl who are admired by the rest of the students, and the nerd. These types are clearly identified by the costumes of the ice skaters: for example, cheerleader outfits with crop tops and short skirts for the cheerleaders, and jeans and a simple striped shirt for the outsider girl, who is thus marked as different from the others. Scenes that drive the plot forward alternate with scenes that allow the performers to showcase their ice-skating talents, with the story of the outsider girl who becomes part of the cheerleader group serving as a frame. The show ends with the prom dance, which is the longest scene and features all the performers. The outsider girl changes her costume on stage so that she now perfectly fits in with the rest of the group and the narrative tension is solved.

While Ice College combines narrative and athletics and is thus reminiscent of classic stage musicals or ice-skating revues, other shows in Phantasialand do not have a coherent plot. JUMP! (2012–2020) at the Silverado Theatre, for example, is a stunt show featuring trampoline stunts, freestyle BMX, a break dancer, and stilts-jumpers. The stage set depicts the construction site of an apartment building in which the trampolinists jump up and down, while the BMX bikers use ramps located on either side of the stage. These two elements of the show are interrupted by the break dancers and the stilts-jumpers, who use boxes standing to the left and right of the house. Here, it is not a narrative that holds the elements of the show together but the visual ensemble of the stage and the costumes.

Miji African Dancers offers yet another type of show experience. The dance performance takes place outside in Phantasialand's "Deep in Africa" section and features four dancers and two drummers. The performance contains African dances and songs, which are described on the park's homepage as "exotic." Like JUMP!, the show has no real narrative but simply showcases seemingly "authentic" African dances—the dancers are in fact part of an existing dance company from that continent. The costumes are kept in brown and orange tones with patterns and furs that a European audience may recognize as African but feature no reference to any specific African countries. The show is mostly made up of drumming, dancing, and singing while the changes in rhythm and volume create dramaturgy and tension. At certain points during the show, individual spectators are invited to take part and join in the dance moves shown by the African dancers. As Jan-Erik Steinkrüger has pointed out (2013, 275), the show stands for vitality and the joy of life, which is the underlying motif of the "Deep in Africa" area and thus brings this theme to life in a performative, embodied way. As visitors may interact with the dancers, the show also provides a way for visitors to participate in the theming. Moreover, the dancers represent not only the joy of life, but also how Africa is seen in Phantasialand in general: it is the traditional European view of a continent "closer to nature than to culture" that is represented here, reproducing problematic colonial stereotypes.

Indeed, theme parks put not only hired performers on stage, but also their own visitors. Drawing on the tradition of the Insanitarium at Coney Island's Steeplechase Park (see Shiffman 1995, 87), theme parks frequently capitalize on the spectacle of visitors enjoying daring thrill rides, and stage this as a performance for other visitors. At Poseidon in Europa-Park, for instance, visitors exchange different gazes that connect them to each other and turn the ride into a theatrical space (Schwarz 2017; see also VISITORS). Poseidon, a water coaster, is the main ride in the "Griechenland" ("Greece") section of the park, and most of the other attractions are arranged around the arena-shaped body of water that belongs to the coaster. The ride's climax—the main drop—is clearly visible from the entire area: along the area's main walkway, which also serves as a passageway to the entrance of the attraction, strategically placed viewpoints offer visitors a spectacular view of the drop. The start of the coaster, housed in a Greek temple, is located on the other side of the arena. On the third, front side, visitors find the Taverna Mykonos restaurant from which the calmer parts of the ride can be observed (see above).

The first gaze one may identify is from the passageway and the viewpoint. It is the gaze of someone who comes to the arena as a passer-by and notices the scenery, but may still be undecided whether to go on the ride or not. The gaze is an identifying gaze: it allows observing visitors to put themselves "into the virtual shoes of the protagonist or to make him understand the emotional meaning of the situation for the character in question" (Tan 2009, 188). Yet this gaze is not solely empathetic, it is also voyeuristic. Hoeppel has argued in the context of the film that empathy and voyeurism do not exclude each other (Hoeppel 1986, 62), and this also applies to these settings. In this broader sense, the voyeur is someone "participating in an action without having his participation agreed upon reciprocally" (198; see SPACE).

In the waiting area of Poseidon, there is another point from which the visitor can see the arena. In contrast to the persons at the viewpoint, the persons waiting in line have already decided to go on the ride and thus already know that they are "about to exchange [the] role as a spectator for the role of an actor on stage" (Domes 2013, 67). Visitors also know that a change of roles will occur and that they will be watched. Warstat's observations on nervousness can help us to understand the situation, as far as the different gazes are concerned: "Here, nervousness is the result of the experience of being seen or of the act of seeing while being aware that one could be seen anytime" (2006, 90). Upon reaching the loading station, visitors finally become riders themselves. The visitor experience resembles that of an actor on stage: they know that they are watched but cannot control or guide the gaze of the other visitors.

The goal of this play with gazes is clear: to make the ride as attractive and suspenseful as possible, to underline its spectacular character, and to play with visitor expectations. But this dynamic exchange is only possible because visitors are physically co-present in one space. As Adam Czirak writes, "producing moments of being seen is an essential feature of physical co-presence" (2012, 144; our translation) and thus the gaze, as it implies the reciprocity of seeing and being seen, constitutes a basic condition of theater. According to Erika Fischer-Lichte, the gaze is "a performative phenomenon par excellence" (2012, 149). Whenever a second person is present, a situation develops in which there is a physical co-presence of two subjects. Hence, we can observe a constant redistribution of the roles of actor and spectator, with the active and passive roles partly converging. Moreover, following Fischer-Lichte's definition of a performance, in the case of the theme park these persons share an experience and take an active part in the "event" of the water coaster and its staging—be it consciously or unconsciously.

Similar—and similarly staged—performances can also be found at other theme parks: the two layered walkways opposite the Talocan suspended top spin ride in Phantasialand's "Mexico" section clearly form a sort of bleachers that allow a maximum number of visitors to watch the ride and the riders—who are aware of the fact that they are being watched. Similarly, the bridge across the big splash of The

Journey of Odyssey at Happy Valley Beijing (China) allows passers-by a spectacular view of their family, their friends, or simply strangers getting soaked, while they themselves are protected from the splashing water by plexiglass sheets (Carlà-Uhink 2020, 144). Such viewpoints turn rides into attractions for non-riders as well.

8 Conclusion

Academic research into theme park attractions is still scarce, as histories and close readings of particularly shows or food and retail locations are few and far between. Williams's recent analysis of Disney's Haunted Mansion has set a new standard for such studies, as it not only considers the ride's paratexts but also investigates how the attraction functions as part of a larger mix (see Williams 2020). However, as this chapter has shown, attractions in the theme park are manifold, and are not—as one may conclude from marketing campaigns, as well as from the focus of academic research—just the rides alone. A theme park's attractions mix has to be carefully calibrated, engaging several different demographics, and often incorporating seasonal as well as larger cultural and historical changes in order to guarantee the success of a theme park. If well balanced, rides (such as thrill rides, dark rides, and flume rides), merchandise and retail, food and beverage, as well as entertainment (shows, character meets, street performers, and other visitors) come together in a park-specific blend that makes the theme park a truly attractive tourist destination.

References

Baedeker, Karl. 1867. *Northern Italy: Handbook for Travellers*. Coblenz: Karl Baedeker.

Bolter, Jay David, and Richard Grusin. 2000. *Remediation: Understanding New Media*. Cambridge: The MIT Press.

Bryman, Alan. 2004. *The Disneyization of Society*. London: Sage.

Bukatman, Scott. 1998. Zooming Out: The End of Offscreen Space. In *The New American Cinema*, ed. John Lewis, 248–272. Durham: Duke University Press.

Carlà-Uhink, Filippo. 2020. *Representations of Classical Greece in Theme Parks*. London: Bloomsbury.

Cartmell, Robert. 1986. *The Incredible Scream Machine: A History of the Roller Coaster*. Bowling Green: Bowling Green State University Popular Press.

Czirak, Adam. 2012. *Partizipation der Blicke: Szenerien des Sehens und Gesehenwerdens in Theater und Performance*. Bielefeld: Transcript.

Domes, Aljoscha. 2013. *Themenparks als inszenierte Flucht(t)räume: Eine kulturwissenschaftliche Untersuchung*. Diploma thesis, University of Hildesheim.

Eco, Umberto. 1986 [1975]. Travels in Hyperreality. In *Travels in Hyperreality: Essays*, trans. William Weaver, 1–58. San Diego: Harcourt Brace Janovich.

Fischer-Lichte, Erika. 2012. *Performativität: Eine Einführung*. Bielefeld: Transcript.

Fjellman, Stephen M. 1992. *Vinyl Leaves: Walt Disney World and America*. Boulder: Westview.

Freitag, Florian. 2017. "Like Walking into a Movie": Intermedial Relations between Disney Theme Parks and Movies. *The Journal of Popular Culture* 50 (4): 704–722.

Freitag, Florian. 2021. *Popular New Orleans: The Crescent City in Periodicals, Theme Parks, and Opera, 1875–2015*. New York: Routledge.
Gottdiener, Mark. 1997. *The Theming of America: Dreams, Visions, and Commercial Spaces*. Boulder: Westview.
Gray, Jonathan. 2010. *Show Sold Separately: Promos, Spoilers, and Other Media Paratexts*. New York: New York University Press.
Gutierrez-Dennehy, Christina. 2019. Taming the Fairy Tale: Performing Affective Medievalism in Fantasyland. In *Performance and the Disney Theme Park Experience: The Tourist as Actor*, ed. Jennifer Kokai and Tom Robson, 65–83. Cham: Palgrave Macmillan.
Hoeppel, Rotraut. 1986. *Psychologie des Filmerlebens*. Frankfurt am Main: Bundesarbeitsgemeinschaft für Jugendfilmarbeit und Medienerziehung e.V.
Janzen, Jack E., and Leon J. Janzen. 1998. Imagineering and the Disney Image: An Interview with Marty Sklar. *The "E" Ticket* 30: 5–11.
King, Margaret J. 2002. The Theme Park: Aspects of Experience in a Four-Dimensional Landscape. *Material Culture* 34 (2): 1–15.
Koenig, David. 2007. *Realityland: True-Life Adventures at Walt Disney World*. Irvine: Bonaventure.
Kokai, Jennifer A., and Tom Robson, eds. 2019. *Performance and the Disney Theme Park Experience: The Tourist as Actor*. Cham: Palgrave Macmillan.
Lonsway, Brian. 2009. *Making Leisure Work: Architecture and the Experience Economy*. New York: Routledge.
Lukas, Scott A. 2008. *Theme Park*. London: Reaktion.
Lukas, Scott A. 2013. How the Theme Park Got Its Power: The World's Fair as Cultural Form. In *Meet Me at the Fair: A World's Fair Reader*, ed. Laura Hollengreen, Celia Pearce, Rebecca Rouse, and Bobby Schweizer, 383–394. Pittsburgh: ETC.
Mangels, William F. 1952. *The Outdoor Amusement Industry: From Earliest Times to the Present*. New York: Vantage.
Marin, Louis. 1984 [1973]. *Utopics: The Semiological Play of Textual Spaces*, trans. Robert A. Vollrath. New York: Humanity.
Mittermeier, Sabrina. 2021. *A Cultural History of the Disneyland Theme Parks: Middle Class Kingdoms*. Chicago: Intellect.
Murray, John. 1838. *A Hand-Book for Travellers on the Continent*. London: J. Murray.
N.N. 1966. *Profile: New Orleans Square: Disneyland*. Glendale: WED Imagineering.
N.N. 2020. Savi's Workshop. *Walt Disney World*. https://www.disneyworld.eu/shops/hollywood-studios/savis-workshop-handbuilt-lightsabers/. Accessed 28 Aug 2022.
Ndalianis, Angela. 2010. Dark Rides, Hybrid Machines, and the Horror Experience. In *Horror Zone: The Cultural Experience of Contemporary Horror Cinema*, ed. Ian Conrich, 11–26. London: I.B. Tauris.
Nelson, Emily. 2016. The Art of Queueing up at Disneyland. *Journal of Tourism History* 8 (1): 47–56.
Online Etymology Dictionary. https://www.etymonline.com/. Accessed 28 Aug 2022.
Rahn, Suzanne. 2000. Snow White's Dark Ride: Narrative Strategies at Disneyland. *Bookbird: A Journal of International Children's Literature* 38 (1): 19–24.
Raz, Aviad E. 2000. Domesticating Disney: Onstage Strategies of Adaptation in Tokyo Disneyland. *The Journal of Popular Culture* 33 (4): 77–99.
Rohde, Joe. 2006. From Myth to Mountain: Insights into Virtual Placemaking. Communication présentée à l'occasion du congrès SIGGRAPH 2006, Thirty-Third International Conference and Exhibition on Computer Graphics and Interactive Techniques, Boston, August 31. http://www.siggraph.org/publications/newsletter/volume/from-myth-to-mountain-insights-into-virtual-placemaking. Accessed 20 Dec 2020.
Schwarz, Ariane. 2016. Spieglein, Spieglein an der Wand: Wer ist das echte Schneewittchen im Land? In *Where the Magic Happens: Bildung nach der Entgrenzung der Künste*, ed. Torsten Meyer, Julia Dick, Peter Moormann, and Julia Ziegenbein, 149–152. Munich: Kopaed.

Schwarz, Ariane. 2017. Staging the Gaze: The Water Coaster Poseidon as an Example of Staging Strategies in Theme Parks. In *Time and Temporality in Theme Parks*, ed. Filippo Carlà-Uhink, Florian Freitag, Sabrina Mittermeier, and Ariane Schwarz, 97–112. Hanover: Wehrhahn.

Schweizer, Bobby, and Celia Pearce. 2016. Remediation on the High Seas: A Pirates of the Caribbean Odyssey. In *A Reader in Themed and Immersive Spaces*, ed. Scott A. Lukas, 95–106. Pittsburgh: ETC.

Shiffman, Dan. 1995. "Let's Go to Coney Island, Old Sport": The Ethnic Sensibility of F. Scott Fitzgerald. *Popular Culture Review* 6 (1): 87–96.

Soto-Vásquez, Arthur. 2021. Mediating the Magic Kingdom: Instagram, Fantasy, and Identity. *Western Journal of Communication* 85 (5): 588–608.

Sotto, Eddie. 2011. Walt Disney Imagineering Who's Who: Ub Iwerks. *ThemedAttraction: A Look Inside Professional Theme Park Attraction Design.* http://www.themedattraction.com/eddiesotto/sotto4-fastpass.htm. Accessed 20 Dec 2020.

Steinkrüger, Jan-Erik. 2013. *Thematisierte Welten: Über Darstellungspraxen in Zoologischen Gärten und Vergnügungsparks.* Bielefeld: Transcript.

Strodder, Chris. 2012. *The Disneyland Encyclopedia*, 2nd ed. Solana Beach: Santa Monica Press.

Surrell, Jason. 2005. *Pirates of the Caribbean: From the Magic Kingdom to the Movies.* New York: Disney Editions.

Tan, Ed. 2009. Wenn wir uns so gut auf die Kunst des Einfühlens verstehen, praktizieren wir sie dann nicht ständig? In *Einfühlung: Zu Geschichte und Gegenwart eines ästhetischen Konzepts*, ed. Robin Curtis and Gertrud Koch, 185–210. Munich: Fink.

Warstat, Matthias. 2006. Vom Lampenfieber des Zuschauers: Nervosität als Wahrnehmungserlebnis im Theater. In *Wege der Wahrnehmung: Authentizität, Reflexivität und Aufmerksamkeit im zeitgenössischen Theater*, ed. Erika Fischer-Lichte et al., 86–97. Berlin: Theater der Zeit.

Williams, Rebecca. 2020. *Theme Park Fandom: Spatial Transmedia, Materiality and Participatory Cultures.* Amsterdam: Amsterdam University Press.

Younger, David. 2016. *Theme Park Design & the Art of Themed Entertainment.* N.P.: Inklingwood.

Authenticity: Theme Parks between Museological, Performative, and Emergent Authenticity

Abstract Since their beginnings, theme parks have employed the notion of authenticity to advertise their offers and the accuracy of their themed areas. Yet authenticity is a complicated and disputed concept. Discarding the idea of authenticity as an attribute of objects, in the sense of the "museological authenticity" of displayed material and of precise, philological reconstructions, recent scholarship has developed a relational, performative notion of authenticity, which views authenticity as the result of an interaction between visitor and object—authentic is what they consider real, believable, and convincing. Hence, obvious mistakes, e.g. in the representation of foreign cultures or anachronisms in the representation of historical themes, do not necessarily contrast with the "perceived" authenticity of a themed area. Furthermore, authenticity cannot be considered a quality that a theme park either has or lacks; instead, it must be understood as a continuum ranging from less to more authentic experiences. Authenticity is conveyed by design choices, for which, among others, budget plays a big role: delivering authentic spaces requires massive investments that are available only to the industry's leaders. This chapter investigates the history of the concept of authenticity in scholarship and discusses its application to theme parks, providing case studies for different meanings and approaches.

1 Introduction

Among the most prized possessions in the personal collection of one of the co-authors of this book is a vinyl leaf that had detached itself from the hedges of Alice's Curious Labyrinth at Parc Disneyland in Paris. The co-author picked up the leaf as a souvenir during his first visit to this park shortly after its opening in 1992—the same year Stephen M. Fjellman published *Vinyl Leaves: Walt Disney World and America*. Like the vinyl leaves from Magic Kingdom's Swiss Family Treehouse which Fjellman describes at the beginning of his study, the leaf in the

This work is contributed by Crispin Paine, Filippo Carlà-Uhink, Florian Freitag, Salvador Anton Clavé, Astrid Böger, Thibault Clément, Scott A. Lukas, Sabrina Mittermeier, Céline Molter, Ariane Schwarz, Jean-François Staszak, Jan-Erik Steinkrüger, Torsten Widmann. The corresponding author is Crispin Paine, Department of Religious Studies, Open University, Milton Keynes, UK.

F. Freitag et al., *Key Concepts in Theme Park Studies*,
https://doi.org/10.1007/978-3-031-11132-7_4

co-author's collection is fake, part of a labyrinth "representing a fake story told in a different medium from, but alluding to, a classic piece of literature" (Fjellman 1992, 1). To the co-author, however, the leaf also constitutes a genuine piece of Parc Disneyland, taken from the original Alice's Labyrinth (Shanghai Disneyland's Alice in Wonderland Maze would not open until 2016), and before the effects of EuroDisney's poor financial performance and the general day-to-day adjustments to operations manifested themselves in the first alterations to the park from its "original" state on opening day. But then, Parc Disneyland itself—in terms of its layout, design, attractions mix, financial structure, and operational procedures—already marked a significant departure from the "original" Disneyland in Anaheim, California, the first and only Disney park created under the supervision of Walt Disney. Moreover, the leaf has no label, stamp, certificate, or anything else that would vouchsafe its provenance from Alice's Curious Labyrinth—except for the story of its owner, whose memory may have attached it to his first visit to Parc Disneyland when he actually gathered it during a subsequent visit and from a different area of the park.

In all its simultaneous fakeness and realness, the vinyl leaf in the co-author's collection emblematizes the debate about authenticity in tourism in general and theme park tourism in particular. Over the course of the last 60 years, this debate has led to fundamental changes in the scholarly conceptualization of "authenticity" and, consequently, to critics' evaluation of (Disney) theme parks as variously fake, real, "totally real" fakes (Eco 1986, 43), and "fake-real" (Pine and Gilmore 2007, 248). Indeed, within academic circles, the notion of authenticity has long been connected to a sense of inaccuracy, discontinuity, and a lack of presence in time and space in terms of the landscapes, material and symbolic design, narrative content, and forms of performance that are representative of themed spaces (Benjamin 1936). Moreover, theme park criticism has often invoked the idea of authenticity in a normative sense, suggesting that a guest who visits a theme park space that may lack authenticity is participating in behaviors or exhibiting ideologies connected with the ills of consumer society and popular culture (Lukas 2016).

Of course, scholars are not the only ones invested in this debate. Within the theme park industry, there is significant discussion of authenticity, and the industry focus on the concept is notably different from that of academics and researchers of theme parks. Among the design communities of the theme park industry, a discussion of authenticity may focus on the very same issues of validity, fidelity, and truthfulness that are also analyzed in academic circles, yet the purpose of such rumination is often for the sake of the visitor who will explore the spaces of theme parks (see also below). The design focus, in turn, differs from the way theme parks have employed the notion of authenticity in their communication with visitors, e.g. in advertisements and other paratexts (see PARATEXTS AND RECEPTION). As theme parks can be conceptualized as a relation between a space representing the theme and a space represented by it ("doubling of landscapes"; see SPACE), from the very beginning, theme park advertisements and self-representations have used the concept of "authenticity" to signify the "overlapping" between these two spaces, their correspondence, and thus also the capacity of the theme park to generate immersion (see IMMERSION). Finally, visitors themselves, and especially fans, may conceive of theme park spaces as authentic, inauthentic, or anywhere in-between for yet other reasons.

Hence, the debate about tourism and theme park authenticity has been not only dynamic, but also multi-perspectival. This chapter seeks to capture significant developments and points of view by first briefly recapitulating the evolving scholarly conceptualization of the term, from more object-centered to subject-centered, and, eventually, performative models of authenticity. In a second step, "New Orleans Square" at Disneyland (Anaheim, California), "Chinatown" at Phantasialand (Brühl, Germany), and "The Wizarding World of Harry Potter" at Universal Orlando Resort (Florida), among others, will serve as case studies for a discussion of the "authentication" strategies used by theme parks in their advertisements and other paratexts. Whereas theme parks' paratextual discourses of authenticity sometimes harken back to more traditional, object-centered notions of the term, the examination of representational theme park spaces in the following subchapter will mainly rely on more recent, subject-centered, and performative models of authenticity. The subsequent section of this chapter will then focus on what has been referred to as "emergent authenticity" (Cohen 1988, 380). In the conclusion, finally, we will broaden our perspective to other tourism spaces as well as museums and heritage sites in order to reflect on the role of theme parks in the authenticity debate.

2 From Authenticity to Authenticities

"Authentic" and "authenticity" belong to those words that we continuously use in daily life without really feeling the need to explain them; we trust that our conversation partners will immediately understand, with semantic nuances determined by the context. When we try to provide a clear, univocal, scholarly definition of the concept, however, we realize that nothing about it is easy or immediate. In its meaning as "real" and "genuine" as opposed to "fake," authenticity was originally, and remained for a long time, mostly if not exclusively understood as a quality pertaining to an object (or a person, for that matter). A Van Gogh painting, in this sense, can be authentic if it has indeed been realized by the famous Dutch artist, or not authentic if it was painted by someone else imitating Van Gogh's style. This is what scholars have referred to as object-centered or "museological" authenticity. Such an authenticity of the object might generate an aura in Benjamin's sense of the term (Benjamin 1936): people seeing and interacting with an object—for instance, when they see Leonardo da Vinci's *Mona Lisa* in the Louvre rather than a copy elsewhere—may perceive such an encounter as an "authentic experience." However, such a connection between "authentic" object and "authentic" experience is not automatic, and the museological authenticity of an item is neither necessary nor sufficient to generate in the visitor an authenticity of experience, which is located in the "feeling subject."

A famous example is provided by German philosopher Martin Heidegger's journeys to Greece in 1962 and 1967. Steeped in philhellenism, Heidegger arrived in Greece full of expectations that the country would immediately spark in him "authentic" inspirational feelings. This was not the case, however, and his visit to Olympia in particular, with the modern village and the touristic structures, proved underwhelming: the museological authenticity of the place and of the archaeological site had failed to generate an authenticity of experience (see Heidegger 2000). By contrast, when in "Travels in Hyperreality" (1986) Umberto Eco compared the "fake New Orleans of Disneyland to the real one," he admitted that some visitors to New Orleans may "risk feeling homesick for Disneyland," for "Disneyland tells us that technology can give us more reality than nature can" (44)—the museological authenticity of the real may prove lacking in the face of the hyperreal recreation.

In these variations and complexities, the concept of "authenticity" has been the topic of critical debates since the 1970s, particularly in the field of tourism studies. Indeed, the "search for authenticity" had already characterized such early forms of travel as the Grand Tour, during which individuals hoped to find the "real" historical, cultural, and natural Southern Europe. With the expansion of the tourism industry in the nineteenth century, "authenticity" also became a trope to distinguish the bourgeois "traveler," looking for "real" people, experiences, and cultures, from the "tourist," who, according to a quite classist and old-fashioned idea of tourism, would not be interested in authenticity but rather pursue, as formulated for example by Daniel Boorstin, "his own provincial expectations" (1961, 106; see also Bendix 1997).

This idea was challenged by Dean MacCannell, who argued that tourists are constantly in search of the authentic, and who thus shifted the attention from what the touristic destinations offer to what the visitors demand. This led him to focus on what he called "staged authenticity," where certain places serve as a "stage for outsiders who are permitted to see details of the inner operation" of their destination (1973, 596). Yet, by focusing on this "front stage" open to the tourist, MacCannell also postulated the existence of a "backstage" to which they would not be admitted and implicitly assumed that the "staged authenticity" was not, in the end, "really authentic": "there is a staged quality to the proceedings that lends to them an aura of superficiality, albeit a superficiality that is not always perceived as such by the tourist, who is usually forgiving about these matters" (595). The tourist remained, in the end, stuck in his role of uneducated visitor. This also means that tourists perceive as authentic what corresponds to their expectations, the latter being shaped by their cultural background. In his analysis of Holy Land tourism, for example, Ron noted that while Catholic visitors include churches and religious sights into their itineraries, Protestant visitors perceive these as inauthentic and prefer visiting strips of seemingly untouched nature to establish a connection with the past (and Jesus) (2010, 115–116).

Indeed, the shift in perspective from an object-centered to a subject-centered authenticity was a long process for scholarship and happened only slowly: the influence of thinkers such as Eco and Baudrillard also played a huge role in reinforcing and perpetuating the idea of an "authentic original": Eco compared "real" cities with their "fake" recreations (see above), and Baudrillard's "simulacrum" distinguishes itself from images precisely by the fact that it does *not* refer to some "profound reality" (Baudrillard 1994, 6). Orvell thus distinguished a "culture of authenticity," which in his opinion had dominated the early twentieth century, from a "culture of the factitious," which expanded after World War II, particularly through the development of mass popular culture and in which "we have a hunger for something like authenticity but we are easily satisfied by an ersatz facsimile" (1989, xxiii). Disneyland, whose opening in 1955 had already led Boorstin to consider it "the example to end all examples" of the fake and inauthentic that would satisfy mass tourists (1961, 103), was an intrinsic part of this discussion. Indeed, "no place before or since its opening in 1955 has provoked more debate on authenticity within modern culture" (Pine and Gilmore 2007, 1).

Yet around the same time, Erik Cohen proposed a constructivist perspective on authenticity that was much more subject-centered: highlighting that different persons have different attitudes toward their touristic experience, he suggested that some visitors might "accept as 'authentic'" what tourists with stricter criteria might "reject as 'contrived'" (1988, 376). This also implies accepting the idea that the "degree of authenticity" derived from an encounter with an item might change over time and thus for example recognizing that Disneyland "will, no doubt, in the future be perceived even by historians and ethnographers, as an 'authentic' American tradition" (380; see also below). It is at this point that the concepts of "authentic" and "original" as well as those of "inauthentic" and "copy," once inextricably linked, were eventually separated.

In the mid-1990s, Bruner (1994, 399–401) thus distinguished four different meanings of (historical) authenticity, which cover both object- and subject-centered perspectives:

(1) Originality, in the sense of "museological authenticity";
(2) Genuineness, which defines as authentic a copy or reconstruction considered historically accurate;
(3) Verisimilitude, which means that the copy or reproduction, even if not accurate, is credible;
(4) Authority, defining who has the power to authenticate.

"Verisimilitude" can be further explained through the concept of "pastness," the quality of "appearing from the past" independent from the actual time of production—this means that while the object may in fact be brand new, it generates an "authentic" feeling of this kind. Holtorf illustrates this with a ruin in Harambe, the African village at Disney's Animal Kingdom (Orlando, Florida):

> The ruin [...] was built in its present state as recently as the mid 1990s. This building is "of the past" because its designer wanted it to be of the past. Pastness is thus an idea that can be built into physical objects. The technique through which this is achieved has been termed "narrative placemaking" by Joe Rohde, a creative executive of Walt Disney Imagineering. (2010, 27)

All of Bruner's criteria, in fact, relate to objects (or persons, or scenarios, or narratives) from the past. Authenticity is indeed commonly perceived—at least in Western societies—as a characteristic that only belongs to "relics" from the past.

As Taylor shows using the example of Maori cultural tourism, this Otherness is associated with an ahistorical origin before (Western) interference:

> Within cultural tourism, and wherever else the production of authenticity is dependent on some act of (re)production, it is conventionally the past which is seen to hold the model of the original. Authenticity in the present must pay homage to a conception of origins. In this way, tourism sites, objects, images, and even people are not simply viewed as contemporaneous productions, or as context dependent and complex things in the present. Instead, they are positioned as signifiers of past events, epochs, or ways of life. In this way authenticity is equated with the "traditional." (2001, 9; see also Steinkrüger 2013)

History before Western interference is thus seen as static and holistic. More generally, Valentin Groebner (2018) has suggested that every form of tourism, aiming at an experience of the visited place in an "original" and "pristine" state, represents a form of time travel. "Authentic" in Bruner's sense thus always refers to a historical dimension—something that also applies when object-centered definitions of authenticity are applied to theme parks: Eco's New Orleans seems particularly authentic because it represents the "true" character of the city as imagined by the visitors, who according to Bruner's "verisimilitude" prefer to ignore the effects of modernization and globalization on the actual city.

The shift away from "museological" authenticity has thus allowed tourism scholars to concentrate on the "tourist gaze" and to recognize that "authentic" is a form of perception by the tourist. In this sense, "authenticity" is composed of two parts: "tourists' preconceptions of the visited culture" and "tourists' perceptions of the actual manifestation of the culture in the attraction" (Yang 2011, 321; see also Bruner 1991, 243; Wang 1999, 355). Hence, "authentic" experiences are those that meet what—based on representations, discourses, and stereotypes available in their country of origin—tourists already know about and expect from the culture, period, etc. they visit (see THEMING). By contrast, as in the case of Heidegger's travel to Greece, simple museological authenticity may fail to generate a feeling of authenticity and thus a satisfactory touristic experience.

Tourism authenticity is thus generated via a process of "authentication," which takes place through a bodily experience of the visitor within the visited spaces. Knudsen and Waade (2010) use the notion of "performative authenticity" to capture this interactive, relational dimension of authenticity; Wöhler calls this the "realization of the virtual" (2011, 72); and Piazzoni speaks of "the physical and social production of space" underlying authenticity (2018, 3). For example, in her study of Thames Town in Shanghai, Piazzoni notes that people within the themed environment change their behavior, seeking to match a perceived authenticity of their performance to the perceived authenticity of the space:

> the Western appearance of the built environment triggers the enthusiasm of the residents and visitors, who willingly modify their behaviors to enhance their own experience of the British atmosphere. At the same time, the sets of aesthetic and moral codes that residents associate with the English theme marginalize those who do not look like they belong or act "appropriately." (2018, 4)

These approaches, then, aim at "bridging the two positions that have emerged in tourism studies with respect to the concept of authenticity, namely object-related (authenticity synonymous with original and trace) and subject-related modes of authenticity (existential authenticity covering bodily feelings, emotional ties, identity construction, and narration related to place)" (Knudsen and Waade 2010; see also Piazzoni 2018, 5).

It is also in the field of tourism (and particularly the field of themed environments) that authenticity has been conceived as a gradable phenomenon: in contrast to those who view authenticity and inauthenticity as an either/or dichotomy, Lukas has noted that within the theme park industry authenticity is discussed as a gradual question on a continuum ranging from less authentic to more authentic experiences (2013, 110–112). Taking the perspective of the producers, Lukas is interested in how to offer visitors the most authentic experience possible within a themed environment. Authenticity, he notes, "is about bringing life to a space in such a way that the guest will see it as real, believable, and worthy of their time and attention" (107). Therefore, he recommends multisensorial and multi-experiential designs full of details and nuances that connect them to a (known) world—real or fictional—and a backstory (107). One challenge in terms of creating authentic theme park spaces is budgetary limitations (132): some of the industry leaders in terms of authentic theme park space design, including Disney and Universal, are able to deliver nuanced, detailed, and deliberately designed spaces within their theme parks because of the massive capital that is available due to the company's size, structure, capital, and transmedia potential.

For Pine and Gilmore, in turn, authenticity constitutes a new imperative not just in tourism but in business in general: consumers, they argue, increasingly see the world in terms of real and fake, seeking "to buy something real from someone genuine, not a fake from some phony" (2007, 1). In an increasingly hyperreal world consumers decide based on their perception of an offering as real or authentic: "Business today, therefore, is all about being *real*. Original. Genuine. Sincere. *Authentic*" (1). Therefore, the authentic experience becomes a costumers' purchase criterion after availability, the reliability of supply, cost, the affordability of a product, quality, and the performance of a product have been fulfilled (5). Pine and Gilmore distinguish five types of authenticity associated with five types of economic offerings:

(1) *Commodities—Natural authenticity*. People tend to perceive as authentic that which exists in its natural state on or of the earth, remaining untouched by human hands; not artificial or synthetic.

(2) *Goods—Original authenticity*. People tend to perceive as authentic that which possesses originality in design, being the first of its kind, never before seen by human eyes; not a copy or imitation.

(3) *Services—Exceptional authenticity*. People tend to perceive as authentic that which is done exceptionally well, executed individually and extraordinarily by someone demonstrating human care; not unfeelingly or disingenuously performed.

(4) *Experiences—Referential authenticity*. People tend to perceive as authentic that which refers to some other context, drawing inspiration from human history, and tapping into our shared memories and longings; not derivative or trivial.

(5) *Transformations—Influential authenticity*. People tend to perceive as authentic that which exerts influence on other entities, calling human beings to a higher goal and providing a foretaste of a better way; not inconsequential or without meaning. (49–50)

As the following section will show, in their design and in their marketing, theme parks have drawn on virtually all of the various notions of authenticity discussed in this section.

3 Authenticities in Theme Parks

3.1 Marketing

For the official dedication of its "New Orleans Street" section in August 1955, Disneyland in Anaheim invited, among others, New Orleans-born actress Dorothy Lamour as a guest star. As the *Disneyland News*—a tabloid-sized, illustrated mock newspaper sold at the park from 1955 to 1957—reported in its article on the official dedication, during the ceremonies the "native of the Southern city" testified that "New Orleans Street" "looks exactly like the wonderful city in which I was born" (N.N. 1955, 11). This early example illustrates how from their beginning, theme park marketing and paratexts (see PARATEXTS AND RECEPTION) have drawn on "experts" or authoritative figures to assert the authenticity of the parks' recreations of both real and fictional cultures and places, thus mixing what Bruner refers to as "genuineness" and "authority."

Indeed, for the opening of "New Orleans Square" eleven years later, Disneyland enlisted Victor Schiro, then Mayor of the Crescent City, whose assessment of the theme park version of the city was readily covered in the local press: in an article entitled "'Looks Just like Home': Mayor of Real New Orleans Praises Disneyland Replica" and published the day after the opening, for example, the *Los Angeles Times* emphasizes how "plainly enchanted" and impressed Schiro was with the "authentic" visuals, smells, foods, and especially sounds offered by "New Orleans Square." Concerning the "highly trained staff," Schiro is quoted as having said "in his soft accent": "They pronounce it New Olyuns. […] Not Noo Orleens" (McPhillips 1966). Apparently not satisfied with Schiro's judgment alone, the *Orange Daily News* had sent Bill Carney, "who is presently Sports Editor and a staff writer for the Daily News, [but] originally from New Orleans and lived there for 22 years," to cover the event (Carney 1966). Carney fully corroborates Schiro's remarks by judging "New Orleans Square" as "another triumph for Disney in factual and delightful portrayal of America's rich history and heritage" (Carney 1966).

Much like Disneyland did in the 1950s and 1960s with its New Orleans-themed spaces, Phantasialand in Brühl (Germany) has regularly invited state representatives from China to events connected to its Chinese-themed spaces—firstly, the park's "China Town" section, opened in 1981 in the presence of bishop Titus Chang and the "virtual" presence of representatives of the Republic of China's ministry of culture, which awarded the park owners a medal for the promotion of Chinese culture; and secondly, the adjacent Ling Bao hotel, whose foundation stone was laid in 2003 by Zhenjiang Mayor Shi Heping and then Counselor of the embassy of the People's

Republic of China Shi Mingde (it is worth noting that the park thus does not consider the People's Republic the only "authenticating authority," and that from a German perspective, even a Christian bishop can have an authenticating function, as long as he is of Chinese origin). This strategy has even been employed with themed spaces based on fictional worlds: according to Waysdorf and Reijnders, the much-publicized involvement of author J. K. Rowling in the design of "The Wizarding World of Harry Potter" at Universal Resort (Orlando, Florida) played an important role in Potter fans' acceptance of the space: "Authenticity for these fans is based on the figure of Rowling and her approval of the park. As the series' 'brand guardian' [...], her approval gives the park a sense of legitimacy" (2018, 179).

At the same time, theme park marketing and paratexts also rely on museological authenticity and the idea of originality to evoke "authentic feelings" and to convince visitors that they are encountering "the real thing." For example, Phantasialand—a park that insists on being "unique and authentic" and paying attention to "the smallest detail" (Phantasialand 2021)—routinely stresses the object-centered authenticity of its offerings, pointing out, for example, that the souvenirs on sale in the "Deep in Africa" area are actually "original" products of Cameroon. In press publications about its Chinese-themed spaces, Phantasialand has also regularly evoked both Bruner's "originality" and Pine and Gilmore's "original authenticity" by stressing that, on the one hand, the park used component parts produced in China for the construction of the "China Town" area and the Hotel Ling Bao, and, on the other hand, the food offered in the area's restaurants "has a unique taste, very authentic, it isn't off-the-shelf. This is what we wanted: a genuine experience" (Kranz 2018).

By contrast, in a 1978 interview with *New West* magazine, then Executive Vice-president and Chief Operating Officer of WED Enterprises (the company responsible for the design of the Disney parks, now known as Imagineering) John Hench claims that Disney's designs are authentic in that they unlock "deeper truths" about what they depict: "You take a certain style, and take out the contradictions that have crept in there through people that never understood it or by accident or by some kind of emergency that happened once and found itself being repeated—you leave those things out, purify the style, and it comes back to its old form again" (Haas 1978, 18; see also Wallace 1985, 33–36)—thus evoking Pine and Gilmore's concept of "natural authenticity."

Another notion of authenticity that has been regularly evoked in theme park marketing is Pine and Gilmore's "exceptional authenticity"; this is particularly relevant in paratexts that offer visitors glimpses behind the scenes such as coffee table books on theme park design, TV documentaries, or guided tours of the park's backstage areas. As Mathew J. Bartkowiak has argued in his analysis of such "behind-the-scenes" material on Disney parks, even though the backstage supposedly revealed in these paratexts may itself be staged (as had already been suggested by MacCannell; see above), there is "cultural capital to be amassed for the curious, the critical, and the Disney fan garnering insider knowledge" (2012, 946–947). Perhaps even more importantly, however, "behind-the-scenes" paratexts routinely stress the enormous efforts that have gone into the design, construction, and maintenance of the parks' landscapes, thus inviting a "revered gaze" (Griffiths 2008, 286) toward and testifying to the "exceptional authenticity" of the parks.

A curious example of this is provided by a 1987 Europa-Park souvenir book, which opens with two double pages on that season's newest attraction, Piraten in Batavia (an imitation of Disney's Pirates of the Caribbean). To be sure, the three paragraphs that accompany the numerous pictures of the ride also briefly tell readers about the setting and the plot of the attraction. Yet the rest of the text is mainly concerned with technical details, listing the size of the ride building, the overall length of the journey, the number of robots and ride vehicles, and the ride's theoretical hourly capacity. Hence, while the pictures convey the content and atmosphere of the ride, the text invites readers to marvel at the complex machinery necessary to produce this spectacle (it is not extraneous to this that Europa-Park is owned and operated by the Mack family, who also owns Mack Rides, and that originally the park was also intended as a sort of showcase for their products). This impression is reinforced by the shorter French version of the text, which contains no descriptions of the ride itself, but instead stresses the quality of the attraction by judging it, five years before the opening of Disneyland Paris and its own Pirates of the Caribbean ride, as "rivalling that of Disney" (Europa-Park 1987, 1; our translation). Hence, Europa-Park asserts the "exceptional authenticity"—as well as the "genuineness" (sensu Bruner)—of Piraten in Batavia by explicitly identifying it as a successful copy.

In early summer 2020, finally, Europa-Park started its "Urlaub in Europa" ("holiday in Europe") campaign, which promoted the park as a surrogate destination when leisure travel to many European countries was restricted due to the COVID-19 pandemic. Posting shots of its various themed landscapes on Instagram—all carefully framed so that no "Europa-Park" logos or other tell-tale signs were visible—and captioning them with questions such as "Are you in [the Netherlands] or in the Dutch-themed area at #Europa-Park?", Europa-Park made potential visitors themselves the arbiters of "verisimilitude" (Bruner) and sought to convince them that a visit to the park could substitute for a visit to one of the European destinations depicted in it. Hence, the campaign explicitly asked potential customers to perform a process of "authentication" based on pictorial representations of the park landscape.

3.2 Design and Performance

Usually, the process of "authentication" takes place within the parks themselves, however, with visitors checking their bodily experience of the representation of a particular theme against their media-based preconceptions and imaginaries of this theme (see above). Tourists, Maria Månsson maintains, have a "mediated imaginary sense of a space and this sense will be negotiated in the physical contact" (2010, 176); similarly, Karlheinz Wöhler (2011, 73) has argued that texts and images about tourism sites distributed through print media, radio, movies, television, and the internet aggregate to form a virtual image of a tourism space—a process he refers to as "virtualization" or "cognitive mapping" (72)—that is then "realized" by the actors involved in the touristic space—or in the latter's representation in a theme park. It is therefore no surprise that "verisimilitude" or "recognizability"—the idea that "[s]omething authentic is simply something that looks as you imagine it might, based on a lifetime of movies and television and glossy advertisements in magazines" (Curtis qtd. in Holtorf 2017, 502)—constitutes one of the cornerstones of theme park design: rather than trying to "authentically" (in the museological sense of the term) represent a particular culture, place, or time, theme parks seek to anticipate their visitors' (historically and culturally contingent) imaginaries of these themes by directly basing their designs on them (see THEMING). The process of "authentication" is turned into a self-fulfilling prophecy.

The "World Showcase" at Epcot (Orlando, Florida) and "Frontierland" at Parc Disneyland (Paris, France) provide intriguing case studies. With respect to the former, Disney had already confidently predicted successful processes of "authentication" in its paratexts. Much like Europa-Park would do decades later in its "Urlaub in Europa" Instagram posts, a 1982 Epcot coffee table book, describing the "France" pavilion in the park's "World Showcase," asks potential visitors: "Are you in Paris?" and immediately gives the answer: "Yes" (Beard 1982, 172). To ensure "verisimilitude" and thus to enable the park to "pass the tourist's test of recognizable authenticity" (Sheppard 2016, 73), Epcot's designers had created not a series of simulacra of the countries represented in the pavilions themselves, but, as Fjellman has argued, a series of "simulacra of the touristic world" (1992, 233), thus turning "the tourist experience of another culture" into "a tourist experience in itself" (Carson 2004, 232). For example, by evoking the "image of a homogenous mestizo Mexican national culture," the "Mexico" pavilion corresponded, as Randal Sheppard has argued, "to images of Mexican geography, history, and culture popularized through the promotion of Mexico as a tourist destination in the United States for roughly four decades prior to the pavilion's opening" (2016, 81). Likewise, regarding the pavilion dedicated to the U.S.'s northern neighbor, Florian Freitag has shown how the amalgamative combination of architectural and other references to Canada's various regions reflected the nation's then-current self-conception and -promotion as a "confederation of regions" (2018, 169). A similar "metatouristic" approach (Köck 2006, 14) would later also be taken by Disney California Adventure in Anaheim, California, which had visitors enter the park through a giant postcard of California.

With respect to "Frontierland" at Parc Disneyland in Paris, Jean Baudrillard had already maintained in *America* that it is "useless to seek to strip the desert [of Death Valley] of its cinematic essence in order to restore its original essence; these features are thoroughly superimposed on it and will not go away. The cinema has absorbed everything—Indians, mesas, canyons, skies" (1988, 69). Seeking to provide "an interpretation that fulfills your expectations of what these romantic areas might be like," Parc Disneyland's Executive Producer Tony Baxter thus turned to the movies in order to create a version of "Frontierland" that a European audience would be able to recognize and successfully authenticate during their visit:

> [W]e noticed the intrigue that the American southwest had for the French and for other Europeans. The Grand Canyon or Monument Valley, the images that have become familiar through John Wayne westerns are symbolic for Europeans of the entire American west, even

> if we feel that in reality these regions are as varied and diverse as Europe is diverse. That is why we created what we feel is going to be a stunning red environment, that is as much in contrast with the Marne Valley here as the greenery of our Disneyland river [at Disneyland's "Frontierland" in Anaheim, California] is with the dry Southern California climate. (Baxter 1992, 71–79)

As both Andrew Lainsbury and Deborah Philips have pointed out, Disney thus sought to "identify and accommodate [Europeans'] intangible perceptions of the United States" (Lainsbury 2000, 62) by "calculatedly inflect[ing]" the design of Parc Disneyland's "Frontierland" "through cinema" (Philips 2012, 231). A 2021 article in *Attractions Magazine* even provides a "close-up on the movies influencing Frontierland at Disneyland Paris" and identifies, among others, Don Siegel's *The Shootist* (1976; starring John Wayne) and Lawrence Kansas' *Silverado* (1985) as having inspired the designers (see N.N. 2021). Bruce Broughton's *Silverado* theme has also been integrated—alongside many other western movie themes—into the "Frontierland" soundtrack, thus making the cinematic basis of the land's design explicit to visitors.

Visitors' preconceptions of and imaginaries about a particular theme are perhaps most easily anticipatable when they are exclusively based on one particular medial artifact, as is the case with IP-based theming (see MEDIA). Alongside economic factors in the shape of synergy effects, their high potential for recognizability and verisimilitude may indeed be one of the reasons behind the recent popularity of IP-themed areas and attractions (see THEMING). The 2013 re-theming of the "Hotel Santa Fé" at Disneyland Paris (France) is a case in point: originally featuring a movie-based, but rather generic and much-criticized southwestern theme—here, too, the cinematic approach was made explicit by having the entrance area evoke a drive-in theater, complete with a giant movie screen showing a painting of Joe, Clint Eastwood's "man with no name" character in Sergio Leone's *A Fistful of Dollars* (1964)—the hotel would eventually switch to referencing the much more broadly recognizable Disney/Pixar animated movie *Cars* (2005) in 2013.

This is not to say that museological authenticity does not play a role in actual theme park landscapes. In fact, their verisimilar approach notwithstanding, all of the themed areas discussed above have also incorporated "authentic" objects into their design—from pre-Columbian artifacts and Mexican artworks at Epcot's "Mexico" pavilion and nineteenth century mining equipment at Parc Disneyland's "Frontierland" to original artwork from *Cars* at "Hotel Santa Fé." Significantly, however, these objects are also often displayed museologically: in "Mexico," for example, the artifacts and artworks are presented in glass cases and dutifully labeled in a separate room at the entrance to the pavilion; in "Frontierland" some of the equipment is free-standing (rather than integrated into "new" structures) behind fences; and at "Hotel Santa Fé" the *Cars* artwork has been carefully framed. Amidst countless credibly accurate replicas, object-centered authenticity needs to be framed as such at theme parks in order to be recognizable to visitors. Museological authenticity can thus even take the shape of exhibitions or small museums, such as the Museo del café at the coffee-themed Parque del café near Montenegro (Colombia; see Fig. 1), the Chasing Rainbows Museum at Dollywood in Pigeon Forge, Tennessee (dedicated to the career of country singer Dolly Parton; see Morales 2014, 123–125), or the Scriptorium in the religious theme park Holy Land Experience in Orlando, which was advertised as a "factual library and research center that houses [...] ancient scrolls, manuscripts, and early printed editions of the Bible" (Holylandexperience 2018). In fact, the Scriptorium worked as a legitimizing root or anchor that added proof, and thereby, authenticity to the otherwise artificial exhibits (Beal 2005, 60; Paine 2019, 117).

Finally, performative authenticity at the theme park is also—one could say, by definition—a matter of performance, namely the "performative" or "emotional labor" of theme park employees. Like other service and experience industries, theme parks require their staff not only to perform specific tasks such as pushing buttons at rides, serving food at restaurants, or ringing up visitor's purchases at shops, but also to convey certain emotions and attitudes that contribute to "the aura surrounding the service" (Bryman 2004, 105), for example via a smile. Visitors, however, are highly attuned to these performances and quick to note glitches. Thus, Helen Morales recalls a visit to Disney California Adventure when she and her daughter

> were dining in Ariel's Grotto and visited by a relay of Disney princesses. When Snow White came up to our table I said to her, "Poor Snow White, you must be fed up with smiling." She replied in character, "Oh no! I *love* smiling," but there was something strained in her eyes,

Fig. 1 In addition to thrill and family rides, Parque del café near Montenegro (Colombia) also offers visitors an "authentic" reproduction of a Colombian coffee village and a "Museo del café." *Photograph* Salvador Anton Clavé

as if she might snap at any moment and pull out an AK-47 from under her petticoats. (2014, 135)

Indeed, lest visitors should perceive employees' emotional labor as disingenuous, Disney and other theme parks usually require what sociologist Arlie Hochschild has referred to as "deep acting" (see Hochschild 1983), which asks performative workers to not only go through the motions of displaying the correct emotional form, but also to really feel the emotions that they are supposed to exhibit (see LABOR). Whether "meticulously scripted or heartfelt," or perhaps a case of "deep acting," the emotional labor at Dollywood proved more convincing to Morales: "everyone I spoke to professed their gratitude to Dolly Parton and their love of their jobs. […] There was no strain in the eyes of anyone I spoke to at Dollywood; the affection for Dolly Parton seemed widespread and genuine" (2014, 135). Like theme parks' designs, the performances of their employees must successfully pass the visitor's "test" of authentication.

3.3 *Emergent Authenticities*

Visitors authenticate theme park spaces not only by measuring them against their virtual images of the respective themes, however. A second standard for the "tourist's test of recognizable authenticity" is provided by visitors' preconceptions and imaginaries about the theme park itself. In order to be perceived as authentic, the theme park must, like any other tourist space, live up to its own virtual image, which is based on images and texts distributed through the media by the park itself and others (see PARATEXTS AND RECEPTION). Obviously, such media-based preconceptions and imaginaries become more powerful over time: a park with a long history—or a park belonging to a well-established park brand—thus possesses, and needs to be aware of and contend with, a much richer, "thicker" virtual image than one that has just recently been opened. As a result, such a park will have to stand a double test of authentication by not only meeting visitors' preconceived notions about its particular theme(s), but also those about the way the park works with these themes. Morales' disappointment in her encounter with Snow White at Disney California Adventure (see above) may have been due to the fact that she measured it not only against her virtual image of Snow White or fairy tale princesses in general, but also against her specific preconceptions about encounters with Disney princesses at Disney parks. Failing to live up to the expectations raised among potential visitors through, e.g. advertisements, a park may thus risk being perceived as inauthentic.

Moreover, repeat visitors and fans in particular may arrive at a park not only with purely media-based preconceptions and imaginaries about their visit, but also with expectations that have been formed during their past interactions with the park (see VISITORS). This is what Cohen has described as "emergent authenticity": referring explicitly to Disney's American parks as examples, he maintains that "it is possible for any new-fangled gimmick, which at one point appeared to be nothing but a staged 'tourist trap,' to become over time, and under appropriate conditions, widely recognized as an 'authentic' manifestation of local culture" (1988, 380). As the result of a growing emotional attachment to a theme park space, "emergent authenticity" constitutes yet another standard for visitors' "test of recognizable authenticity" against which the park in its current state and especially alterations to it are measured. Indeed, both Kiriakou and Williams have examined fans' reactions to such changes as the updating, replacement, and closure of "classic" theme park attractions as rooted within a "nostalgic attachment to a version of the resort that no longer exists" (Kiriakou 2017, 100) and a perceived threat to fans' sense of "ontological security" and "home" (Williams 2020, 217). At the same time, however, "emergent authenticity" has also provided theme parks with a rationale for keeping and even bringing back certain elements that may have otherwise fallen victim to theme parks' constant striving for novelty in the interest of economic competitiveness.

A case in point is the Carousel of Progress. Originally designed by Disney for the General Electric Pavilion at the 1964–65 New York World's Fair, this revolving theater show about the constant improvement of domestic life through electric appliances was subsequently added to the "Tomorrowland" section of Disneyland (Anaheim, California) before being relocated, in 1975, to the "Tomorrowland" of Magic Kingdom (Orlando, Florida). The decision to keep the Carousel during the 1994 renovation of "Tomorrowland" could be explained by the fact that the by then almost 30-year-old ride perfectly fit the section's new retro theme of the "future that never was." As Robson has argued, however, as "one of the few attractions at Walt Disney World that Walt Disney himself actually helped design and execute," the attraction "holds a special place in the heart of many long-time Disney fans" and thus (also) remains in the park "in memory of Walt Disney himself" (2019, 36). In fact, the park has even sought to heighten the nostalgic appeal of the Carousel by restoring its original soundtrack and by adding "Walt Disney's" to its official name and marquee, thus enlisting Disney himself as an authoritative figure to present the attraction as an "authentic" Disney experience.

It was also largely due to nostalgia that Europa-Park's Piraten in Batavia (see above) was immediately rebuilt after it had spectacularly burned to the ground in 2018. According to designer Marc Heinzelmann, for the park "it was pretty clear straight away that we would rebuild the Pirate ride as a reminder of the old ride" (see Ralph 2020). Indeed, as discussions among fans on social media and in fan fora attest, Europa-Park's copy of Disney's Pirates of the Caribbean had become, in the eyes of some, an original that could never be replaced. Fan attachment supposedly not only dictated the decision to reconstruct the ride, but also the way it would be rebuilt: in addition to toning down the original ride's colonialism (see WORLDVIEWS), the designers sought to develop a version that was "creative enough on its own merit" but that would also respect the "many emotions attached to the old ride" (Ralph 2020).

Not all theme park aficionados can rely on "their" park bringing back fan favorites (or leaving them in their original state in the first place), however—some feel inspired to take action themselves. Examining fans' reactions to the closure of selected Walt Disney World rides and attractions, Williams has highlighted how, for example, the replacement of the Maelstrom ride at Epcot's "Norway" pavilion with the *Frozen*-inspired Frozen Ever After in 2016 met with criticism as fans felt the switch to an IP-based ride was incompatible with the park's original educational mission and thus hurt Epcot's authenticity: "The idea that a *Frozen* attraction was 'inauthentic' was common in Tweets posted with the #savemaelstrom hashtag" (2020, 222). Rather than an emergent authenticity, fans noted and were concerned about a "decreasing" authenticity. Moreover, Williams documents fans' attempts to virtually preserve long-gone Walt Disney World attractions such as the River Country water park (1976–2001) through "online memorialization" on unofficial websites (2020, 229–31). In fact, there are numerous fan websites and YouTube channels commemorating and virtually recreating "extinct" theme park rides and attractions, with e.g. Werner Weiss's *Yesterland* (www.yesterland.com) conceptualizing itself as a "theme park on the web" composed of closed Disneyland attractions (see Weiss 2012) and Kevin Perjurer's "Defunctland" (www.youtube.com/c/Defunctland) offering digital recreations of bygone rides.

Hence, in response to the fact that "Disneyland is not a museum"—a dictum sometimes ascribed to Disney designer Tony Baxter and frequently quoted in fan discussions about changes at (Disney) theme parks (see Koehler 2017, 140)—fans have created their own virtual museums. These and other practices that attest to fans' attachment to a particular ride, shop, restaurant, or park—from debates on social media to farewell rituals such as taking one final visit (see Williams 2020, 220–221)—point to the central role of museological authenticity in these emotional relationships with theme park spaces: for fans, "emergent authenticity" often takes the shape of a canonization of specific theme park elements that they consider fundamental to the identity of the place. As a result, they may experience the constant physical evolution of theme parks not as a necessary part of theme parks' economic strategy (see ECONOMIC STRATEGY), but rather as "nerve-wrecking" (Kiriakou 2017, 105) or even as threatening their sense of ontological security (see Williams 2020, 216–217) and may thus resist any changes. From the perspective of the parks, in turn, "emergent authenticity" constitutes a thoroughly ambivalent phenomenon: while visitors' attachment to the theme park ensures repeat visits and even provides the source of events and marketing campaigns such as ride anniversaries or "farewell seasons," it may also cause highly emotional reactions and resistance to even minor alterations, with theme park operators finding themselves in the role of heritage guardians rather than that of entrepreneurs and entertainers.

4 Conclusion

There has long been a popular view that, unlike theme parks which deal in simulacra and illusion, museums (and perhaps also heritage sites) guard "the real thing." Even theme park enthusiasts sometimes accept this. As one post in the *Theme Park Insider* blog put it on April 15, 2019, the day a fire had broken out at Notre-Dame de Paris that would leave the famous cathedral severely damaged, "Go see things if you can.

Theme parks are great, but there is REAL beauty and wonder in this world that could be gone in an instant" (Niles 2019). This easy dichotomy between "fake" theme parks and the "REAL," however, can be challenged from several perspectives.

Firstly, the notion of authenticity has come to be as strongly problematized and disputed in the museum and heritage world as it is elsewhere. If the 1964 *Venice Charter* of the International Council on Monuments and Sites (ICOMOS) claimed that "[i]t is our duty to hand [the historic monuments of generations of people] on in the full richness of their authenticity" (ICOMOS 1964) without any attempt to define the term, fifty years later Mary Brooks has argued "that authenticity has shifting meanings and that, in the museum, this resides both in the object and in the experience of viewing the object" (2014, 8). Moreover, museums have developed strategies to make the visitors' experience "authentic"—in all the senses discussed above—well beyond simply caring for the age and production of the exposed objects. Like theme parks, they readily embrace new technology to interpret their collections and tell their stories, from photographs and plaster casts in the 1860s to AR and VR today. Titanic Belfast, for example, a museum opened in 2012 on the site where the Titanic was built, includes a reproduction of the staircase of the Titanic and a short trip on a car that moves—very much alike a theme park dark ride—along a replica of the Titanic's rudder, past scenes representing the various phases of shipbuilding.

Secondly, it is not only museums' adoption of certain (re)presentational strategies commonly associated with theme parks, but also certain economic practices that contribute to making strict distinctions between theme parks and heritage sites increasingly difficult. During a research trip to E-Da, a theme park in Kaohsiung (Republic of China) that features a themed area inspired by the Greek island of Santorini, one of the contributors to this chapter took a selfie and sent it to a friend who lives in Greece. The trick worked and he promptly received a text from said friend who asked the contributor why he had not let her know that he was coming to Greece and visiting Santorini. The joke relied as much on the park's ability to create a convincing version of Santorini that could be used in selfies as on the contrast between the "real" (Greek) and the "fake" (Taiwanese) Santorini. But there is a further twist to E-Da's metatouristic offering: the shops in the "Santorini" area sell fridge magnets and other souvenirs of Santorini and Greece. As many souvenirs sold to tourists in the "real" Santorini are made in Taiwan, what visitors to E-Da can purchase is, in fact, the very same object they would bring back from an actual trip to Greece (see Carlà-Uhink 2020, 159–161).

Thirdly, and perhaps most importantly, at least some people have come to relate to theme parks as to other, "real" places. Indeed, the devastated reactions of fans to the 2018 fire that destroyed Europa-Park's Piraten in Batavia (see above) were surprisingly similar to the responses to the Notre-Dame fire: "Terrible. I have been crying the entire day," one post in the *EP-Board* blog noted, "It sounds silly, I know, but for me an entire world has broken down" (EP-Board 2018; our translation). Likewise, in the introduction to a book chapter Brian Lonsway writes:

> I want to start with the obvious, or at least what is obvious to me: themed environments are authentic. [...] My experience in a themed environment, constructed and narratively framed

> as it may be, is real, sensorial, and personally meaningful; as a result of such an experience, have I not in fact become more empowered with the complexities of my reality? (2016, 239)

The somewhat apologetic and defiant tone of the blog post and Lonsway's statement may indicate that popular views on the "fakeness" of theme park spaces are still lingering, but that they are very well changing.

References

Bartkowiak, Mathew J. 2012. Behind the Behind the Scenes of Disney World: Meeting the Need for Insider Knowledge. *The Journal of Popular Culture* 45 (5): 943–959.

Baudrillard, Jean. 1988. *America*. Trans. Chris Turner. London: Verso.

Baudrillard, Jean. 1994 [1978]. The Precession of Simulacra. In *Simulacra and Simulation*, trans. Sheila Faria Glaser, 1–42. Ann Arbor: The University of Michigan Press.

Baxter, Tony. 1992. Euro Disneyland [Interview]. In *Euro Disney: Special Issue of Connaissance des Arts*, ed. Philip E. Jodidio, 65–84.

Beal, Timothy. 2005. *Roadside Religion: In Search of the Sacred, the Strange, and the Substance of Faith*. Boston: Beacon Press.

Beard, Richard R. 1982. *Walt Disney's Epcot: Creating a New World of Tomorrow*. New York: Harry N. Abrams.

Bendix, Regina. 1997. *In Search of Authenticity: The Formation of Folklore Studies*. Madison: University of Wisconsin Press.

Benjamin, Walter. 1936. L'œuvre d'art à l'époque de sa reproduction mécanisée. *Zeitschrift für Sozialforschung* 5: 40–68.

Boorstin, Daniel J. 1961. *The Image; Or, What Happened to the American Dream*. New York: Atheneum.

Brooks, Mary. 2014. "Indisputable Authenticity": Engaging with the Real in the Museum. In *Authenticity and Replication: The "Real Thing" in Art and Conservation*, ed. Rebecca Gordon, Erma Hermens, and Frances Lennard, 3–12. London: Archetype.

Bruner, Edward M. 1991. Transformation of Self in Tourism. *Annals of Tourism Research* 18 (2): 238–250.

Bruner, Edward M. 1994. Abraham Lincoln as Authentic Reproduction: A Critique of Postmodernism. *American Anthropologist* 96 (2): 397–415.

Bryman, Alan. 2004. *The Disneyization of Society*. London: Sage.

Carlà-Uhink, Filippo. 2020. *Representations of Classical Greece in Theme Parks*. London: Bloomsbury.

Carney, Bill. 1966. Disneyland Adds New Orleans Charm. *Orange Daily News*, July 28: A4.

Carson, Charles. 2004. "Whole New Worlds": Music and the Disney Theme Park Experience. *Ethnomusicology Forum* 13 (2): 228–235.

Cohen, Eric. 1988. Authenticity and Commoditization in Tourism. *Annals of Tourism Research* 15: 371–386.

Eco, Umberto. 1986 [1975]. Travels in Hyperreality. In *Travels in Hyperreality: Essays*, trans. William Weaver, 1–58. San Diego: Harcourt Brace Janovich.

EP-Board. 2018. Brand bei den Piraten in Batavia. https://www.ep-board.de/viewtopic.php?f=5&t=9102&sid=66e055bf63fb2049d11e0f015909b2a2&start=80. Accessed 27 Oct 2021.

Europa-Park. 1987. *Europa-Park*. N.P.: N.P.

Fjellman, Stephen M. 1992. *Vinyl Leaves: Walt Disney World and America*. Boulder: Westview.

Freitag, Florian. 2018. "Who Really Lives There?": (Meta-)Tourism and the Canada Pavilion at Epcot. In *Gained Ground: Perspectives on Canadian and Comparative North American Studies*, ed. Eva Gruber and Caroline Rosenthal, 161–178. Rochester, NY: Camden House.

Pine II, B. Joseph, and James H. Gilmore. 2007. *Authenticity: What Consumers* Really *Want*. Boston: Harvard Business School Press.

Griffiths, Alison. 2008. *Shivers Down Your Spine: Cinema, Museums, and the Immersive View*. New York: Columbia University Press.

Groebner, Valentin. 2018. *Retroland: Geschichtstourismus und die Sehnsucht nach dem Authentischen*. Frankfurt: S. Fischer.

Haas, Charlie. 1978. Disneyland Is Good for You: Charlie Haas on the Magic Kingdom's Master Manipulator. *New West*, December 4: 13–19.

Heidegger, Martin. 2000. *Zu Hölderlin—Griechenlandreisen. Gesamtausgabe*, Vol. 75, ed. Curd Ochwadt. Frankfurt: Vittorio Klostermann.

Hochschild, Arlie. 1983. *The Managed Heart: Commercialization of Human Feeling*. Berkeley: University of California Press.

Holtorf, Cornelius. 2010. The Presence of Pastness: Themed Environments and Beyond. In *Staging the Past: Themed Environments in Transcultural Perspectives*, ed. Judith Schlehe, Michiko Uike-Bormann, Carolyn Oesterle, and Wolfgang Hochbruck, 23–40. Bielefeld: Transcript.

Holtorf, Cornelius. 2017. Perceiving the Past: From Age Value to Pastness. *International Journal of Cultural Property* 24: 497–515.

Holylandexperience. 2018. The Scriptorium. *Holylandexperience.com*, April 20. https://holylandexp.com/2018/04/20/the-scriptorium/. Accessed 6 Sep 2021.

ICOMOS. 1964. International Charter for the Conservation and Restoration of Monuments and Sites (The Venice Charter, 1964). *ICOMOS*. https://www.icomos.org/charters/venice_e.pdf. Accessed 27 Oct 2021.

Knudsen, Britta Timm, and Anne Marit Waade, eds. 2010. *Re-Investing Authenticity: Tourism, Place and Emotions*. Bristol: Channel View.

Köck, Christoph. 2006. Die Konstruktion der Erlebnisgesellschaft: Eine kurze Revision. In *Erlebniswelten: Herstellung und Nutzung touristischer Welten*, ed. Karlheinz Wöhler, 3–16. Münster: LIT.

Koehler, Dorene. 2017. *The Mouse and the Myth: Sacred Art and Secular Ritual of Disneyland*. East Barnet: John Libbey.

Kiriakou, Olympia. 2017. "Ricky, This Is Amazing!": Disney Nostalgia, New Media Users, and the Extreme Fans of the WDW Kingdomcast. *Journal of Fandom Studies* 5 (1): 99–112.

Kranz, Rebecca. 2018. Eine Küchencrew—vier Länder. Phantastische Vielfalt. https://magazin.phantasialand.de/eine-kuechencrew-vier-laender-phantastische-vielfalt/. Accessed 20 Dec 2021.

Lainsbury, Andrew. 2000. *Once Upon an American Dream: The Story of Euro Disneyland*. Lawrence: University Press of Kansas.

Lonsway, Brian. 2016. Complicated Agency. In *A Reader in Themed and Immersive Spaces*, ed. Scott A. Lukas, 239–248. Pittsburgh: ETC.

Lukas, Scott A. 2013. *The Immersive Worlds Handbook: Designing Theme Parks and Consumer Spaces*. New York: Focal.

Lukas, Scott A. 2016. Judgments Passed: The Place of the Themed Space in the Contemporary World of Remaking. In *A Reader in Themed and Immersive Spaces*, ed. Scott A. Lukas, 257–268. Pittsburgh: ETC.

MacCannell, Dean. 1973. Staged Authenticity: Arrangements of Social Space in Tourist Settings. *American Journal of Sociology* 79 (3): 589–603.

Månsson, Maria. 2010. Negotiating Authenticity at Rosslyn Chapel. In *Re-Investing Authenticity: Tourism, Place and Emotions*, ed. Britta Timm Knudsen and Anne Marit Waade, 169–180. Bristol: Channel View.

McPhillips, William. 1966. "Looks Just Like Home": Mayor of Real New Orleans Praises Disneyland Replica. *Los Angeles Times*, July 25: 6.

Morales, Helen. 2014. *Pilgrimage to Dollywood: A Country Music Road Trip through Tennessee*. Chicago: The University of Chicago Press.

N.N. 1955. Mardi Gras Dedicates New Orleans Street. *The Disneyland News*, 1 (3): 11.

N.N. 2021. Close-Up on the Movies Influencing Frontierland at Disneyland Paris. *Attractions Magazine*, January 18. https://attractionsmagazine.com/close-up-movies-influencing-frontierland-disneyland-paris/. Accessed 20 Dec 2021.

Niles, Robert. 2019. The Awful Lesson of the Impermanence of History. *Theme Park Insider*, April 15. https://www.themeparkinsider.com/flume/201904/6723/. Accessed 20 Dec 2021.

Orvell, Miles. 1989. *The Real Thing: Imitation and Authenticity in American Culture, 1880–1940*. Chapel Hill: The University of North Carolina Press.

Paine, Crispin. 2019. *Gods and Rollercoasters: Religion in Theme Parks Worldwide*. London: Bloomsbury.

Phantasialand. 2021. Phantastische Themenwelten. https://www.phantasialand.de/de/themenpark/phantastische-themenwelten/. Accessed 25 Oct 2021.

Philips, Deborah. 2012. *Fairground Attractions: A Genealogy of the Pleasure Ground*. London: Bloomsbury.

Piazzoni, Maria Francesca. 2018. *The Real Fake: Authenticity and the Production of Space*. New York: Fordham University Press.

Ralph, Owen. 2020. Pirates in Batavia at Europa-Park: From Destruction to Resurrection. https://blooloop.com/theme-park/in-depth/pirates-in-batavia-europa-park/. Accessed 20 Dec 2021.

Robson, Tom. 2019. "The Future Is Truly in the Past": The Regressive Nostalgia of Tomorrowland. In *Performance and the Disney Theme Park Experience: The Tourist as Actor*, ed. Jennifer A. Kokai and Tom Robson, 23–42. Cham: Palgrave Macmillan.

Ron, Amos S. 2010. Holy Land Protestant Themed Environments and the Spiritual Experience. In *Staging the Past: Themed Environments in Transcultural Perspectives*, ed. Judith Schlehe, Michiko Uike-Bormann, Carolyn Oesterle, and Wolfgang Hochbruck, 111–133. Bielefeld: Transcript.

Sheppard, Randal. 2016. Mexico Goes to Disney World: Recognizing and Representing Mexico at EPCOT Center's Mexico Pavilion. *Latin American Research Review* 51 (3): 64–84.

Steinkrüger, Jan-Erik. 2013. *Thematisierte Welten: Über Darstellungspraxen in Zoologischen Gärten und Vergnügungsparks*. Bielefeld: Transcript.

Taylor, John. 2001. Authenticity and Sincerity in Tourism. *Annals of Tourism Research* 28: 7–26.

Wallace, Mike. 1985. Mickey Mouse History: Portraying the Past at Disney World. *Radical History Review* 32: 33–57.

Wang, Ning. 1999. Rethinking Authenticity in Tourism Experience. *Annals of Tourism Research* 26 (2): 349–370.

Waysdorf, Abby, and Stijn Reijnders. 2018. Immersion, Authenticity and the Theme Park as Social Space: Experiencing the Wizarding World of Harry Potter. *International Journal of Cultural Studies* 21 (2): 173–188.

Weiss, Werner. 2012. Welcome to Yesterland. https://www.yesterland.com/welcome.html. Accessed 20 Dec 2021.

Williams, Rebecca. 2020. *Theme Park Fandom: Spatial Transmedia, Materiality and Participatory Cultures*. Amsterdam: Amsterdam University Press.

Wöhler, Karlheinz. 2011. *Touristifizierung von Räumen: Kulturwissenschaftliche und soziologische Studien zur Konstruktion von Räumen*. Wiesbaden: VS.

Yang, Li. 2011. Cultural Tourism in an Ethnic Theme Park: Tourists' Views. *Journal of Tourism and Cultural Change* 9 (4): 320–340.

Economic Strategy: Conceptual, Customer-Based, and Environmental, Social, and Governance Strategies in the Theme Park Economy

Abstract This chapter introduces a selection of economic strategies that theme parks use to achieve their economic objectives. The conceptual design of any theme park is of utmost importance as initial investments in the immovable property are high and risky, and conceptual errors are hard to correct at a later stage. Hence, conceptual strategies relate to the investment decisions taken during the design phase of a park and the continuous investment decision-making by operators during the life cycle of the venue. Customer-based strategies, in turn, point to the extent to which the economic success of a theme park depends on the average daily number of visitors, the length of the season, and the average spends per capita. They seek to increase the visitor and expenditure volumes, including expanding the catchment area, utilizing the visitor potentials, and stimulating secondary spends. Hence, customer-based strategies relate to the theme park product mix development, management, and marketing. Finally, Environmental, Social, and Governance (ESG) strategies address the current major societal global challenges from the perspective of the theme park industry. This includes risk management practices which, as drivers of innovation to create long-term value and respond to the quest for sustainability, are of key importance to this industry. Operators arrange their organizations in such a way as to encourage the increase of positive impacts (and mitigate negative impacts) on their surrounding areas and the environment, while increasing positive financial results.

1 Introduction

The rise and growth of the theme park industry in the last quarter of the twentieth century is closely connected to a societal shift, for which Schulze (2005) coined the term *Erlebnisgesellschaft* ("event-driven society/experience society"). Schulze describes how Western societies have changed in the post-WWII period, when basic needs were increasingly saturated, enabling the individualization of consumption,

This work is contributed by Torsten Widmann, Salvador Anton Clavé, Jan-Erik Steinkrüger, Astrid Böger, Filippo Carlà-Uhink, Thibaut Clément, Florian Freitag, Scott A. Lukas, Sabrina Mittermeier, Céline Molter, Crispin Paine, Ariane Schwarz, Jean-François Staszak. The corresponding author is Torsten Widmann, Fakultät Wirtschaft, DHBW Ravensburg, Ravensburg, Germany.

F. Freitag et al., *Key Concepts in Theme Park Studies*,
https://doi.org/10.1007/978-3-031-11132-7_5

self-identification by means of consumption, and the necessity for an additional emotional benefit, which he sees in experience. In the long run, this also changed the whole economy from a Fordist mass production system to a customer-based economy of scope, based on individualized production and often described as post-Fordism. Nevertheless, it must be acknowledged that this term is highly contested since the very idea of mass production is not contradicted (Schulze 2005) and mass production continues to play a role in post-Fordism.

Pine and Gilmore (2011) consider Disney's theming approach as the blueprint for the rising experience economy. In 1955, Disney founded what is often considered the world's first modern theme park (see ANTECEDENTS), which immerses visitors in an unfolding story rather than just thrilling or entertaining them. Based on the example of Disneyland, Bryman (1999a, 2004) identifies four characteristics of theme parks that can be considered a reaction to the societal change toward an experience society and at the same time defines the core of the theme park as a product: theming (see THEMING), dedifferentiation of consumption, merchandizing, and emotional labor (see LABOR). To refer to these strategies, Bryman (1999a, 2004) introduces the term "Disneyization." Hence, to paraphrase Steinecke (2002; 2009), according to Bryman's model, a theme park might be defined as a mixed-use-center that offers different usages, including rides, shops, restaurants, animals, exhibitions, etc. Theming, blurring the spheres of consumption, merchandizing, and engaging employees all contribute to individualizing the experience of customers and catalyzing an infinite collection of unique experiences.

However, whereas theme parks enable the individualization of and identification with the product by creating an additional experience benefit, they are also simultaneously products for the masses based on the Fordist principles of Ritzer's *McDonaldization of Society* (1983, 2008). According to Ritzer, Fordist principles of rationalization—ideally visible at McDonald's restaurants—have changed not only the fast food business but society in general and the theme park industry in particular, with queuing, time planning, fast food-based dining, and mass-produced merchandise sold throughout the park as its clearest illustrations (Bryman 1999b; Ritzer and Liska 2005; Pine and Gilmore 2011).

Following these general principles, three types of economic strategies will be discussed in the various sections of this chapter: conceptual, customer-based, and ESG (Environmental, Social, and Governance) strategies. Of course, the management of a theme park as a whole, but also of each of its single area components (attractions, rides, retail, food and beverage outlets, transport, etc.), is highly complex. With customers at the center of the stage, economic success depends on the ability to achieve the economic and financial objectives of the park, from the spatiality of the proposed experience to the appropriate use of management tools. Moreover, in the case of some of the most popular parks, the investment in and/or operation of a theme park is part of the corporate strategy of an industrial conglomerate that seeks to increase its brand value rather than just the viability of the theme park facility. The so-called brand theme parks make the brand tangible, be it those of the two megaparks companies, Disney and Universal, or those of the main Chinese theme park corporations Chimelong, Fantawild, and OCT (Happy Valley), the Lego venues with different types of parks distributed worldwide under the management of Merlin Entertainment (see INDUSTRY), or individual corporate visitor centers such as Autostadt Wolfsburg. In all those cases, the economic strategy of theme parks is part of the global economic and financial strategy of such brands.

Therefore, a large variety of factors—corporate, contextual, and location factors, the park's life cycle, marketing efforts, the length of the season, the value of the park brand, visitor spending, corporate social responsibility and sustainability issues, the impact of the digital transformation in the management, and product offering of theme parks—will be discussed in the following sections on conceptual, customer-based, and ESG strategies in the theme park industry. Making sure that the theme park delivers a unique customer experience, attracts an adequate volume of visitors, and is socially responsible in terms of management—respectively, conceptual, customer-based, and ESG strategies—is ultimately determined by the economic viability of the theme park as it is related to the investment, the attractions mix composition (see ATTRACTIONS), the proposed customer experience, and the operations organization.

Operating a theme park has been compared to "operating a small city" (Vogel 2001, 154). However, compared to regular urban places, a theme park property cannot be utilized for other purposes. To decrease the risk of under-utilization, selecting a location with a sufficient catchment area regarding its actual surface, as well as choosing the number, type, and location of attractions, are crucial factors (Wenzel 1998). Indeed, location is a key issue as very few parks in the world can attract a significant number of visitors from further away than a two-hour ride by car (see below). Likewise, the product definition of a theme park and the transformations of its attractions mix through investments in new products and experiences over the course of its lifecycle is of utmost importance when defining economic strategies. Finally, theme parks are increasingly adaptive and innovative when it comes to sustainable trends and practices that create a long-term value for society and, even more importantly, higher efficiencies for the parks' economic viability. Hence, the industry is re-thinking its current business strategies in order to survive in a competitive, multi-faceted global entertainment market where customers are increasingly committed to environmental and social causes and ESG-responsible investment practices, with the result that processes consistent with the United Nations Sustainable Development Goals (SDGs) are becoming more and more relevant.

Europa-Park, owned and run by the Mack family and Germany's largest theme park, provides a range of examples of the three types of economic strategies that influence theme park profitability. Opened in 1975 with 15 attractions on an area of 16 ha, the park has been steadily expanding and currently offers over 100 attractions on an area of 95 ha. In order to satisfy the demands of repeat visitors, who accounted for 80% of visitors in 2018 and thus constitute an extremely important segment, the park continually makes new investments and creates new attractions (N.N. 2020, 2019). A key point in this development was in the early 1990s when the park became a multi-day destination due to its growth in size. Even though many sector experts did not consider the approach of having in-house accommodation too promising at the time, the company nevertheless decided to build the first themed hotel, El Andaluz, in 1995. Between 1999 and 2012, four more hotels with themed restaurants and shows were added, as well as low-budget overnight accommodation for price-sensitive guest segments (e.g. camping and guesthouse). The year 2019 saw the opening of the most recent hotel project, Hotel Krønasår, which is themed in a Scandinavian style and thus connects to the park's second gate, the Nordic-themed water park Rulantica, where visitors can enjoy 25 water attractions located in nine themed areas (N.N. 2020). The aim of having this extra park open all year-round plus the extensive accommodation options (Europa-Park resort now has 5,800 beds in total) is to significantly increase the number of overnight guests, extend the average length of stay (2018: 1.4 days; N.N. 2019), and turn the theme park into a short vacation destination. In addition, Europa-Park has increased its ESG commitment and is taking a holistic approach to the environmental, economic, and social elements of sustainability (see Europa-Park 2022).

2 Conceptual Strategies

Conceptual design is key to the economic success of a theme park. It should incorporate the characteristics of the place, the storytelling of the experience, as well as critical indicators of the performance of the investment, such as an estimate of the development costs, the forecast revenues, the expected profit margin, and the proposed attractions mix for the park (see ATTRACTIONS). Hence, the design of a theme park is a strategic component of its economic viability and has an impact on several strategic decisions, such as investment models, funding and financing mechanisms, target groups and visitor selection, park operations and management, as well as make-or-buy decisions. To define the size requirements of the park, the process of design must also consider the potential revenue sources when the park will be opened, the potential visitor mix depending on the length of stay, the catchment area and visitors' propensity to spend, and the expected customer perception of the resulting product. Moreover, park design also has crucial effects on management strategies throughout the entire lifecycle of the theme park. The conceptual design of the park thus constitutes a key basic economic strategy that must be strongly integrated within the business model of the project and the experience that will be proposed to the customer.

Following Cornelis (2017), budget decisions related to conceptual design form the basis of the economic success of a park, and both initial and subsequent investments over the years are key in terms of the increase or decline in attendance and revenue (see THEMING). Other factors that influence the return of a theme park investment are as follows: a proper recognition of the characteristics of the region where the park is located, a proper monitoring of potential markets for the park, and a proper understanding of the propensity of potential customers to visit the park according to its more or less unique position within the entertainment industry (see Cornelis 2017).

Design and layout factors influence the position of each park and shape its entertainment value (see SPACE). This includes the mix of attractions, events, and facilities;

the theming and storytelling; the use of the intellectual property and the associated brand qualities; the size of the park; the ticket entrance price; the quality of the design; the efficiency of service; and the operational control of processes and details. The relevance of these factors may vary over the lifecycle of the park, depending on the adoption of new strategies by the operating managers. Hence, the conceptual economic strategy depends on the uniqueness of the park and the capability of its contents, the attractions mix, and ride capacity to create sufficient drawing power and ensure long stays, repeatability, and customer value. Moreover, the conceptual economic strategy of a park should properly analyze not only direct development costs such as land, basic infrastructure development, construction, attractions, taxes, and design, but also asset depreciation and the financial and operational costs associated with the investment. In this context, the design of the components that will define the guest experience is key—in particular, the number of attractions and entertainment opportunities that the park will offer, the investment per attraction according to variables such as innovativeness, technological disruption, or theming, and the coherence of the proposed visitor experience for the targeted audience. Additionally, scholars have shown that the effects of theatrical elements such as the attractiveness of scripts, the charm of the settings, and the park's ability to generate consumer immersion, surprise, participation, or fun relate positively to customer satisfaction and thus to loyalty intentions (Kao et al. 2008).

As it can determine the limits and effects of the investment, a park's potential for growth is a key factor in its conceptual economic strategy (see Cornelis 2017, 46). Besides the design elements described above, the potential for growth is also directly related to the size of the market catchment area and the existing theme park competition. The potential frequentation of a given park is usually considered to be related to the rate of the park's penetration among the existing population in the different concentric catchment areas that surround it, as well as its ability to directly attract tourists. More precisely, the conceptual economic strategy of the park depends on its potential geographical reach, which may be local, regional, national, or even global. In each of the scales of reference, there is only a certain percentage of the population that visits theme parks. This percentage represents the limit of growth of each particular park within its catchment area.

Cornelis (2017, 56–59) uses an analysis of the market potential of the regional park Toverland (Sevenum, Netherlands) to illustrate this. Rough estimates indicate that Toverland's market potential amounts to over 1.5 million people within a two-hour catchment area, including tourists within a one-hour car ride from the park. Since the park was, at the date of analysis, attracting less than half of this market, its potential for growth was around 800,000 visitors. Changing the "imagescape" of the park (Cornelis 2017, 143), transforming the attractions mix, applying aggressive price policies, or creating new sources of revenue such as accommodation or retail might change the capacity of the park to reach new markets and overcome its asymptotic limit of growth. As the size of a park's potential market depends on a multiplicity of factors, however, the potential response of the market is not easy to predict. Uncertainty and risk are common factors when new investment for park additions and changes are needed in order to increase attendance.

Obviously, the size of the investment and resources needed for the operation of a park depend directly on the size of the potential market it can cater to. This is defined as the effective market, i.e. the number of people who visit each park either occasionally or recurrently. Hence, in order to develop and build the economic strategy of a park, planners, investors, and operators need to have a clear idea of what the park's target market is, how many people are going to visit it, what the park's sphere of influence is, and when it can expect its visitors.

Assessing the potential impact of a new investment in an existing park is a complex matter, but there is, as Cornelis (2017, 75) acknowledges, one commonly accepted certainty: "if a park does not invest in new attractions, shows, parades, etc. 'once every few years' the attendance will decrease, and the park will gradually disappear from the market." Hence, there is a consensus that parks need continuous and regular investments over the course of their product lifecycle. The key questions are as follows: how much and when to invest; how to combine the existing data about park performance, visitor satisfaction, and financial opportunity to decide the optimal timing for the allocation of resources in order to renew the park; and how to determine the extent to which the investment will achieve the expected results both in terms of visitor numbers and revenues. This also depends on the phase within the product lifecycle in which the park currently is and the type of investment in terms of expected market and revenue growth. Besides this, there is also a need to assess and evaluate the following: the capacity of the park to manage the resulting capacity and demand changes, the expected role of the investment in terms of stabilization or growth of the park performance, and the duration of the effects on the financial performance of the park over the years.

Other contextual factors lie outside the immediate influence of budget decisions and are less controllable, but are nevertheless key to the process of defining the economic strategy of the park during the phase of planning and design as well as throughout its lifecycle. These include the following: potential changes to the park's reputation and image; the evolution of competitors and of "imagescape" uniqueness and differentiation; support from public administrations; improvements to surrounding transportation networks and bottle-necks; short- and long-term weather variability in the case of open-air facilities; the evolution of the urban, social, and economic landscape surrounding the park; the general economic situation; and major unforeseeable area risks such as earthquakes, terrorist attacks, pandemics, or industrial hazards (Anton Clavé 2007, 337–340).

3 Customer-Based Strategies

The theme park industry is a service industry targeting different customer groups with different needs that need to be fulfilled to ensure their satisfaction. Defining the appropriate strategies to attract target customer groups, offer them a satisfying experience, and maximize customer spending on entrance fees and secondary spends (e.g. food and beverage, souvenirs, accommodation, etc.) in accordance with the product mix and the design concept of the park is crucial to success. Theme parks compete not just with other theme parks, but also with other classic out-of-home competitors like casinos, shopping malls, leisure facilities adopting themed entertainment—all of which are strong competitors within the respective catchment areas or markets—as well as with the diverse forms of passive entertainment practices derived from the digital entertainment at home.

As the economic success of a theme park largely depends on the average number of daily visitors, the length of the season, and the average spending per capita, theme parks deploy multiple management and marketing strategies to achieve or increase the planned visitor and revenue volumes and fit the economic objectives to the investment. These strategies are as follows:

- **Expanding the catchment area:** The introduction of new attractions (see ATTRACTIONS) can help to switch visitor-flows from one theme park to another. New attractions must be related to the creation of new experiences and the fulfillment of the expectations and needs of the targeted population. For this reason, new attractions should be related to specific information about how large the market areas are, where the new potential visitors live, from where they travel to the park, and how accessible it is for them. The creation or enhancement of transportation infrastructure is also a suitable measure to expand the catchment area (for example, the construction of a superhighway exit especially for Europa-Park in 2002) and the transformation of the station of Ringsheim into a stop for long-distance trains by German railways DB. In this sense, the connection between a park and major transportation routes constitutes a strong and important success factor. To target customer groups within the catchment area, strategic approaches for market segmentation might be designed, including psychographic segmentation (e.g. lifestyle groups and *milieus*), product-related segmentation (e.g. thrill-seekers), demographic segmentation, and geographic segmentation.
- **Utilizing the visitor potential:** For theme parks, it is vital to have a well-balanced attractions mix for all possible target groups based on a general theme. Depending on its financial capabilities, a park may seek to reach new target groups but needs to carefully consider their desire and ability to spend, their available leisure time, and their motivations. Increasing the impact of the actual visitors can be accomplished by creating new spending opportunities or exercising a variety of revenue management techniques and by expanding the key characteristics of the park, e.g. opening further themed areas or even different forms of parks on the same site. The latter refers to the so-called "second gate strategy," a concept that Walt Disney World first developed in Orlando with the opening of Epcot in 1982, with two more theme parks and other entertainment venues and tourism facilities eventually following. More recently, this strategy has been employed, for example, by Disneyland Paris with the addition of Walt Disney Studios Park in 2001 and by Europa-Park's addition of the Rulantica water park (see above). The second gate strategy seeks to increase the number of attractions and therefore stimulate overnight stays and short-term vacations.

- **Increasing the share of repeaters:** While there are local and regional differences, in general and irrespective of its size, the attendance rate of a theme park significantly decreases after 2 h of driving time (Stiftung für Zukunftsfragen 2013). Within the core catchment area, potential customers have to be stimulated to repeat a visit to a theme park on a regular basis, for example by a new attraction. Ideally, new attractions would be provided on a yearly basis, although this may be problematic for smaller parks with spatial or financial restrictions as annual investments require a high volume of visitors. Additionally, parks need a clear vision of what customers are looking for, what they wish to experience, and what the subsequent additions will be. This means that parks need to know who to prioritize among the existing loyal markets, depending on how much they spend and how they will ensure the best return of investment. Research results from Ali, Kim, Li, and Jeon (2018) also suggest that theme park operators need to pay attention to maintaining a good physical park setting, managing human resources well, and managing the behavior of other customers in order to ensure that visitors enjoy delightful experiences and have the propensity to be repeaters.
- **Enticing customers away from other parks:** Measures to gain competitive advantage in a regional system with several theme parks can be described with the four fields of marketing policy: firstly, the product policy should be characterized by a wide range of attractions in order to reach a broad range of target groups. Secondly, the pricing policy needs to establish a good cost/performance ratio for the customers by also providing a high density of attractions and amenities for repeaters. Annual season-tickets are a popular instrument to stimulate visits to theme parks and to increase the secondary spends (food and beverage, souvenirs, etc.) per capita. Thirdly, measures in the field of communications policy need to constantly keep the park interesting by creating special events (e.g. evening shows such as Night.Beat.Angels at Europa-Park), having their own fan-clubs (see VISITORS), and interacting with theme park aficionados. Newsletters, print material, and Public Relations are common fields of communication. Social media, from YouTube to Facebook and Instagram, offer a wide range of possibilities for self-expression (see PARATEXTS AND RECEPTION); this may include cooperating with social media influencers and opinion leaders. Finally, for distribution purposes, it is favorable if the park is located in an attractive tourist destination. Therefore, theme parks can be part of cooperative selling and a network of tourist attractions. In this vein, Cheng et al. (2016, 1) conclude that "there are seven factors influencing visitor brand-switching behavior: 'visitor variety-seeking,' 'visitor satisfaction,' 'switching cost,' 'perceived value,' 'competitor attraction,' 'theme park image,' and 'visitor involvement.' More precisely, the perceived value and visitor satisfaction strongly influence brand-switching behavior as intermediate variables. Visitor variety-seeking and competitor attraction are positively related to visitor brand-switching behavior, whereas the other five factors are negatively related to visitor brand-switching behavior."
- **Increasing the length of stay:** The spatial expansion and the agglomeration of attractions of a theme park (including second gate developments) can lead to an increase in the length of the average visit and, accordingly, to higher spends per

visit. Larger theme parks offer themed accommodation and encourage short trips of two to three days in the park instead of same-day visits. Parks like Disneyland Paris even create Urban Entertainment Centers, where they simulate urbanity in a theme park environment (Widmann 2006). Smaller parks cooperate with local accommodation providers. Even if maintaining hotels on theme park grounds can pose problems (closing some hotels while the park is not open; high operating costs throughout the year, including energy costs; high-pressure situations among theme park operators and hotel operators, e.g. high seasonal demand), the opening of hotels by theme parks companies is a strategic tool to stimulate multi-day and repeat visits. Therefore, it constitutes a state-of-the-art, industry-specific benefit.

- **Stimulating secondary spends:** Running a theme park involves high fixed costs. To make the theme park business more profitable, it is of great importance not only to have high attendance numbers, but also to sell goods with high profit margins, such as food and beverages, merchandise, or upgrades like entrance fees to shows. Other secondary expenditure sources of income are games, parking, accommodation, corporate meetings and events, parties (birthdays or weddings, for instance), advertisements, sponsorship and naming rights, or temporary exhibitions. Such measures help to decrease the fixed costs, increase the variable costs, and increase the contribution margin, thereby increasing the operational profitability. It is important that parks have a clear idea of what products, experiences, and dreams are sold to which customer segments and for what reasons, and how to use digital media in order to enhance the selling proposition of the park.
- **Expanding the core business:** Theme parks acquire new business segments as they become destinations for short holidays with their themed accommodations (see above). Resort-like theme park hotels often have high-quality standards and amenities like spas, conference infrastructure, and signature restaurants and also provide the framework for family celebrations. Some parks expand into the MICE (meetings, incentives, conferences, and exhibitions) or "confertainment" tourism segment by combining their offers such as conference infrastructure, dinner shows, and banquets with overnight stays and theme park visits to make customized packages for business tourists (e.g. Phantasialand's Business2Pleasure package). Business tourists are encouraged to visit the theme park in private with their families. Further on, larger theme parks offer their infrastructure as a location for media productions. For example, Europa-Park has developed into a well-established media center with approximately 200 TV productions per year, and the hotel infrastructure plays a crucial role in this context. Another field of action in the business-to-business segment is to offer the possibility of customer-company presentations. Theme parks offer the theming of attractions according to the companies' core competencies or values (for example, Adidas Soccerhall at Europa-Park).
- **Expanding the season:** Especially for parks located in moderate-to-cold climates, it is of great importance to expand the season and find business segments for year-round attendance. Accordingly, parks increase their share of indoor attractions,

shows, and rides; offer Christmas markets; and expand their offerings to the aforementioned MICE or confertainment market. Evening entertainment, banquets, and dinner shows (e.g. Phantasialand's Fantissima) can also be offered the whole year-round. The major theme parks' hotels and resorts attract tourists not only during the park's opening season, but also off-season.

4 Environmental, Social, and Governance Strategies

Theme park operators are increasingly implementing sustainability initiatives such as recycling programs (e.g. water for irrigation or food waste), LED lighting, and green energy and electricity (Hung 2019) as part of a key economic strategy to ensure sustainability in all senses, both within the facility and in its surrounding area. This is in line with the increasing desire among customers to do business with organizations that operate with social and environmental responsibility. As early as 2006, Porter and Kramer pointed out that in the future a proper understanding of the mutual dependence of business and society would be a key economic strategy to stimulate value creation for both companies and communities (see PGAV Destinations 2021). Hence, currently, theme park operators are sensitive to include a "social dimension to the value proposition" to their visitors (PGAV Destinations 2021). As an industry that welcomes large numbers of people from all around the world, such issues are therefore of key importance to the theme park industry. Moreover, as visitors tend to be open to new experiences when visiting theme parks, the industry also has the opportunity to educate customers about sustainability practices and inspire them to mitigate their impact on the world (see WORLDVIEWS).

For example, during the Earth Day celebrations in 2021, 425 attractions industry professionals from around the world gathered online for "Greenloop: Sustainability in Visitor Attractions," a conference dedicated to exploring sustainability issues in the visitor attractions sector organized by Blooloop (see METHODS). Among many other topics, participants discussed the rapidly accelerating move to socially responsible investing, how sustainability impacts businesses, and ways to financially benefit from making attractions sustainable. It also highlighted why having a favorable ESG rating will be essential for raising finance or divesting (Coates 2021).

In fact, sensitivity to environmental, social, and governance (ESG) issues has also been climbing on the priority lists of boards and stakeholders, as failing to meet compliance obligations for environmental impact, diversity, equity, or other governance issues can put companies at legal or reputational risk. It is for this reason that for theme parks, sustainability, social responsibility, and circularity have been increasingly institutionalized as part of their economic strategy. More and more, they have started to not only develop environmental and social programs to benefit people, planet, and profit, but also measure and report the degree to which their economic value is at risk due to ESG factors. And it is also for these reasons that theme park companies have increasingly rated themselves through third-party agencies, reporting research organizations, and analysis firms in order to determine their ESG scores for investment decisions (Farnham 2020). These scores can be a significant factor in attracting capital and maintaining transparent stakeholder communication (Holcomb et al. 2010).

This process of re-thinking the current business strategies of theme parks with a high sensitivity to ESG issues is consistent with the United Nations Sustainable Development Goals (SDGs) and the United Nations Global Compact initiative, a voluntary global movement based on multisector CEO commitments to implement universal sustainability principles and to take steps to support the UN SDGs (García Alba et al. 2015). Consequently, the adoption of ESG considerations in private investments is evolving from a risk management practice to a driver of innovation and new opportunities. ESG reports and ratings include a list of key indicators and metrics in each ESG category.

Common evaluation criteria metrics on Environment include climate change, soil and water contamination, renewable energy, and environmental policy indicators. The Social category examines business relationships with employees, suppliers, partners, shareholders, and other groups throughout the supply chain. Social scores may also reflect charitable contributions, customer interactions, community impact, and policy influence. Finally, Governance criteria evaluate legal and compliance issues and board operations, including the representation of diverse backgrounds and perspectives on company boards or how executive and non-executive compensation compare to the company's peers (Farnham 2020). A well-known example in this vein is the CSR report of The Walt Disney Company. It includes priority ESG topics related to the many industries and geographies where the Disney corporation operates, including the following: (1) diversity, equity, and inclusion in the workplace; (2) issues concerning protecting the planet; and (3) support for communities, and especially children (The Walt Disney Company 2021; see Rodríguez 2021).

In the same sense, in 2021, the theme park PortAventura World was recognized and awarded by the IAAPA with the first "IAAPA EMEA Award for Extraordinary Excellence," which acknowledges their firm strategic commitment to sustainability. The PortAventura World corporate responsibility strategy is based on the SDGs and the UNWTO Global Code of Ethics for Tourism and focuses on ESG. More precisely, at the Environmental level, the park controls its carbon footprint and encourages the saving and reuse of water, circularity, energy transition, and the minimization of plastic use, in addition to promoting electric modes of transportation and training its staff in environmental awareness. At the Social level, besides employee safety, health, and wellbeing programs as well as gender and diversity equality initiatives, one of the park's most challenging projects is PortAventura Dreams, an initiative promoted by the PortAventura Foundation that offers unique experiences to children and young people who suffer from serious illnesses, as well as to their families (see Fig. 1). At the Governance level, the company prioritizes the hiring of local services and suppliers and invests in innovation and digitalization (PortAventura World 2020, see Fernández 2021).

In 2017, the Swedish theme park Liseberg (Gothenburg) was already using a materiality analysis to identify five areas where the operators believe they can make the biggest difference in regards to sustainability according to what stakeholders

Fig. 1 Founded in 2019, PortAventura Dreams Village offers free stays to families whose children suffer from serious illnesses. *Photograph* PortAventura World

think is important, what impact the business has on people and the environment, and what challenges and opportunities this entails. The five most important sustainability areas are as follows: (1) a safe and secure environment; (2) job satisfaction, service, and equal treatment; (3) resource and climate efficiency; (4) responsible purchasing; and (5) contribution to local community development (Liseberg 2017).

In the same vein, the French SNLAC (Syndicat National des Espaces de Loisirs et Culturels) has created the label "divertissement durable" (sustainable entertainment) to certify, through a process of continuous improvement, parks' commitment to social responsibility issues at the environmental, social, and economic levels (www.divertissement-durable.fr/). The certification acknowledges the engagement of park operators in developing an awareness for and a training policy on social responsibility issues among their employees. To be awarded the label, parks should also be committed to the preservation of local heritage and biodiversity. Purchasing policies take into account the origin, materials, and environmental impact of the products while promoting shorter material loops and local producers. Additionally, environmental impact is taken into account for each new project in order to promote eco-design. Finally, in terms of waste management, certified parks set targets to minimize packaging, to reuse, sort, and recover waste, and are committed to a policy of optimizing water and energy resources. At the beginning of 2022, DéfiPlanet' (Dienné), Micropolis (Besançon), Le PAL (Saint-Pourçain-sur-Besbre), Puy du Fou (Les Epesses), and La grotte de Clamouse (Saint-Jean-de-Fos) were already certified.

Finally, it is instructive to examine the role of ESG issues in the economic strategy of theme parks from a different geographical context and with a different use of the concept of environment. Zhang and Jin (2010), discussing the case of the Overseas Chinese Town in Shenzhen (People's Republic of China), underline how valuing the environment can be conceived as a resource that can empower the product brand, lead to environmental protection, attract customers aligned with the values of the company, and thus increase its business and short, mid-, and long-term value creation.

5 Conclusion

The economic strategies of theme parks manifest themselves in the conceptual planning and design, in marketing and management of customer attraction strategies, and in the Environmental, Social, and Governance (ESG) practices. Theme parks employ these strategies to attract as many visitors as possible and deliver a valuable, memorable, and satisfactory experience. As highlighted by Kim and Kim (2016), customer satisfaction is the result of "the pleasant fulfillment of consumption experience" and the "evaluation process on the degree of consistency between pre-experience expectation and post-experience performance." To minimize the gap between expectation and performance is the main purpose of the entire operationalization of the three types of economic strategies of theme parks summarized in this chapter. Nonetheless, besides all the strategic considerations and opportunities, theme park managers must not forget their core business, which is, as Walt Disney put it with respect to Disneyland, to create nothing less than "the happiest place on earth" (Thomas 1994, 246).

This purpose means designing the park according to the customer experience that will be offered, to be realistic about financial goals and objectives, attendance and in-park spending, as well as about the anticipated margin profit, and to align the proposed experience and the theme park operation as a whole with the creation of social, environmental, and community value as it is expected by visitors. Last but not least, the ongoing digital transformation—beyond the use of technology in order to enhance and personalize the customer experience—is already changing and will deeply transform the ability of parks to achieve their goals in terms of their economic strategy. Service automation and especially the digital data gathered from ticketing, sales, customer feedback, etc. can be used to gain insights into the continuous process of decision-making. Digital technology is, in fact, the new key player in the process of defining the economic strategy of theme parks and especially of reinforcing their commitment to sustainable and responsible development objectives and ESG practices.

References

Anton Clavé, Salvador. 2007. *The Global Theme Park Industry*. Wallingford: CABI.

Ali, Faizan, Woo Kim, Jun Li, and Hyeon-Mo Jeon. 2018. Make It Delightful: Customers' Experience, Satisfaction and Loyalty in Malaysian Theme Parks. *Journal of Destination Marketing & Management* 7: 1–10

Bryman, Alan. 1999a. The Disneyization of Society. *The Sociological Review* 47 (1): 25–47.

Bryman, Alan. 1999b. Theme Parks and McDonaldization. In *Resisting McDonaldization*, ed. Barry Smart, 101–115. London: Sage.

Bryman, Alan. 2004. *The Disneyization of Society*. London: Sage.

Cheng, Qiang, Ruoshi Du, and Yunfei Ma. 2016. Factors Influencing Theme Park Visitor Brand-Switching Behaviour as Based on Visitor Perception. *Current Issues in Tourism* 19 (14): 1425–1446

Coates, Charlotte. 2021. Show Me the (Green) Money: The Financial Benefits of Making Attractions Sustainable. *Blooloop*. https://blooloop.com/theme-park/in-depth/sustainable-attractions-financial-benefits/. Accessed 28 Aug 2022.

Cornelis, Pieter C.M. 2017. *Investment Thrills: Managing Risk and Return for the Amusement Parks and Attractions Industry*. Nieuwegein: NRIT Media.

Europa-Park. 2022. Sustainability at Europa-Park. *Europa-Park*. https://corporate.europapark.com/en/company/sustainability/. Accessed 28 Aug 2022.

Farnham, Kezia. 2020. ESG Scores and Ratings: What They Are, Why They Matter. *Diligent*. https://www.diligent.com/insights/esg/esg-risk-scores/. Accessed 28 Aug 2022.

Fernández, Choni. 2021. Una estrategia transversal e integradora. *Compromiso RSE*. https://www.compromisorse.com/responsabilidad-social/turismo-y-restauracion/portaventura-world/. Accessed 28 Aug 2022.

García Alba, Jaume, Anthony Miller, William Speller, Helene Winch, Karin Malmberg, Careen Abb, and Elodie Feller. 2015. *Private Sector Investment and Sustainable Development*. N.P.: United Nations Global Compact.

Holcomb, Judy, Fevzi Okumus, and Anil Bilgihan. 2010. Corporate Social Responsibility: What Are the Top Three Orlando Theme Parks Reporting? *Worldwide Hospitality and Tourism Themes* 2: 316–337.

Hung, Linda. 2019. Sustainability in Theme Parks. *Forrec*. https://www.forrec.com/blog/sustainability-in-theme-parks/. Accessed 28 Aug 2022.

Kao, Yie-Fang, Li-Shia Huang, and Cheng-Hsien Wu. 2008. Effects of Theatrical Elements on Experiential Quality and Loyalty Intentions for Theme Parks. *Asia Pacific Journal of Tourism Research* 13 (2): 163–174.

Kim, Changhee, and Soowook Kim. 2016. Measuring the Operational Efficiency of Individual Theme Park Attractions. *Springerplus* 5. https://doi.org/10.1186/s40064-016-2530-9. Accessed 28 Aug 2022.

Liseberg. 2017. *Sustainability Report 2017*. https://www.liseberg.com/globalassets/om-liseberg/arsredovisningar/liseberg_sustainability_report_2017.pdf. Accessed 25 March 2022.

N.N. 2019. *Europa-Park. Meeresschlange auf Gästefang. Allgemeine Hotel- und Gastronomie-Zeitung* 119, June 24.

N.N. 2020. Rulantica: Der neue Wasser-Themen-Park in Rust. *Euro Amusement Professional* 1: 32–38.

PGAV Destinations. 2021. Destinations and Social Responsibility. *PGAV Destinations Quarterly Publications*. https://pgavdestinations.com/quarterly-publications/destinations-and-social-responsibility/. Accessed 28 Aug 2022.

Pine II, B. Joseph, and James H. Gilmore. 2011. *The Experience Economy*. Updated ed. Boston: Harvard Business Review Press.

PortAventura World. 2020. Informe de Responsabilidad Social Corporativa del Grupo PortAventura. *PortAventura World*. https://www.portaventuraworld.com/ca/responsabilitat-corporativa. Accessed 28 Aug 2022.

Porter, Michael, and Mark Kramer. 2006. Strategy and Society: The Link between Competitive Advantage and Corporate Social Responsibility. *Harvard Business Review*, December. https://hbr.org/2006/12/strategy-and-society-the-link-between-competitive-advantage-and-corporate-social-responsibility. Accessed 28 Aug 2022.

Ritzer, George. 1983. The "McDonaldization" of Society. *Journal of American Culture* 6 (1): 100–107.

Ritzer, George. 2008. *The McDonaldization of Society*. Thousand Oaks: Pine Forge.

Ritzer, George, and Allan Liska. 2005 [1997]. "McDisneyization" and "Posttourism": Complementary Perspectives on Contemporary Tourism. In *Touring Cultures: Transformations of Travel and Theory*, ed. Chris Rojek and John Urry, 96–109. London: Routledge.

Rodríguez, Sara. 2021. ESG Case Study: The Walt Disney Company. *EFT Trends*. https://www.etftrends.com/2021/12/esg-case-study-the-walt-disney-company/?utm_source=Yahoo&utm_medium=referral&utm_campaign=ReadMore. Accessed 28 Aug 2022.

Schulze, Gerhard. 2005 [1992]. *The Experience Society*. London: Sage.

Steinecke, Albrecht. 2002. Kunstwelten in Freizeit und Konsum. *Geographie Heute* 23 (198): 2–7.

Steinecke, Albrecht. 2009. *Themenwelten im Tourismus: Marktstrukturen—Marketing—Management—Trends*. Munich: Oldenbourg.

Stiftung für Zukunftsfragen. 2013. *Forschung aktuell* 246. Hamburg: Stiftung für Zukunftsfragen.

The Walt Disney Company. 2021. *2021 Corporate Social Responsibility Report*. https://impact.disney.com/app/uploads/2022/02/2021-CSR-Report.pdf. Accessed 28 Aug 2022.

Thomas, Bob. 1994 [1976]. *Walt Disney: An American Original*. New York: Hyperion.

Vogel, Harold L. 2001. *Travel Industry Economics: A Guide for Financial Analysis*. Cambridge: Cambridge University Press.

Wenzel, Carl-Otto. 1998. Freizeitimmobilien. In *Spezialimmobilien: Flughäfen, Freizeitimmobilien, Hotels, Industriedenkmäler, Reha-Kliniken, Seniorenimmobilien, Tank- und Rastanlagen/Autohöfe*, ed. Bernd Heuer and Andreas Schiller, 85–159. Cologne: Müller.

Widmann, Torsten. 2006. *Shoppingtourismus: Wachstumsimpulse für Tourismus und Einzelhandel in Deutschland*. Trier: Geographische Gesellschaft Trier.

Zhang, Hazhoe, and Guangjun Jin. 2010. Depending on Marketing to an Ecological Vision: A Case Study of Overseas Chinese Town's Developing Management in Shenzhen, China. *IEEE*. https://ieeexplore.ieee.org/stamp/stamp.jsp?tp=&arnumber=5577924&tag=1. Accessed 28 Aug 2022.

Immersion: Immersivity, Narrativity, and Bodily Affect in Theme Parks

Abstract Theme parks are distinct from other immersive media in that the mediated space and the space of reception are one and the same. This does not necessarily or automatically lead to the "total" immersion of theme park visitors, as immersion depends on a large variety of factors that include not only the specific buildup of the immersive space, but also the recipients' current disposition (i.e. their willingness to let themselves be immersed) and the general context of the experience. This chapter, therefore, discusses immersion in terms of theme park's immersivity, i.e. their capacity to engender or induce an immersed state of mind in visitors. After offering a brief overview of the various theoretical conceptualizations of immersion and their application to the theme park space, the chapter focuses on the relationships between immersion and narrativity, as well as immersion and bodily affect. Finally, the chapter discusses how theme parks encourage visitors to accept their offers of immersivity.

1 Immersion Versus Immersivity

In taglines, advertising copy, and other paratexts (see PARATEXTS AND RECEPTION), theme parks frequently promise to transport customers to other worlds. The famous plaque at the entrance to Disneyland, which matter-of-factly tells visitors "Here You Leave Today and Enter the World of Yesterday, Tomorrow and Fantasy," is only one of many examples. What visitors may expect from the parks according to these promises are experiences of immersion. Derived from Latin *immergo* (to plunge, dip, or sink into liquid), the English term "immersion" is still used literally—e.g. for the administration of Christian baptism—but has also developed a transferred or figurative sense and has been used to refer to both a mental state of absorption within a condition, task, or medial representation, as well as the capacity of a condition,

This work is contributed by Ariane Schwarz, Florian Freitag, Thibaut Clément, Salvador Anton Clavé, Astrid Böger, Filippo Carlà-Uhink, Scott A. Lukas, Sabrina Mittermeier, Céline Molter, Crispin Paine, Jean-François Staszak, Jan-Erik Steinkrüger, Torsten Widmann. The corresponding author is Ariane Schwarz, Institut für Medien, Theater und Populäre Kultur, Universität Hildesheim, Hildesheim, Germany.

F. Freitag et al., *Key Concepts in Theme Park Studies*,
https://doi.org/10.1007/978-3-031-11132-7_6

task, or medial artifact to engender or produce such a state ("immersive"). In order to distinguish between the two figurative senses, immersion as a state of mind or a certain form of reception has sometimes been described as "mental immersion" (Sherman and Craig 2018, 10), or has been replaced by such terms as "aesthetic illusion" ("the impression of being imaginatively and emotionally immersed in a represented world," Wolf 2013a, v) or "presence" (the "perception of non-mediation or immediacy," Holtorf 2009, 36). By contrast, the property of medial and other systems to facilitate or "produce" mental immersion, aesthetic illusion, presence, etc. has been referred to as "physical immersion" (Sherman and Craig 2018, 10) or "immersivity" (Freitag et al. 2020).

As a state of mind or form of reception, immersion has been described as a gradable phenomenon of "diminishing critical distance to what is shown and increasing emotional involvement in what is happening" (Grau 2003, 13)—a "feeling, *of variable intensity*, of being imaginatively and emotionally immersed in a represented world" (i.e. in the mediated environment) "and of experiencing this world in a way similar (but not identical) to real life" (i.e. to the space and time of reception; Wolf 2013b, 51; emphasis added). Indeed, while some scholars have speculated about "total immersion" (see Lukas 2016, 85), there seems to be a critical consensus that there is no "simple relationship of 'either-or' between critical distance and immersion" (Grau 2003, 13) and that some residual awareness of the outside world—some "latent rational distance" (Wolf 2013b, 51) or a "split loyalty" (Ryan 2015, 68)—is even constitutive of the experience. Thus, media scholar Britta Neitzel (2008, 147) and theater scholar Doris Kolesch (2016) have both considered total immersion to be a myth; and to Janet H. Murray's assertion that "[w]e seek the same feeling from a psychologically immersive experience that we do from a plunge in the ocean or swimming pool" (1997, 98), Marie-Laure Ryan has responded that "[t]he ocean is an environment in which we cannot breathe; to survive immersion, we must take oxygen from the surface, stay in touch with reality" (2015, 68). The idea of immersion as entailing a "split loyalty," a divided consciousness, or a meta-awareness (Wolf 2013b, 16) also and especially applies to the theme park: what makes theme parks attractive is, among other things, the constant awareness that there is an outside world that is completely different, from which the visitor has only temporarily been separated (see Kolb 2008, 122; Carlà-Uhink 2020, 12–13; see also THEMING). By contrast, what distinguishes the theme park experience from other (potentially) immersive situations is that the mediated space and the space of reception are one and the same.

In the sense of physical immersion or immersivity—i.e. from a technological or production rather than from a phenomenological or reception perspective—immersion has often been conceptualized as one of several factors that may induce an "immersed" state of mind. Werner Wolf, for instance, has argued that "like all reception effects," immersion (in the sense of mental immersion, aesthetic illusion, or presence) is

> elicited by the conjunction of factors that are located (a) in the representation themselves, which tend to show certain characteristic features and follow certain illusion-generating principles, (b) in the reception process and the recipients, as well as (c) in framing contexts, e.g. cultural-historical, situational, or generic ones. (Wolf 2013b, 52)

Several different characteristics of medial artifacts have regularly been identified as being particularly conducive to states of immersion, including the multimodality of a medium (i.e. the number of senses addressed; see Wirth and Hofer 2008, 165), the "vividness and credibility" or the "believability" of the represented world (Therrien 2014, 451, 453; see also Murray 1997, 15), and the degree of interactivity it allows (see Grau 2003, 7; Ryan 2015, 257). Two characteristics of medial artifacts that have been considered central to fostering states of immersion in theme parks are the use of (fictional) narratives and bodily affect, which will be discussed in the second and third sections of this chapter.

Critics have been divided, however, on whether the presence of some sort of framing device (for example, the picture frame of a painting, the proscenium arch at the theater, the paratexts of a literary work) enhances or diminishes the immersive potential of a given artifact. For Freitag and Schwarz as well as for Carlà-Uhink, for example, theme park entrances use careful staging to dramatize and ritualize the separation of customers from the world outside and to thus prepare them for the liminal experience of the park visit (see Freitag and Schwarz 2015/16, 22; Carlà-Uhink 2020, 12–13; see also SPACE and THEMING). Grau agrees, arguing that framing devices contribute to "stage symbolically the aspect of difference," but since he focuses on two-dimensional images, he concludes that they "leave the observer outside and are thus unsuitable for communicating virtual realities in a way that overwhelms the senses" (Grau 2003, 13).

At the same time, Grau is one of very few scholars who subscribe to a mechanical, deterministic conception of the relationship between physical immersion/immersivity and mental immersion/aesthetic illusion/presence. According to this model, the presence of certain medial features—in Grau's case, "360° images, such as the fresco rooms, the panorama, circular cinema, and computer art in the CAVE: media that are the means whereby the eye is addressed with a totality of images" (Grau 2003, 5–6)—somewhat automatically leads to the experience of immersion. Most other scholars allow various other factors to play a role as well, most notably the motivation of the recipients, i.e. Coleridge's "willing suspension of disbelief for the moment" (Coleridge 1989, 52) or "the willingness to ignore [those elements] that might destroy the illusion of another reality" (Lonsway 2009, 125). In this sense, "the process of immersion cannot work if the visitors are not 'willing' to enjoy it and bring to the theme park the 'correct' predisposition" (Carlà-Uhink 2020, 13; see also Wolf 2013b, 52).

What this ultimately means for theme parks is that contrary to their promises, they simply cannot offer experiences of immersion (see Pinzer 2012, 118). What they can offer, however, is immersivity in the shape of immersive environments that may, given the right disposition on the part of the visitor (which may, of course, also be influenced by the parks; see below), temporarily lead to a more or less intensive degree of the experience of immersion. Hence, Disneyland's famous plaque should rather read "Here You *May* Leave Today and Enter the World of Yesterday, Tomorrow and Fantasy." The immersivity of theme parks is mainly characterized by their multimediality (see MEDIA), the role of narratives, and the role of bodily affects, which will be discussed in the following subchapters.

2 Immersivity and the Role of Narratives

2.1 Theme Parks as Fiction

In their efforts to operate as vehicles to narratives, theme parks—especially those in the Disney tradition, whose design draws heavily on filmmaking (see MEDIA)—invite the same type of interpretive attitudes as works of fiction. More specifically, theme parks often meet some of the very same criteria of "narrativity" that make narratives immersive in the first place—namely their capacity to suspend disbelief, invite the audience's presence inside the story and its storyworld, and, by absorbing their (almost) undivided attention, keep the real world at bay (see Clément 2016).

Yet, before they may even be called "narrative" (that is, identify protagonists and events arranged into a plot), theme parks rest on a foundation of fiction, meaning they represent an effort to cut reality's ties to the "here and now" and relocate visitors to an imaginary time-place location (Odin 2000, 107) that often ties the park's lands to canonical worlds of fiction: the Wild West, storybook villages and fairytale castles, etc. (see Philips 2012). Parks thus primarily serve escapist functions—a literal change of scenery made evident by the parks' externality, as notably supported by theming (see THEMING). Additionally, fiction—and, consequently, immersion in fiction—relies not only on the shift in one's "deictic center" (with audiences substituting real life for imaginary or represented space-time coordinates; Segal 1995), but also on a shift in modes of reference and representation.

As notably explained by philosopher John Searle, the logical status of fictional discourse is one of "pretended reference," meaning that fiction rests on operations by which the narrator feigns to merely describe events, persons, or things which, in reality, are entirely made up (Searle 1979). Critical to fiction are framing devices or conventional cues (e.g. the "once upon a time" formula) that allow it to explicitly designate itself as such, momentarily suspend language's ordinary uses (such as truth-telling), and "break the connection between words and the world" (Searle 1979, 66), with the audience's full cognizance (in contrast to fiction, meanwhile, outright lies aim to conceal the disconnect between language and extralinguistic reality). In theme parks, theming constitutes one such form of "pretended reference" in that it allows themed areas to present themselves as "copies without an original" or "signs without references," much in keeping with Eco's and Baudrillard's definitions of simulacra and hyperreality (Eco 1975; Baudrillard 1983). In other words, themed areas "feign" to reproduce time-space locations which, in reality, are wholly invented.

This mode of representation consists of the application of strategies of selection and abstraction and results in "something that is entirely new but that still feels oddly familiar" (Wright 2009, 23; see THEMING). The somewhat comical or parodical appearance of the park's environment thus provides it with a touch of unreality that allows it to be readily identified as distinct from the outside world, even when imaginary space-time locations refer to specific places or historical periods, as with Disneyland's "New Orleans Square" or "France" in Europa-Park in Rust, Germany. Such explicit marks of fictionality or "meta-messages" (Bateson 1973) cue visitors to the same interpretive attitudes as novels, films, or folktales—most crucially, their "willing suspension of disbelief." One might even say that fiction's "pretended reference" invites "pretended belief" in the audience. Such a response is further assisted by the park's idealized environment, whose generally "reassuring" appearance, in which even danger is domesticated, feeds into a form of reverie most conducive to the suspension of criticality (Haas 1978).

2.2 *Immersive Narratives*

Moreover, narratives themselves are immersive or, to reprise Marie-Laure Ryan's expression, there is a "worldness" to narrative representation which makes immersion into narratives possible in the first place (Ryan 2019). This fact is particularly captured by the notion of story world (or "diegesis," in the language of classical narratology), that is, the complex, dynamic image that spectators form of the world presented in a work of fiction. According to Etienne and Anne Souriau, the diegesis is "the world established by a work of art"—one not so much "contained" or represented as implied by or inferred from it (Souriau 1990, 581–582). As one critic explains, narrative interpretation requires audiences "to reconstruct not just what happened but also the surrounding context," much of which "explain[s] narratives' immersiveness, their ability to 'transport' interpreters into places and times that they must occupy for the purposes of narrative comprehension" (Herman 2004, 579).

Design wise, such attitudes are also supported in the theme park by the use of "backstories," namely the "behind-the-scenes legend" (The Imagineers 2010, 26) or origin story that justifies a themed area or attraction's overall appearance, helps make it especially cohesive, and allows it to present the outward image of a world captured *in medias res*—the manifest end-result of past causes and ongoing processes (see Grice 2017). As a result, the audience is invited to piece together an attraction's many fragments into a coherent whole (with the fragments including founders' dates, names, portraits, tools of the trade, and other tell-tale details scattered throughout a themed area). In their overt appeal to the audience's deductive and inference-making capacities, backstories call on audiences to actively reconstruct the rules and circumstances that govern the environment at hand (its "mental model," in the language of cognitive psychology) in much the same way that diegeses or story worlds are deduced from traditional narratives, providing the theme park environment with the same kind of immersive "worldness."

On a more perceptual level, the resulting environment tends to suggest that it exceeds visitors' limited perception of it and exists independently of them, as if moved by a life entirely its own: thus imbued with the "imprint [...] of past action," such landscapes help "create the impression that the story did not begin the moment you arrived, that you have joined a story that has momentum, and action, and is moving forward," as one Disney designer explains (Rohde 2009). The resulting detail-rich, internally coherent, and vivid environments contribute to a sensory overload that increases their reality effect and overall believability. Whether they appeal to conscious or unconscious responses, e.g. by encouraging visitors to explore the environment and go on the hunt for tell-tale details, or by simply overwhelming their capacity to absorb information, backstories serve to let the audience's critical guard down and elicit acceptance, as well as to make the park's storyworlds welcoming and intriguing—an invitation for the audience to step inside.

One additional way to invite visitors' presence inside the storyworld is to make them the protagonists of their own first-person narratives, allowing them to reenact stories and perform roles within them: as one Disney theme park guidebook has it, "Disneyland is essentially a movie that allows you to walk right in it and join in the fun" (Wright 2009, 6). To this extent, theme park environments might be said to convey at least two narratives at once: the themed area's origin story (often implicit and merely glimpsed at by visitors) and the story of the visitor's own visit. Though a matter of debate among park designers and critics (some early Disney designers, most notably Marc Davis, deny that attractions might serve as vehicles to stories; see Surrell 2006, 29–30), it might be argued that all parks, lands, and attractions more or less build on the same basic motif and narrative—that of a voyage into unknown territory. The parks' backdrops serve to set the stage for the patrons' own role as visitors in the literal sense of tourists, subjecting their progression through space to a sequence, each step of which adds up to a minimal, three-act narrative along a canonical structure: departure, progression, and return. Such stories conform to Hollywood storytelling canons for conventional, three-act narratives, as per requirements codified and rigidified in how-to manuals like Syd Field's 1979 *Screenplay* and fully absorbed by studio executives eager to streamline production through tried-and-true formulas (Thompson 1999, 339; Thompson 2003, 43; see also VISITORS).

Such demands for "well-told stor[ies], with a beginning, a middle, and an end" (in the words of then-Disney CEO Michael Eisner; see Eisner and Schwartz 1999, 201) have, at least in Disney's case, since been found to extend to its theme park operations (see Younger 2017, 76–81; see also ATTRACTIONS and THEMING). More importantly, they also meet the basic criteria for narrativity, as identified by structural linguist and narratologist A.J. Greimas. Disney attractions, for example, reveal a whole economy of interactions between the protagonist or "subject" (here, the visitor) and the environment (the "object"), by which both end up mutually transformed: narrative discourse is essentially concerned with "transformations" and thus moves from "an initial state to a final state" marked by changing relationships between subjects and objects, which Greimas describes as "operations of conjunction and disjunction" (Greimas and Courtès 1982, 350). Put simply, thrill rides—and not only thrill rides—operate as a "hero's journey," with riders typically emerging victorious from the perils sent their way by antagonistic forces out to chase them away (see, for example, Expedition Everest at Disney's Animal Kingdom in Orlando, Florida), or alternatively hold them captive (see, for example, the Haunted Mansion at Disneyland in Anaheim, California, or Monsieur Cannibal at Efteling, Kaatsheuvel, the Netherlands), achieve their adventurous goal (in El Rescate de Ulises at Terra Mítica, Benidorm, Spain, this is freeing Ulysses, who is a prisoner of Poseidon), or find the hidden treasure they were seeking (as in Abenteuer Atlantis in Europa-Park). But whether met with hostile or friendly forces, the intrepid travelers only ever briefly reach their destination—the object of their quest—before their inevitable return to earth: each step of the journey thus stages transformations in the visitors' relationship with their fictional environment, from initially far apart (departure) to temporarily conjoined (progression) and, finally, back to separated (return; see Clément 2016, 84–86).

Much in keeping with the latent animism of many animated films (Disney's among them), this relationship might even take on an enchanted, marvelous character, by which the environment takes on the features of a sentient being—when it is not directly personified by a *genius loci*: a Ghost Host (Haunted Mansion), a fabled guardian deity (Fluch der Kassandra at Europa-Park), etc. The visitor's intrusion into the story world thus often serves as an inciting incident and allows it to gradually come to life, suggesting some degree of interaction between them: sometimes called "the Sleeping Beauty effect" (Gardies 1993), this effect is key to enhancing visitors' perceived agency over and, therefore, immersion into their fictional environments. Of course, visitors' agency over the park's physical (if not social) environment is largely a pretense, though this should not be taken to mean that the theme park experience is not authentically interactive (for the immersive effects of social interactions between visitors and staff, see below). As some definitions suggest, while interactive experiences do rely on one's sense of agency, such agency need not mean agency over one's physical environment (for example, by altering its course) so much as agency over oneself—one's sensory and mental states—through one's actions *in*, not *on*, the environment. As one critic explains: "we should understand interactivity as the creation of experiences that appear to flow from one's own actions, including the ability to devote attention to phenomena" (Grodal 2009, 175). In other words, simply using, navigating, or even inspecting one's environment for one's gratification is enough to make the experience interactive.

One additional, rarely discussed aspect of narratives' immersive potential is their capacity to captivate attention and engross audiences. Simply put, it is hard to become immersed in stories that fail to compel. As psychologist Jerome Bruner observes, stories build on incidents—departures from the world's expected course—that warrant their retelling in the first place and constitute their "tellability" (Bruner 1991). The more serious the breach and the higher its stakes, the greater the story's "tellability"—i.e. the greater the audience's interest and engagement with the story become. Such a phenomenon helps identify the socio-cognitive role of narratives (which Bruner calls "the narrative construction of reality"), namely their "normativeness": by staging the disruption of situations otherwise recognized as "normal" or simply "ordinary," narratives help build consensus on the world's expected course and reiterate cultural groups' core norms and values.

That theme parks and theme park narratives serve ideological functions has been widely remarked upon (see WORLDVIEWS), though the fact has seldom been as overtly acknowledged by their creators as at Disney's Animal Kingdom. According to its lead designer, "the underlying value system upon which a story is built" is so essential to the park that, in a departure from prevailing definitions, it provides Animal Kingdom's theme: "the intrinsic value of nature" (Rohde 2008). The violation of nature (by loggers, poachers, and others who prey on nature's resources) provides the park's attractions with their basic story arc: by staging the "conflict between materialistic goals and the intrinsic value of nature" (Rohde 2009), Animal Kingdom helps present nature's sanctity as a norm to be restored in the face of man's destructive actions. To the extent that it increases its appeal and interest in audiences, the normativity of theme park narratives is also key to their immersivity: immersion in narratives calls for a stakeholder stance rather than that of a coolly detached observer; the story resonates with and engrosses the audience so that it keeps the outside world at bay.

3 Immersivity and the Role of Bodily Affect

As defined above, immersion requires a residual awareness of the outside world: "we must take oxygen from the surface" in order to be able to breathe. Nevertheless, theme parks can sometimes be breathtaking: roller coasters such as Black Mamba in Phantasialand (Brühl, Germany) or Silver Star in Europa-Park can literally take your breath away. For a short period of time, visitors experience extreme bodily sensations that make it difficult to adopt a reflective pose. Moreover, roller coasters confront visitors with their own bodies. Phenomenologically, one may distinguish between the material body (German: "Körper") and the body as a medium or a carrier of sensations (German: "Leib"; Husserl qtd. in Kasprowicz 2019, 37). When one rides a fast ride, this (synthetic) separation between the outer body and the carrier of the mental and sensual sensations collapses; visitors are thrown back on the physical components of their bodies, which makes it difficult to distinguish between the two entities. However, visitors experience a loss of distance not only in relation to their ability to reflect, but also in their relation to their bodies. They are thus placed in an "experiential presence" ("Erlebnisgegenwart"; see Szabo 2006, 169) with their bodies and minds.

According to Husserl's phenomenological approach, it is also difficult for roller coaster riders to perceive their surroundings, as the latter are more like an "optical illusion." In *Ding und Raum* (1991), Husserl notes that in a car, one feels the surroundings moving rather than one's own body (282). The movement of the vehicle notwithstanding, the body remains "without kinaesthetic change" (282); it rests, so to speak, while everything around the person in the vehicle is moving. This also applies to theme park visitors on a roller coaster. However, the visitors' difficulty to perceive their surroundings not only increases the focus on their own physicality, but it also makes it harder for them to truly immerse themselves into the park environment. Just as the roller coaster ride could also be seen as the end point of total immersion, the lack of opportunities for reflection, the difficulty to perceive one's surroundings, and the accompanying focus on one's own physicality could also be said to make it difficult for immersion to take place on a roller coaster—in stark contrast to the narrative rides, as well as the underlying narratives of the theme park discussed in the previous section.

Still, other rides combine the theme park's underlying narrative with the kinetic elements of the roller coaster, so that there is an effect on the body here as well. The water coaster Chiapas, located in the "Mexico" section of Phantasialand, for example, casts visitors as the heroes of a journey (see above); thrill sections alternate with calmer parts of the ride that provide different story elements. Having passed several excavation sites, the ride vehicles suddenly change direction and move backwards, taking riders back in time to the Mayas, the remains of whose civilization had been excavated in the previous scene. Roughly in the middle of the ride, visitors come to a grotto that anachronistically juxtaposes different temporal layers: skulls and the depiction of a divine figure refer to Mayan culture, but the dance music and disco ball that reflects the light of numerous colored spotlights and lasers feel contemporary (see TIME). Visitors nevertheless understand that they have traveled through time, not least due to the kinetic effect of moving backwards. Here, then, the ride's kinetics underline and reinforce the narrative. Unlike a book or a movie, a theme park allows visitors to not only physically enter the story, but also "feel" it with their senses. The "willing suspension of disbelief" thus always also depends on the physical disposition of the visitors. They are not only mentally involved in the environment of a park, but also have to be physically open to enjoying the experience.

Riding an attraction like Chiapas obviously also impacts the visitors' perception of the rest of the park, as the speed and thrill change the atmosphere of the surroundings. The atmosphere may further help us understand the perception of immersive environments like theme parks, as the concept focuses on the body: according to German philosopher Gernot Böhme, atmosphere "is the reality of the perceived as the sphere of its presence and the reality of the perceiver insofar as he or she, in sensing the atmosphere, is bodily present in a particular way" (2018, 23–24). As Böhme goes on to state, the atmosphere can produce a certain mood in the perceiver (24), which is used by the producers of such atmospheres to "impact the unconscious" of the perceivers (28): when walking through a theme park, artificially produced atmospheres (unconsciously) influence the visitors, put them in a certain mood, and allow them to feel immersed in the surroundings (see Fig. 1).

For example, different parts of Phantasialand produce different atmospheres. The park's "Deep in Africa" section thus represents a rural African village, with winding narrow paths and wood and clay buildings in brown and beige. By contrast, the "Berlin" section, which depicts the city at the turn of the last century, features one wide cobblestone street, comparatively massive buildings, and pastel shades. Visitors do not perceive each of these details separately and in an analytic way, however, but rather, due to the atmosphere produced by them, unconsciously perceive the spaces in their entirety. And while an atmosphere "can produce a [...] mood in the perceiver," the visitor also has to be attuned to the different atmospheres, because these are part of "a shared reality" (Böhme 2018, 24; see also SPACE).

Fig. 1 A simple statue of Confucius sets the mood at OCT East's Tea Valley park in Shenzhen (People's Republic of China). *Photograph* Salvador Anton Clavé

4 From Immersivity to Immersion

High levels of immersivity on the part of the theme park, achieved through a combination of medial multimodality, believability of the represented world, interactivity, narrative, and bodily affect, do not necessarily lead to the immersion of theme park visitors—the latter need to "play along" as well.

Indeed, theme parks put considerable effort into encouraging visitors to suspend their disbelief. For example, using a politics of selection, they create a stark contrast between the "real," external world on the one hand and the carefree world of the theme park on the other, so that visitors may experience the latter as a place where they need not worry about violence, war, politics, money, sickness, or death, and may thus "let go." The contrast between the theme park and the surrounding world is maintained both on the level of representation as well as that of the lived experience of the theme park visit. Thus, whereas turn-of-the-century amusement parks such as those on Coney Island included controversial topics "ranging from the Boer War and the Galveston Flood to the Fall of Pompeii and the gates of hell" (Lukas 2015, 50–51), and whereas other themed environments, including restaurants, museums, and even cruise ships, have increasingly addressed such serious and politically relevant themes as social justice (see Lukas 2013, 254–255), theme parks have from their very beginning rigorously excluded certain themes or certain aspects of a theme. Death, war, and sickness are a case in point: although prominently featured in spaces that use "dark theming" (see Lukas 2007, 276–280), these issues are usually silenced in theme parks, or at least only used in contexts where they seem far removed from the life world of the visitors. Carlà-Uhink and Freitag argue, for instance, that deaths in theme park representations of ancient Greek myths "are not perceived by the public as 'tragic' and are indeed quite 'aseptic' because they seem detached from daily life and 'historical reality'" (2015, 149; see also THEMING).

Sickness and death are likewise excluded—as far as possible—from the lived reality of the theme park visit. Drawing on interviews with Walt Disney World employees—and putting a venerable urban legend into academic print—Jane Kuenz has maintained that "no one actually dies on Disney property" and that dying persons are kept "'alive by artificial means until they're off Disney property'" lest anyone should die at a Disney park (1995, 115). In fact, in case of a medical emergency, most theme parks will transport suffering visitors or employees to the backstage areas, not only (or even mainly) to maintain the atmosphere of the theme park as a carefree environment, but also to protect patients from being stared at and because the backstage areas offer much better access for medical personnel and equipment than the crowded park. Sick persons may be a more common sight at theme parks in countries that place special emphasis on the social inclusion of chronically ill or recovering patients such as the Netherlands, but here the sight of a sick person may also be perceived as less "unsettling" by a local audience and may thus not keep them from suspending their disbelief (see INCLUSION AND EXCLUSION).

Moreover, theme parks also work hard to discipline uncooperative, "nay-saying" guests, whose questions on backstage logistics "ruin the show" or "ruin the fun."

Some theme park critics have encouraged visitors to positively resist and undermine theme parks' efforts to create an atmosphere of "letting go": Klugman invites "anyone who," like herself, "in the presence of constructed joviality, feels like a cultural misfit" to go on what she calls the "Alternative Ride"—an experience that turns "even the most ordinary situations into ironic social commentary," as, for instance, when the Disney characters are "beheaded by mouths that are eating them as ice cream and candy" (1995, 164, 167). Similarly, Gottdiener and Giroux suggest that "analysts should search for alternate and possibly resistant *behaviors*" in immersive spaces such as theme parks (Gottdiener 1997, 158; emphasis original) and that the parks "should be mined for the spaces of resistance they provide and for the progressive possibilities they offer" (Giroux 1999, 26), respectively (see INCLUSION AND EXCLUSION).

Parks should also be mined, however, for how they respond to such resistant behaviors. With visitors given a role to play, much of their conduct and even their emotions take on the appearance of a loosely scripted performance. Parkgoers are socially expected to engage in at least superficial "facework," i.e. the outward display of emotions and attitudes socially appropriate to the context—in the case of theme parks, an impression of "mock belief" (Clément 2016, 151–152). To this extent, social interactions in the parks follow the same principles as "gaming encounters" (Goffman 1961) in that they create self-contained worlds and require participants to narrow their field of vision, disregard events not directly relevant to the situation at hand, and play by specialized rules that apply only within the game world. Breaches to the park's preferred mode of engagement in particular evince the collective enforcement of the parks' "willful suspension of disbelief"—suggesting that immersion is "a process of mutual engagement in a space" (Lukas 2016, 117).

Rules to maintain attitudes of mock belief and keep interfering behaviors in check are notably evident in the way Disney employees discipline uncooperative guests. For example, they will typically answer questions like "How many Mickey Mouses are there?" with answers that are demonstrably false yet coherent with the park's storylines: "There is only one Mickey Mouse," as per the sanctioned, management-approved answer. The very same dynamics play out in the following exchange, extracted from a message board for employees exasperated with "stupid guests" intent on breaking the park's tacit norms:

> I LOVE messing wit[h] the minds of SG's [stupid guests] when they bring stuff like that up, and it's fully sanctioned! This is one of my favorite conversations with a guest.
>
> SG: "So, how many Mickeys are there?"
>
> Me: "Sir, there is only one Mickey Mouse."
>
> SG: "No, how many people play Mickey Mouse?"
>
> Me: "Sir, it's really Mickey Mouse, there's only one of him."
>
> SG: How many people wear [Mickey Mouse's] costume?
>
> Me: (in a fake sad/shocked voice) "*Gasp* You mean that's a costume? I always thought Mickey was real!" (quoted in Clément 2016, 154)

In a "fully sanctioned" retribution for such breaches, the cast member's ironic self-parodical expression of shock not only pokes fun at the visitor's "superior" perceptiveness but cues him into the desired interpretive attitude—one of "mock belief." This suggests that parks operate as "double inhabitations," i.e. the "doubl[ing] of participating and staging" by which visitors "are both immersed in and staging in the experience": as David Kolb explains, in theme parks "[w]e are involved in both the story being performed and in its staging; we enjoy their interplay" (Kolb 2008, 124–128). This likewise suggests that immersion and rule-compliance are of course neither uncritical nor involuntary, but rather self-conscious and even, to some extent, self-elicited. Such role play evident in "pretended belief" offers one instance of the games that sociologist Roger Caillois calls "mimicry," a form of play that "consist[s] not only of deploying actions or submitting to one's fate in an imaginary milieu, but of becoming an illusory character oneself, and of so behaving," to the effect that "the subject makes believe or makes others believe that he is someone other than himself" (Caillois 1961, 19).

Of course, theme park employees and theme parks themselves sometimes deliberately play with visitors' willingness or ability to suspend their disbelief—for instance, such "behind the scenes" tours as Walt Disney Studios Paris's Studio Tram Tour: Behind the Magic. The various "movie sets" that form the tour feature cameras and lighting equipment and thus look as if the movie crew had simply left for a break. Later, visitors see stunts and special fire and water effects, and it is always clear that the elements will not reach the visitor and are man-made effects normally used in films. Hence, during this tour visitors are deliberately invited to disbelieve and to look "behind the magic"—not the magic of the theme park, of course, but merely that of the movies. "Real" behind-the-scenes tours such as those described by Karen Klugman (1995, 100–103), by contrast, may perhaps even enhance rather than distract from the magic: as Boorstin noted as early as 1961, "[i]nformation about the staging of a pseudo-event simply adds to its fascination" (38)—and thus perhaps to the immersion of the visitor.

References

Bateson, Gregory. 1973. A Theory of Play and Fantasy. In *Steps to an Ecology of Mind: Collected Essays in Anthropology, Psychiatry, Evolution and Epistemology*, ed. Gregory Bateson, 150–166. London: Paladin Granada.

Baudrillard, Jean. 1983. *Simulations*. Los Angeles: Semiotext(e).

Böhme, Gernot. 2018. *Atmospheric Architectures: The Aesthetics of Felt Spaces*. Trans. Anna Christina Engels-Schwarzpaul. London: Bloomsbury.

Boorstin, Daniel J. 1961. *The Image: Or, What Happened to the American Dream*. New York: Atheneum.

Bruner, Jerome Seymour. 1991. The Narrative Construction of Reality. *Critical Inquiry* 18 (1): 1–21.

Caillois, Roger. 1961. *Man, Play and Games*. Trans. Barash Meyer. Champaign: University of Illinois Press.

Carlà-Uhink, Filippo. 2020. *Representations of Classical Greece in Theme Parks*. London: Bloomsbury.
Carlà-Uhink, Filippo, and Florian Freitag. 2015. The Labyrinthine Ways of Myth Reception: Cretan Myths in Theme Parks. *Journal of European Popular Culture* 6 (2): 145–159.
Clément, Thibaut. 2016. *Plus vrais que nature: Les parcs Disney, ou l'usage de la fiction de l'espace et le paysage*. Paris: Presses de la Sorbonne Nouvelle.
Coleridge, Samuel Taylor. 1989 [1817]. *Biographia Literaria*. Cambridge: Cambridge University Press.
Eco, Umberto. 1986 [1975]. Travels in Hyperreality. In *Travels in Hyperreality: Essays*, trans. William Weaver, 1–58. San Diego: Harcourt Brace Janovich.
Eisner, Michael D., and Tony Schwartz. 1999. *Work in Progress: Risking Failure, Surviving Success*. New York: Hyperion.
Freitag, Florian, and Ariane Schwarz. 2015/16. Thresholds of Fun and Fear: Borders and Liminal Experiences in Theme Parks. *OAA Perspectives* 23 (4): 22–23.
Freitag, Florian, et al. 2020. Immersivity: An Interdisciplinary Approach to Spaces of Immersion. *Ambiances: International Journal of Sensory Environment, Architecture and Urban Space* 6. https://doi.org/10.4000/ambiances.3233. Accessed 28 Aug 2022.
Gardies, André. 1993. *L'espace au cinéma*. Paris: Méridiens Klincksieck.
Giroux, Henry A. 1999. *The Mouse That Roared: Disney and the End of Innocence*. Lanham: Rowman & Littlefield.
Goffman, Erving. 1961. *Asylums: Essays on the Social Situation of Mental Patients and Other Inmates*. New York: Doubleday.
Gottdiener, Mark. 1997. *The Theming of America: Dreams, Visions, and Commercial Spaces*. Boulder: Westview.
Grau, Oliver. 2003. *Virtual Art: From Illusion to Immersion*. Trans. Gloria Custance. Cambridge: MIT Press.
Greimas, Algirdas J., and Joseph Courtès. 1982. *Semiotics and Language: An Analytical Dictionary*. Bloomington: Indiana University Press.
Grice, Gordon. 2017. Temporality and Storytelling in the Design of Theme Parks and Immersive Environments. In *Time and Temporality in Theme Parks*, ed. Filippo Carlà-Uhink, Florian Freitag, Sabrina Mittermeier, and Ariane Schwarz, 241–257. Hanover: Wehrhahn.
Grodal, Torben Kragh. 2009. *Embodied Visions: Evolution, Emotion, Culture, and Film*. Oxford: Oxford University Press.
Haas, Charlie. 1978. Disneyland Is Good for You: Charlie Haas on the Magic Kingdom's Master Manipulator. *New West*, December 4: 13–19.
Herman, David. 2004. Storyworld. In *Routledge Encyclopedia of Narrative Theory*, ed. David Herman, Manfred Jahn, and Marie-Laure Ryan, 569–570. London: Routledge.
Holtorf, Cornelius. 2009. On the Possibility of Time Travel. *Lund Archeological Review* 15: 31–41.
Husserl, Edmund. 1991. *Ding und Raum: Vorlesungen 1907*. Ed. Karl-Heinz Hahnengreß and Smail Rapic. Hamburg: Felix Meiner.
Imagineers, The. 2010. *Walt Disney Imagineering: A Behind the Dreams Look at Making More Magic Real*. New York: Disney Editions.
Kasprowicz, David. 2019. *Der Körper auf Tauchstation: Zu einer Wissensgeschichte der Immersion*. Baden-Baden: Nomos.
Kolb, David. 2008. *Sprawling Places*. Athens: University of Georgia Press.
Kolesch, Doris. 2016. Theater und Immersion. *Berliner Festspiele Blog*, September 22. https://blog.berlinerfestspiele.de/theater-und-immersion/. Accessed 25 March 2021.
Klugman, Karen. 1995. Under the Influence. In *Inside the Mouse: Work and Play at Disney World*, ed. Jane Kuenz, Susan Willis, Shelton Waldrep, and Stanley Fish, 98–109. Durham: Duke University Press.
Kuenz, Jane. 1995. Working at the Rat. In *Inside the Mouse: Work and Play at Disney World*, ed. Jane Kuenz, Susan Willis, Shelton Waldrep, and Stanley Fish, 110–162. Durham: Duke University Press.

Lonsway, Brian. 2009. *Making Leisure Work: Architecture and the Experience Economy*. New York: Routledge.

Lukas, Scott A. 2007. A Politics of Reverence and Irreverence: Social Discourse on Theming Controversies. In *The Themed Space: Locating Culture, Nation, and Self*, ed. Scott A. Lukas, 271–293. Lanham: Lexington.

Lukas, Scott A. 2013. *The Immersive Worlds Handbook: Designing Theme Parks and Consumer Spaces*. New York: Focal.

Lukas, Scott A. 2015. Controversial Topics: Pushing the Limits in Themed & Immersive Spaces. *Attractions Management* 20: 50–54.

Lukas, Scott A. 2016. Questioning "Immersion" in Contemporary Themed and Immersive Spaces. In *A Reader in Themed and Immersive Spaces*, ed. Scott A. Lukas, 115–124. Pittsburgh: ETC.

Murray, Janet H. 1997. *Hamlet on the Holodeck: The Future of Narrative in Cyberspace*. New York: Free Press.

Neitzel, Britta. 2008. Facetten räumlicher Immersion in technischen Medien. *Montage AV* 17 (2): 145–158.

Odin, Roger. 2000. *De la fiction*. Paris: De Boeck Université.

Philips, Deborah. 2012. *Fairground Attractions: A Genealogy of the Pleasure Ground*. London: Bloomsbury.

Pinzer, David. 2012. Erlebniswelten und Technikphilosophie. In *Erlebnislandschaft—Erlebnis Landschaft? Atmosphären im architektonischen Entwurf*, ed. Achim Hahn, 97–120. Bielefeld: Transcript.

Rohde, Joe. 2008. From Concept to Reality: Animal Kingdom's Tenth Anniversary. http://www.laughingplace.com/Lotion-View-640.asp. Accessed 28 Aug 2022.

Rohde, Joe. 2009. Story Structure and the Design of Narrative Environments. http://www.andoh.org/2009/12/siggraph-asia-2009-joe-rohde.html. Accessed 28 Aug 2022.

Ryan, Marie-Laure. 2015. *Narrative as Virtual Reality 2: Revisiting Immersion and Interactivity in Literature and Electronic Media*. Baltimore: Johns Hopkins University Press.

Ryan, Marie-Laure. 2019. From Possible Worlds to Storyworlds: On the Worldness of Narrative Representation. In *Possible Worlds Theory and Contemporary Narratology*, ed. Marie-Laure Ryan and Alice Bell, 62–87. Lincoln: University of Nebraska Press.

Searle, John R. 1979. *Expression and Meaning: Studies in the Theory of Speech Acts*. Cambridge: Cambridge University Press.

Segal, Erwin M. 1995. Narrative Comprehension and the Role of Deictic Shift Theory. In *Deixis in Narrative: A Cognitive Science Perspective*, ed. Judith F. Duchan, Gail A. Bruder, and Hewitt E. Lynne, 3–17. London: Routledge.

Sherman, William, and Alan Craig. 2018. *Understanding Virtual Reality: Interface, Application, and Design*, 2nd ed. San Francisco: Morgan Kaufmann.

Souriau, Etienne. 1990. *Vocabulaire d'esthétique*. Paris: Presses Universitaires de France.

Surrell, Jason. 2006. *Pirates of the Caribbean: From the Magic Kingdom to the Movies*. New York: Disney Editions.

Szabo, Sacha. 2006. *Rausch und Rummel: Attraktionen auf Jahrmärkten und in Vergnügungsparks. Eine soziologische Kulturgeschichte*. Bielefeld: Transcript.

Therrien, Carl. 2014. Immersion. In *The Routledge Companion to Video Game Studies*, ed. Mark J.P. Wolf and Bernard Perron, 451–458. New York: Routledge.

Thompson, Kristin. 1999. *Storytelling in the New Hollywood: Understanding Classical Narrative Technique*. Cambridge: Harvard University Press.

Thompson, Kristin. 2003. *Storytelling in Film and Television*. Cambridge: Harvard University Press.

Wirth, Werner, and Matthias Hofer. 2008. Präsenzerleben: Eine medienpsychologische Modellierung. *Montage AV* 17 (2): 159–175.

Wolf, Werner. 2013a. Preface. In *Immersion and Distance: Aesthetic Illusion in Literature and Other Media*, ed. Werner Wolf, Walter Bernhart, and Andreas Mahler, v–vi. New York: Rodopi.

Wolf, Werner. 2013b. Aesthetic Illusion. In *Immersion and Distance: Aesthetic Illusion in Literature and Other Media*, ed. Werner Wolf, Walter Bernhart, and Andreas Mahler, 1–63. New York: Rodopi.

Wright, Alex. 2009. *The Imagineering Field Guide to Magic Kingdom at Walt Disney World*. New York: Disney Editions.

Younger, David. 2017. Traditionally Postmodern: The Changing Styles of Theme Park Design. In *Time and Temporality in Theme Parks*, ed. Filippo Carlà-Uhink, Florian Freitag, Sabrina Mittermeier, and Ariane Schwarz, 63–82. Hanover: Wehrhahn.

Inclusion and Exclusion: Marginalization in Theme Parks

Abstract Theme parks are highly complex spaces, and this is also true when it comes to dealing with marginalization. Far from being practiced only at their main entry gates, exclusion from the theme parks' environments extends into their very realms and manifests itself at both physical or spatial as well as symbolic levels—with both visitors and employees who are subjected to these practices at times seeking to counter them. This chapter thus traces the history of how theme parks have excluded visitors based on class, race, gender, sexuality, or able-bodiedness on two levels: firstly, through active, overt gatekeeping via price limits, bans, and prohibitions; and secondly, on a more symbolic level via representations in theming. Finally, the chapter will not only show how both employees and visitors have reacted to these inequities in order to demonstrate that, thanks to these reactions, theme parks have become more inclusive spaces, but also to highlight areas where exclusions nevertheless persist.

1 Introduction

Theme parks present themselves as inclusive spaces ("for everyone," "for the whole family") but at the same time often claim to offer unique and therefore exclusive experiences (Europa-Park's homepage describes the park as the place where you can "experience Europe in one day"; N.N. N.D.). This already indicates that the dynamics of inclusion and exclusion within the theme park are much more complex than they might first appear. The economic pressure to draw as many visitors as possible to the park pulls toward a greater inclusivity with respect to audiences, which is partly accomplished by excluding controversial or potentially alienating aspects of themes such as poverty or violence (see THEMING). Yet all of this must be developed in combination with the parks' strict safety rules and their ideologically toned representations of the world (see WORLDVIEWS). Thus, prominent signs exclude people who do not comply with the safety requirements of certain attractions, while at the same time the parks have developed special offers for disabled people. In the

This work is contributed by Sabrina Mittermeier, Florian Freitag, Céline Molter, Salvador Anton Clavé, Astrid Böger, Filippo Carlà-Uhink, Thibaut Clément, Scott A. Lukas, Crispin Paine, Ariane Schwarz, Jean-François Staszak, Jan-Erik Steinkrüger, Torsten Widmann. The corresponding author is Sabrina Mittermeier, Fachgruppe Geschichte, Universität Kassel, Kassel, Germany.

F. Freitag et al., *Key Concepts in Theme Park Studies*,
https://doi.org/10.1007/978-3-031-11132-7_7

end, this complex bundle makes mechanisms of inclusion and exclusion far more visible in theme parks than in the outside world that surrounds them.

Indeed, theme parks are complex spaces, and this is also the case when it comes to dealing with marginalization. Far from being practiced at their main entry gates only, where it is reified by the ticket office and security control, exclusion from the theme parks' environments extends into their very realms and manifests itself at both physical or spatial as well as symbolic levels—with both visitors and employees who are subjected to these practices at times seeking to counter them. This chapter thus traces the history of how theme parks have excluded visitors based on class, race, gender, sexuality, or able-bodiedness on two levels: firstly, through active, overt gatekeeping via price limits, bans, and prohibitions; and secondly, on a more symbolic, systemic level via representations in theming. Finally, it will turn to the reactions of both employees and visitors to these inequities to not only show how, thanks to these reactions, theme parks have become more inclusive spaces, but also to highlight areas where exclusions nevertheless persist.

2 Racialized-Turned-Economic Exclusion

Historical perspectives can shed some light on amusement and theme parks' social functions as seats of power (or contestations thereof). In contrast to urban parks and universal expositions (whose promoters hoped they would discipline and educate the public), the carnival-like atmosphere of mid- and late-nineteenth century American amusement parks made them a site of contestation and subversion of dominant norms (see ANTECEDENTS): much to the dismay and concern of the era's cultural arbiters, they "constituted the counterculture of Victorian America; their activities, an inversion of genteel cultural norms" (Kasson 1978, 29). Providing a new arena for spontaneity and abandon rather than self-restraint, Coney Island in New York City expanded public opportunities for romantic licentiousness (some attractions were designed to encourage physical intimacy) and, through the spectacle of variegated crowds freely intermingling, offered a suggestion of social mobility. Revelry even occasionally came laden with anarchic impulses, with attractions inviting patrons to throw objects at imitation china dishes and proclaiming: "If you can't break up your own home, break up ours!" (Kasson 1978, 59). Kasson has correctly noted, however, that the "mechanized, standardized character" of amusements was a mere replication of the urban-industrial life that participants sought to escape, suggesting that Coney Island "affirm[ed] the existing culture" and acted "as a safety valve, a mechanism of social release and control that ultimately protected existing society" (1978, 109).

Yet the most popular ball-throwing game in Coney Island was not "Smash the Dishes," but "Kill the Coon," where patrons threw objects at the faces of African Americans (or Caucasians in blackface; see Sterngass 2001, 105). Indeed, another important way in which American amusement parks perpetuated existing power structures was through racial and class-based exclusion (and of course their intermingling). Spaces of recreation not only reflected wider patterns of urban segregation;

the very escapism of amusement parks was premised on racial exclusion. Originally, American amusement parks, just like other leisure spaces, were open to all; starting from the 1930s, however, many of them would begin to charge an entrance fee on top of charging for rides and other amusements individually. This was nothing but a deliberate means of classist exclusion with inevitable consequences for the "racial profiling" of the desired guests (Morris 2019, 215). The dismantlement of public transport was also instrumental in this regard, cutting off inner-city minorities from suburban places of public recreation. Whereas early amusement parks had often been constructed at the end of trolley lines by transportation companies who sought to increase ridership figures on weekends, by the late 1970s, as Victoria Wolcott remarks, American parks

> no longer needed to define themselves explicitly as white spaces; therefore, there was little noticeable conflict at the new theme parks. [...] This was a public accommodation's version of the "color-blind meritocracy" historians have found in public discourse around housing and education. Those who could afford to go to the theme park in their private cars belonged there. (Wolcott 2012, 9)

Almost invariably then, the presence of Black people at amusement parks (and later theme parks) was perceived as intrusive, suggestive of crime and miscegenation. Not only were spaces of leisure increasingly open to both sexes but some activities (especially dancing and swimming) were associated with physical intimacy and thus perceived as being not suitable for mixed-race interaction.

From the 1930s to the 1960s, recurring episodes of violence caused by white resistance to Black people's desegregation efforts in such spaces ("recreational riots"; see Wolcott 2012) provided additional excuses for the continued exclusion of Black people and sometimes even helped justify the closure of parks. Despite legal advances, the segregation of spaces of leisure thus went on in both the North and South of the U.S., until the final demise of local parks sometime in the late 1960s and early 1970s. To this end, white patrons and operators resorted to the privatization of facilities, with the creation of "clubs" whose membership could be refused on a purely arbitrary basis (Wolcott 2012, 66, 198).

The context of racial segregation in the U.S., and the tensions that accompanied attempts to gradually abolish it, are essential for understanding the success of the Disney theme park model and its imitators around the world, as the development of private parks would come full circle with the opening of Disneyland in 1955: the Walt Disney Company positioned their park (and each of their parks to follow) as targeted at the white middle to upper class (see Mittermeier 2021). Following Wolcott, the hypersanitized and overly controlled environment, the location in the outer periphery of cities (which guaranteed that the parks could only be reached by private cars or regulated transportation), the absence of facilities like dancing venues or swimming pools that might provoke racial tensions, and the integration of security forces into the friendly, costumed employee "cast" to minimize provocation, are all lessons Disney learned from the tensions caused by the segregation practiced in other amusement parks (2012, 155–156). Yet, this segregation was always implicit rather than explicit—Disneyland was even listed in the *Green Book* for Black Americans in

1963, denoting a safe space (Morris 2019, 214). Effectively, then, these measurements allowed Disney to become "color-blind" in its customer policy (in itself a problematic concept) and to definitively shift the issue of accessibility from race to class.

Class-based exclusion is in turn a marker that has remained in place until today. Not only are entrance prices for theme parks, particularly those run by Disney or Universal, rising steadily; theme parks are also increasingly abandoning their original model of all-inclusive pricing and are offering more and more so-called "VIP experiences." Because of this, many areas of the parks, including table service restaurants, certain shops, or specific viewing areas for shows and performances, have become exclusively accessible to visitors who pay an additional fee. This applies, for instance, to restaurants that offer character appearances such as the Plaza Inn at Disneyland in Anaheim, California; shops that offer "transformation" experiences such as the Bibbidi Bobbidi Boutique at Walt Disney World's Magic Kingdom in Orlando, Florida, where children are "transformed" into "princesses" (which is further restricted to certain age groups, though, at least formally, not to a specific gender; see Gutierrez-Dennehy 2019); or certain viewing areas for Disney California Adventure's World of Color show, which are reserved for those who purchase a "dining package." Although some theme parks may allow non-paying customers to take a brief tour of the restricted premises (e.g. for the purpose of taking pictures), these places and offerings nevertheless belie the "democratic" promise of the one-ticket policy of most theme parks, according to which anyone who can afford the entry fee is welcome to enjoy the park in its entirety. More and more theme parks also offer so-called VIP guided tours that offer access to rides, shows, and dining experiences without waiting times and which can cost up to several hundred dollars per visit, such as Universal Orlando's VIP Tour Experience ($189–$419 per person for a seven-hour tour in 2020). It may come as somewhat of a surprise that Disney's Fastpass system, which grants faster access to specific attractions, has remained without an additional cost for as long as it has, although Disneyland's paid Maxpass (in Anaheim) is now falling in line with Universal Studios' Universal Express or Sea World's Quick Queue, which are only available for an additional fee.

Restricting certain areas to more affluent visitors becomes especially controversial when the additional fees are (perceived as) extremely high, when upcharge extras are aggressively advertised in the park, and/or when experiences and services formerly included in the park admission are turned into sources of ancillary revenue. A case in point is Disneyland's Club 33, to which all of the above has applied, particularly since its 2013–14 expansion. Inspired by the executive lounges at the 1964–65 New York World's Fair, Club 33 had welcomed members to its elegantly designed bar and restaurant since the opening of the park's "New Orleans Square" area in 1966 and was the only place at Disneyland to serve alcohol. Membership was limited to a few hundred, yet open to anyone who could afford the membership fee of $5,000 (in 1984) and the additional annual dues (see Johnson 1991). Occasionally, the club also served as a location for special events, such as the opening of the Haunted Mansion in 1969. Members would enter the club through an inconspicuous door located next to the Blue Bayou Restaurant and access the main dining room via a small overhead passageway that was hidden behind a street corner.

When in 2013–14 Disney decided to increase the club's capacity (and the revenues from club membership fees and dues), however, the company no longer tried to hide Club 33: the small passageway was significantly widened and equipped with large picture windows; similar picture windows appeared elsewhere in the club, offering impressive views of "New Orleans Square" as well as of the Fantasmic! Nighttime spectacular, which takes place on and across the neighboring Rivers of America. Especially at night, however, the new picture windows also allow "regular" visitors glimpses into the new facilities from down below. Online reviewers have angrily commented on this, complaining that Disney now dangles the Club 33 experience in front of visitors like a carrot, trying to get them to spend even more money at the park (see DeCaro 2014). Moreover, Club 33 also received a new, highly prominent, and ostentatious entrance: the Court of Angels, a small area formerly accessible to all visitors and a popular photo spot, was closed off with stained-glass screens to serve as the club's new entrance and lobby. In a strange case of what could be called the simulated privatization of simulated public space, the courtyard, a private space originally designed to evoke a (semi-)public place, was redesigned to represent a private space, accessible only to the most affluent among Disneyland's visitors.

Located right next to Club 33, the Disneyland Dream Suite constitutes an even stranger case of exclusion. Originally conceived as a private apartment for Walt Disney and his family, the space above the entrance to the Pirates of the Caribbean ride had been used, following Disney's death, for hosting VIP receptions and as office space. In 1987, it was opened to all visitors as a retail space before it was, in the context of a marketing campaign in 2007, once again turned into a private apartment, open to contest winners, Disney managers, celebrities, and other "special guests." In 2017, the space began accepting reservations for dinner, although at a price that once again ensures that almost all visitors will be excluded (see Freitag 2021, 178–185). That same year, the Walt Disney Company also announced the construction of its recently opened Star Wars: Galactic Starcruiser resort at Walt Disney World, which due to its extremely high prices constitutes yet another step toward ever-more exclusionary, niche-targeted offerings in the experience economy.

3 Between Exclusion and Integration

Other forms of exclusion at theme parks are openly and directly based on physical aspects of visitors' bodies. Certain age groups, very small, tall, or large persons, pregnant women, photo- and audiosensitive people, visitors with cardiac pacemakers, as well as visitors with physical and other disabilities, those who require a wheelchair, a mobility scooter, or the assistance of accompanying people or service animals, are frequently excluded from specific attractions at theme parks. Here, however, the parks have become increasingly attentive to the needs of these audiences and have striven to include them by way of specific design choices. These range from the use of wider queue lines or virtual queuing areas and the installation of benches, ramps, and elevators to the creation of appropriately modified, separate loading stations and ride vehicles, the use of audio guides and subtitles, and the offering of so-called alternative experiences that allow visitors to watch point-of-view videos of specific rides.

In recent years, the range of attractions accessible to people with disabilities has thus significantly increased, especially with the help of new technologies. The Dutch amusement park De Efteling (Kaatsheuvel), for example, offers a virtual reality alternative to Droomvlucht, a suspension gondola ride through a fairyland for which riders usually need to transfer from their wheelchairs. In a special room located near the exit, visitors can experience the ride through VR glasses, including 360-degree vision, sound, and the scent of flowers. Using a headset, they even have the additional option of experiencing the ride at the same time as their companions and communicating with them (Efteling 2017). Of course, the construction of attractions suitable for the disabled is complicated by the high demands of visitor safety: in case of emergency, it must always be possible to leave the attraction smoothly and quickly. And while newly built attractions already often take the needs of physically impaired visitors into account during the design process, it is sometimes difficult to convert older rides.

Particularly over the past few years, food offerings have also been greatly diversified—dietary restrictions for religious (such as Halal or kosher food) or personal reasons (vegetarianism or veganism), as well as food allergies of all varieties, are something that the restaurants at Disney parks, for instance, have become especially attentive to. However, this only applies to their American theme parks. Disney's Asian resorts and Disneyland Paris still lack the variety to accommodate these needs, which suggests that some forms of inclusion in theme parks are mostly due to cultural context. German theme parks rarely cater to any of these needs and the food is often more of an afterthought in these spaces.

4 Symbolic Exclusion: Dealing with Representation

With respect to symbolic or representational exclusion, it has been argued that nostalgia—the stock and trade of many, if not most theme parks—is by nature exclusionary in that its appeal is limited to the section of the public likely to find some comfort in the memory of times gone by; in Europe and North America, therefore, middle- to upper-class whites (see THEMING). Sometimes, the nostalgia that U.S. theme parks seek to indulge in represents pre-desegregation America itself: for example, replacing Cincinnati's Coney Island (which closed in 1971), Paramount's King Island opened the same year, beyond the reach of public transport. In a nod to its predecessor, however, the new park incorporated a Coney Island section—one now free of black visitors and playing to memories of the pre-desegregation park (Wolcott 2012, 220–221). The Disney parks' nostalgic take on American history was brought to the fore with the corporation's once-projected historical theme park, Disney's America, which got caught in the larger crossfire of the so-called Culture Wars raging in the U.S. in the early 1990s; it also highlighted the underlying issues of whitewashed history (Mittermeier 2016; Carlà-Uhink 2020, 11–15). An essential ingredient to Disney's nostalgic and reassuring version of history—or "Distory," as some critics have called it (Fjellman 1992, 59)—is the systematic suppression of conflicts in the parks' environment. As a result, issues of class and race are frequently "skirted over" (Bryman 1995, 131), and marginalized groups are generally erased from view (see THEMING and WORLDVIEWS).

At the same time, racist representations persist in today's theme parks. When they do occur at all, theme park representations of Black people, Indigenous people, or People of Color still often conform to negative or demeaning stereotypes. Perhaps the most astonishing example was the "teacup" (or rather "cooking pot") ride Monsieur Cannibale in Efteling (Netherlands), advertised by the park as a ride "for all the family," which featured a French soundtrack about a European visitor caught by a tribe of cannibals and placed into the "harem" by the "big chief"—it was only after much public criticism that the ride was closed in 2021 (see Fig. 1). However, less blatant examples of stereotypes can also be found in Disneyland's "Adventureland" and "Frontierland" (or many of the themed areas in other parks built in their image), whose themes of conquest, exploration, and colonialism tap directly into well-established racial clichés. "Adventureland"'s exoticism, for instance, is predicated on the "othering" of distant and "primitive" tribes—as notably demonstrated by the Jungle Cruise's Trader Sam character, a headhunter whose appearance varies from vaguely threatening (at Disneyland in California) to frankly grotesque (at Magic Kingdom in Florida). Fjellman has rightly called this "'cute' colonial racism" (1992, 225–226), while Mittermeier speaks of "armchair colonialism" (2021, 174; see WORLDVIEWS).

Somewhat reflecting the U.S. government's policy of confinement to reservations, the presence of Native Americans in "Frontierland" was, at least originally, only "peripheral"—that is, mostly restricted to the land's outer edge, beyond the railroad tracks that encircle the park. The role of Native Americans in the history of the Western settlement was highlighted by the script recited aboard the Santa Fe and Disneyland Railroad in 1962: "watch for Indians and wild animals along the riverbanks … some Indians are hostile, and across the river is proof … a settler's cabin afire. The pioneer lies in his yard … victim of an Indian arrow" (Francaviglia 1999, 175–176). This episode of "Indian" savagery—here evident in their association with the animal world—builds on a popular motif already on display in Buffalo Bill's Wild West Show during the late nineteenth century, namely the "Attack on the Settler's Cabin" narrative (see ANTECEDENTS). Owing to the symbolic meaning of the home, which "presupposed the presence of a woman, particularly a wife," the piece "tapped into a set of profound cultural anxieties" and made the raid "an attack on whiteness, on family, and on domesticity itself" (Warren 2005, 238). While the threat of war-like "Indians" was initially played off against "a friendly Indian village with the inhabitants active in their daily tribal chores" (still as per the railroad's spiel), all descriptions of Native Americans as hostile have since been removed, and the fire that consumes the cabin is now explained as the product of lightning (Francaviglia 1999, 175–176). As Lantz has argued, however, this has de facto "create[d] indexical absence of Nativeness" in Disneyland and "relegated representations to the

Fig. 1 Monsieur Cannibale at Efteling (Netherlands). *Photograph* Pro Shots/Alamy Stock Photo

back areas of Frontierland" (2019b, 48). Hence, while racist stereotyping of Native Americans is largely gone, so are the more positive representations of them.

Another case in point is sexist depictions of (white, straight, cis) women as well as, more generally, the portrayal of stereotypical gender roles in theme park attractions. At the Carousel of Progress at Magic Kingdom's "Tomorrowland," for example, "the position of women is presented in terms of their gradual emancipation from the kitchen by the virtue of the creation of labour saving goods" (Bryman 1995, 131). Created in 1964, the attraction has been updated five times since, as a result of which the attraction's homemaker was promoted to working mother and the father was allowed to cook the family's dinner. As Alan Bryman remarks, though, his incompetence in the kitchen does little to mitigate the attraction's gender stereotypes (1995, 132).

Perhaps even more spectacular were the successive changes brought to the depiction of women in Disney's Pirates of the Caribbean rides. Much of those efforts have been directed to expunging all references to the pirates' sexual exploitation of women and depicting women not as victims, but as protagonists. In 1997, the "pooped pirate" scene was altered so the pirate would no longer appear to be chasing a woman hiding in a barrel behind him: trading the woman's petticoat for a turkey leg (at Disneyland in California) or for a magnifying glass (at Magic Kingdom in Florida), the pirate is now in search of food or riches rather than sex. In the same year, the Magic Kingdom's chase scene was reversed, to the effect that women—brooms and rolling pins in hand—were now in pursuit of pirates, who, judging by their loot, had just raided their pantries. But while the new version removes any reference to assault, it is hard to see how women's representation as kitchen harpies breaks from existing stereotypes. In 2017, the bride auction scene was altered at Disneyland Paris, Magic Kingdom, and Disneyland, so the women would no longer be auctioned off ("Take a wench for a bride," a banner proclaimed) but have become pirates themselves, presiding over the sale of their plunder. While the most recent changes have been generally welcomed by the public, the modifications introduced in 1997 earned the attraction's new version the moniker of "politically correct rehab"—one that, by softening the ride's originally rowdy, sexist edge, runs against the ride's and its creators' original intentions. As one of the designers, X Atencio, says: "The show's called Pirates of the Caribbean, not Boy Scouts of the Caribbean" (Surrell 2006, 99; see also Freitag 2021, 169–170).

Internationally, the case of Asian theme parks provides additional perspectives on issues of representation in theme parks, for example by shedding light on East–West power relations or the nationalist agenda of governments. In the case of "reverse othering" or "reverse orientalism" (Hendry 2000), the Japanese indulge visions of the West as exotic at so-called *gaikoku mura*—theme parks recreating the atmosphere of foreign countries. Themed after such destinations as Germany, Spain, Canada, or Russia, over two dozen of them have opened since the inauguration of Tokyo Disneyland (1983)—with the largest, Huis ten Bosch, being a recreation of Holland (see Schlehe and Uike-Bormann 2010; see ANTECEDENTS). Much in keeping with the Japanese's generally positive attitudes toward imitation, such reconstructions are best understood as efforts to domesticate, "'wrap', or 'tame' foreign countries" (Hendry 2000, 190). Asian countries may conversely try to capitalize on their perceived exoticism to better appeal to Western tastes: such attitudes, already apparent in the "Japan" and "China" pavilions at the World Showcase at Epcot in Orlando, Florida (as well as those of other, non-Asian countries), are probably best illustrated by nearby Splendid China in Four Corners, Florida: a copy of its sister park in Shenzhen (China), the park's miniature replicas of China's natural or man-made wonders failed to draw crowds, however, and closed in 2003 after 10 years.

Also, while clearly in the service of nationalist ideology (see WORLDVIEWS), the People's Republic of China's efforts at self-representation in ethnic theme parks help counter Beijing's homogenizing and hegemonic ambitions. In its celebratory yet unified vision of China's fifty-six ethnic groups, Beijing's Ethnic Park serves as "an important site for reframing and incorporating Tibetans and Tibetan Buddhism within the post-Mao PRC nation-state," making it "an ever-involving attempt, in the fact of the actual unruliness of real people, to structure and render pleasurable citizens' and visitors' participation in the core categories, hierarchies, and thus values of the ascendant PRC" (Makley 2010, 128–129).

5 Visitors Reacting to and Participating in In- and Exclusion

While there may be increased awareness of accessibility issues on the part of the parks, along with legislation working in favor of making theme parks more accessible and inclusive spaces, it is not only the park offerings themselves that determine how inclusive a theme park is, but also the behavior of theme park actors—visitors, fans, and employees. For example, there are cases where the protests of theme park actors against symbolic and structural exclusions have actively contributed to removing them, as theme parks have sought to avoid negative publicity. Over time, many have thus seen their representations in and access to the theme park spaces change and improve, in a response to both individual actions as well as broader cultural changes. At the same time, people with differences may still experience exclusion or difficulties that can ultimately be traced back to the abusive behavior of other, particularly heterosexual and non-disabled, visitors.

Heteronormativity is one area in which activism from visitors, fans, and employees has brought about a change in the relationship of the LGBTQIA+ community to theme parks, at least in the Western world. Critics such as Kuenz (1995, 69–70) or Griffin (2000, 121) have argued that the Disney parks' emphasis on family has made it a gendering, heteronormative institution. Given this heteronormativity, it comes as no surprise that while California's Great America (Santa Clara, California) had organized a "Gay Night" as early as 1979 (see Stryker and van Buskirk 1996, 82), Disney as well as many other theme parks have long remained explicitly discriminatory toward visitors living what officials once called "alternative lifestyles" (Davies 2008): in 1980, Andrew Exler sued Disneyland for encroachment on his civil rights when he and his date Shawn Elliott were escorted off the premises by security guards for dancing together at a Disney After Dark event—as noted above, dancefloors were originally excluded to avoid intimacy that was unwanted by the parks' owners, and it seems pertinent that this occurred in this context. Four years later, a Superior Court judge ruled in their favor and Disneyland had to abolish its ban on same-sex dancing; however, the ruling did not extend to further cases of "homosexual touching" (Davies 2008).

Whereas the exclusion of the LGBTQIA+ community was initially approved even by some queer Disneyland staff members (see Griffin 2000, 126), fans eventually responded with a larger self-organized effort: the "Gay Days." Disneyland had long allowed private parties on property, and in 1978 an organization of gay bar owners rented out the park using the moniker "Los Angeles Bar and Restaurant Organization" (Griffin 2005, 126). Disney's officials realized too late to cancel but hired extra security and barred all public dancing (Griffin 2005, 126). This kind of "take over" of the park has been framed by Griffin as a measure of so-called "zapping" akin to kiss-ins during the post-Stonewall years (2005, 127–128). While no such events re-occurred in the wake of the AIDS Crisis in the 1980s, Disneyland did become an official site of AIDS benefits events; these were, however, intentionally purged of any labeling related to the LGBTQIA+ community (Griffin 2005, 132). In 1991, Gay Days came to Walt Disney World in Florida, here starting out as a small gathering called the "Gay and Lesbian Day." For this event, participants wore red t-shirts, both to recognize each other (Griffin 2005, 134) and blend in with the crowd (rather than stand out in rainbow-colored outfits; Swartz 2017). Also during the early 1990s, Disneyland and Walt Disney World workers started forming queer employee groups (LEAGUE/Anaheim and ALLIANCE, respectively; see Griffin 2000, 116).

Today, Gay Days are a yearly event at Disneyland Park in Anaheim as well as Magic Kingdom in Florida, drawing an estimated 200,000 visitors to several pride events taking place around the same time in Orlando (Swartz 2017).

These events are notably "apolitical" (Griffin 2005, 135), likely one reason why the Walt Disney Company has started to welcome them, even if they are still not officially sanctioned. For example, the parks offer a telling amount of rainbow merchandise around the scheduled dates, as well as during national pride month in June—and recently, items such as Pride Mouse Ears have become available year-round. Because of such tacit endorsement, the company had seen boycotts from evangelical Christian groups in the past, such as the Southern Baptist Convention that upheld their boycott for the first eight years after Gay Days began in Orlando (Griffin 2005, 139). Since 2016, the Pride Event has only increased in importance: after the attack on the gay night club Pulse in Orlando, which also saw an outpouring of support from Walt Disney World and Universal Studios employees (Koman 2016), the visibility of LGBTQIA+ people in the city had gained a certain urgency. Disney itself also donated $1 million to the victims' families in its aftermath (Smith 2016). Meanwhile, Disneyland Paris became the first Disney theme park to host an officially sponsored Pride Event on June 1, 2019 ("Magical Pride"), and it seems only a matter of time before such celebrations are also held in the U.S. Yet, as Lantz rightly argues, "even in moments that celebrate queer aesthetics, such as the unofficial Gay Days at Walt Disney World, gender presents as rigid or separate" (2019a, 1348) and is thus still mostly enforcing a rather homonormative view of queerness. One of Germany's biggest theme parks, Phantasialand, has itself offered pride celebrations since 2004: "Fantasypride," touted as Europe's biggest such event in a theme park, is still one of the few ones around the world and is in line with other generally "more party than politics" focused pride celebrations in larger German cities.

Theme park visitors have also begun to counteract the lack of representation of different groups in theme park attractions in ways that are similar to how consumers of other media or pop cultural products have, namely through "fannish" activities. Disney theme parks in particular have enjoyed a strong and ever-growing fandom (see Williams 2020) that not only includes the fan activism behind the Gay Days, but also cosplaying from a diverse fanbase within the parks (see VISITORS). Through such practices, fans reinsert themselves into spaces that do not directly reflect them, such as, for instance, the participation of People of Color in Disney's Dapper Day. The Dapper Day is an unofficial event that stretches beyond the theme park and features cosplayers in retro outfits of all periods, and thus allows for the reinsertion of diverse audiences into Disneyland's "Main Street, U.S.A." section, counteracting its white-washed and heteronormative image of turn-of-the-century America. As Lantz argues, while "the styling encourages a more varied participant self-expression that may defy the official rules," it simultaneously counteracts the more obviously gendered and heteronormative practices of Disney's own Bibbidi Bobbidi Boutique (2019a, 1341, 1347; see above).

At the same time, people with differences may still experience exclusions or difficulties that can be ultimately traced back to the abusive behavior of other visitors. The exclusion of same-sex couples, for example, has not only been practiced by theme parks themselves, but also by their visitors, who reportedly police behavior, as illustrated by the following report by a female visitor to Disneyland in California:

> I was at Disneyland with my girlfriend having a fun time. During the line for the Haunted Mansion, I gave my girlfriend a kiss. It was nothing serious, just a small kiss. All of a sudden, I felt somebody tapping my shoulder. I turned around and a man was yelling at me and saying I should not be doing things like that at Disneyland and that they were disgusted. I felt so angry and embarrassed that my girlfriend and I got out of the line and onto another ride. (qtd. in Camara et al. 2012, 324)

The perpetrator apparently considers Disneyland a (hetero-)normative space, where any—even the lightest and smallest—form of public expression of LGBTQIA+ life is so inappropriate that it legitimizes setting aside any form of socially expected aloofness toward the other or at least politeness in such a setting. Likewise, in a recurring case amounting to a "niche cottage industry," the structures put in place for disabled people at Disney parks were exploited by non-disabled visitors: people not willing to pay the fees for a VIP tour (see above) have hired patrons with disability passes in order to avoid long waiting lines. In response, Disney switched from unlimited to per-attraction disability passes around 2013, which in turn resulted in longer waiting times that have proven particularly difficult for patrons affected by autism, for example (see St. Clair 2015).

Employees have in turn similarly pushed back against the exclusionary practices of theme parks. A case in point is the conflicts at the opening of Disneyland Paris in France, where the unionization of workers helped push back against Disney's strict grooming rules that include forbidding tattoos, nail polish for women, or long hair for men (Lainsbury 2000). And yet, even though the public image of theme parks may suffer if employees are perceived as being suppressed and underpaid, the situation of many theme park employees is dire (see LABOR). In 2018, a survey by Occidental College called "Working for the Mouse" created a stir by reporting that a high percentage of Disneyland's employees in California were paid poverty wages, causing homelessness, food insecurity, and a lack of health insurance. When U.S. Senator Bernie Sanders attended a protest rally in Anaheim, Disney was quick to announce a proposed wage offer "that they said would amount to a 36 percent increase over a three-year span for hourly workers" and to blame Sanders' engagement on his need for publicity, while stating that "we continue to support our cast members through investments in wages and education" (Stateman 2018).

Hiring practices and typecasting (see LABOR) are another pitfall for theme park publicity and belie more recent strides to make them more inclusionary. In the case of the creationist theme park Ark Encounter in Grant County, Kentucky (in and of itself an exclusionary space), a dispute on hiring policies had to be resolved in court: employees for Ark Encounter must, along with their application, sign a "statement of faith" issued by the park operator Answers in Genesis, in which they certify their belief in the creationist worldview and creationist values, including the rejection of homosexuality and pre-marital sex (Bielo 2018, 9; see WORLDVIEWS). When Ark Encounter applied for a tax incentive program under the Kentucky board of tourism, secular and counter-creationist groups threatened to sue the state for violating the First Amendment establishment clause because of discriminatory employment. When the park's application was rejected, they successfully sued the state for violating their First Amendment rights of free exercise and were finally approved the rebate.

All of these cases illustrate the continued efforts toward making theme parks more inclusive spaces. But as theme parks, like all products of popular culture, reflect an increasingly inclusionary zeitgeist, they also have to account for the conservative backlashes to progressive social movements and the convergence of a diverse audience at any moment in time. It remains interesting to see how employees and visitors alike will continue to push back against the lack of inclusivity in theme parks, and whether or how these spaces will change in turn. While theme park companies like Disney have shown some willingness to go with the times, such changes have often been successful not due to activist pressure, but due to a fear of losing face and thus important revenue.

References

Bielo, James S. 2018. *Ark Encounter: The Making of a Creationist Theme Park*. New York: New York University Press.

Bryman, Alan. 1995. *Disney and His Worlds*. London: Routledge.

Camara, Sakile, Amy Katznelson, Jenny Hildebrandt-Sterling, and Todd Parker. 2012. Heterosexism in Context: Qualitative Interaction Effects of Co-cultural Responses. *Howard Journal of Communications* 23 (4): 312–331.

Carlà-Uhink, Filippo. 2020. *Representations of Classical Greece in Theme Parks*. London: Bloomsbury.

Davies, Erica. 2008. Dancing at Disneyland: Gays in the Fast Lane. *Outhistory.org*. http://outhistory.org/exhibits/show/queer-youth-campus-media/facing-discrimination/dancing-at-disneyland. Accessed 28 Dec 2020.

DeCaro, Dave. 2014. Kool Aid at Club 33, Pt. 1. *Daveland*, July 29. https://davelandblog.blogspot.com/2014/07/kool-aid-at-club-33-pt-1.html. Accessed 28 Aug 2022.

Efteling. 2017. Efteling komt met een virtuele beleving van Droomvlucht. https://www.efteling.com/nl/pers/efteling-komt-met-een-virtuele-beleving-van-droomvlucht/. Accessed 28 Dec 2020.

Fjellman, Stephen M. 1992. *Vinyl Leaves: Walt Disney World and America*. Boulder: Westview.

Francaviglia, Richard V. 1999. Walt Disney's Frontierland as an Allegorical Map of the American West. *The Western Historical Quarterly* 30 (2): 155–182.

Freitag, Florian. 2021. *Popular New Orleans: The Crescent City in Periodicals, Theme Parks, and Opera, 1875–2015*. New York: Routledge.

Griffin, Sean P. 2000. *Tinker Belles and Evil Queens: The Walt Disney Company from the Inside Out*. New York: New York University Press.

Griffin, Sean P. 2005. Curiouser and Curiouser: Gay Days at the Disney Theme Parks. In *Rethinking Disney: Private Control, Public Dimension*, ed. Mike Budd and Max H. Kirsch, 125–150. Middletown: Wesleyan University Press.

Gutierrez-Dennehy, Christina. 2019. Taming the Fairy Tale: Performing Affective Medievalism in Fantasyland. In *Performance and the Disney Theme Park Experience*, ed. Jennifer A. Kokai and Tom Robson, 65–84. Cham: Palgrave Macmillan.

Hendry, Joy. 2000. *The Orient Strikes Back: A Global View of Cultural Display*. Oxford: Berg.

Johnson, Kevin. 1991. No Mickey Mouse Club Here. *Los Angeles Times,* September 8: OC B1-OC B6.

Kasson, John F. 1978. *Amusing the Million: Coney Island at the Turn of the Century*. New York: Hill & Wang.

Koman, Tess. 2016. How Disney Employees are Responding to the Orlando Massacre. *Cosmopolitan*, June 15. https://www.cosmopolitan.com/lifestyle/news/a59974/disney-employees-post-love-pictures-for-orlando-shooting-victims/. Accessed 28 Aug 2022.

Kuenz, Jane. 1995. It's a Small World After All. In *Inside the Mouse: Work and Play at Disney World*, ed. Jane Kuenz, Susan Willis, Shelton Waldrep, and Stanley Fish, 54–78. Durham: Duke University Press.

Lainsbury, Andrew. 2000. *Once Upon an American Dream: The Story of Euro Disneyland*. Lawrence: University Press of Kansas.

Lantz, Victoria. 2019a. Reimagining Tourism: Tourist-Performer Style at Disney's Dapper Days. *The Journal of Popular Culture* 52 (6): 1334–1354.

Lantz, Victoria. 2019b. What's Missing in Frontierland? American Indian Culture and Indexical Absence at Walt Disney World. In *Performance and the Disney Theme Park Experience*, ed. Jennifer A. Kokai and Tom Robson, 43–63. Cham: Palgrave Macmillan.

Makley, Charlene. 2010. Minzu, Market, and the Mandala: National Exhibitionism and Tibetan Buddhist Revival in Post-Mao China. In *Faiths on Display: Religion, Tourism, and the Chinese State*, ed. Tim Oakes and Donald S. Sutton, 127–156. Plymouth: Rowman & Littlefield.

Mittermeier, Sabrina. 2016. Disney's America, the Culture Wars, and the Question of Edutainment. *Polish Journal for American Studies* 10: 127–146.

Mittermeier, Sabrina. 2021. *A Cultural History of the Disneyland Theme Parks: Middle Class Kingdoms*. Chicago: Intellect.

Morris, Jill Anne. 2019. Disney's Influence on the Modern Theme Park and the Codification of Colorblind Racism in the American Amusement Industry. In *Performance and the Disney Theme Park Experience*, ed. Jennifer A. Kokai and Tom Robson, 213–227. Cham: Palgrave Macmillan.

N.N. N.D. Europa-Park Resort. https://www.europapark.de/en?msclkid=5e081f1720f21699fb8a7c1e9b56c0bc. Accessed 28 Aug 2022.

Schlehe, Judith, and Michiko Uike-Bormann. 2010. Staging the Past in Cultural Theme Parks: Representations of Self and Other in Asia and Europe. In *Staging the Past: Themed Environments in Transcultural Perspectives*, ed. Judith Schlehe, Michiko Uike-Bormann, Carolyn Oesterle, and Wolfgang Hochbruck, 57–92. Bielefeld: Transcript.

Smith, Thomas. 2016. Walt Disney Company Donates $1 Million to OneOrlando Fund. *Disney Parks Blog*, June 14. https://disneyparks.disney.go.com/blog/2016/06/walt-disney-company-donates-1-million-to-oneorlando-fund/. Accessed 28 Aug 2022.

St. Clair, Katy. 2015. How Fakers with Wheelchairs Ruined Disneyland's Disabled Line. *Vice*, November 4. https://www.vice.com/en_us/article/jpy3vy/how-fakers-with-wheelchairs-ruined-disneylands-disabled-line. Accessed 28 Aug 2022.

Stateman, Alison. 2018. Bernie Sanders Takes on Disneyland for 'Poverty Wages' at Anaheim Rally. *Commercial Observer Online*, June 2. https://commercialobserver.com/2018/06/bernie-sanders-takes-on-disneyland-for-poverty-wages-at-anaheim-rally/. Accessed 28 Aug 2022.

Sterngass, Jon. 2001. *First Resorts: Pursuing Pleasure at Saratoga Springs, Newport & Coney Island*. Baltimore: Johns Hopkins University Press.

Stryker, Susan, and Jim van Buskirk. 1996. *Gay by the Bay: A History of Queer Culture in the San Francisco Bay Area*. San Francisco: Chronicle Books.

Surrell, Jason. 2006. *Pirates of the Caribbean: From the Magic Kingdom to the Movies*. New York: Disney Editions.

Swartz, Anna. 2017. How Disney GayDays Quietly Became a Massive Pride Event. *Mic*, June 5. https://www.mic.com/articles/178951/how-disney-gaydays-quietly-became-a-massive-pride-event#.uYk35bIcA. Accessed 28 Aug 2022.

Warren, Louis S. 2005. *Buffalo Bill's America: William Cody and the Wild West Show*. New York: Alfred A. Knopf.

Williams, Rebecca. 2020. *Theme Park Fandom: Spatial Transmedia, Materiality and Participatory Cultures*. Amsterdam: Amsterdam University Press.

Wolcott, Victoria W. 2012. *Race, Riots, and Roller Coasters: The Struggle Over Segregated Recreation in America*. Philadelphia: University of Pennsylvania Press.

Industry: Global Trends, Players, and Networks in the Theme Park Industry

Abstract The theme park industry is expanding worldwide, merging with other entertainment and attractions firms, and diversifying toward other areas of amusement, leisure, and consumption. In addition to large transmedia corporations with clear global positioning, the theme park industry also comprises a diverse array of regional companies, including independent attractions, family-owned parks, high-tech facilities, cultural/heritage/environmental-based parks, or small entertainment facilities. Moreover, connections between theme park operators and real estate corporations, shopping outlet companies, and other entertainment activities such as zoos, water parks, or corporate centers have been increasing. This chapter seeks to disentangle the theme park industry system, outlining the various types of theme park organizations, the importance of the main park operators and brands in the evolution of the industry, and the role of the many players that participate in the value chain of the theme park experience delivery. Mainstream trends in the current dynamics of the industry concerning sustainability, social responsibility, digital transformation, and safety/security are discussed, as is the role of influential organizations such as the International Association of Amusement Parks and Attractions (IAAPA) or Themed Entertainment Association (TEA) in setting standards for the industry. Finally, the chapter reflects on the impact of COVID-19 on the theme park industry in different local contexts and depending on the type and characteristics of each venue.

1 Defining the Field

In a context of increasing urban concentration of the population, digital innovation, growth of the middle class, accelerated mobilities, globalization, and greater environmental awareness—and setting aside year-to-year fluctuations and the disruption of activities due to COVID-19 in 2020—in the first two decades of the twenty-first

This work is contributed by Salvador Anton Clavé, Astrid Böger, Filippo Carlà-Uhink, Thibaut Clément, Florian Freitag, Scott A. Lukas, Sabrina Mittermeier, Céline Molter, Crispin Paine, Ariane Schwarz, Jean-François Staszak, Jan-Erik Steinkrüger, Torsten Widmann. The corresponding author is Salvador Anton Clavé, Departament de Geografia, Universitat Rovira i Virgili, Vila-seca, Spain.

F. Freitag et al., *Key Concepts in Theme Park Studies*,
https://doi.org/10.1007/978-3-031-11132-7_8

century the global theme park industry has experienced marked expansion worldwide. New parks, new attractions, new types of facilities, new pricing strategies, new business opportunities, new digital and sustainable operational possibilities, and unexpected economic developments and social transformations in certain locations play a role in the development of the industry. Yet, the theme park industry is one of the clearest manifestations of contemporary global corporate capitalism and symbolizes the potential of entertainment to catalyze social, cultural, spatial, environmental, and economic changes.

As described by Schmid (2009, 1), in a society where urban decision-making processes "are increasingly transforming shopping malls into urban entertainment centers, pedestrian zones into festival market squares, and conventional residential neighborhoods into master-planned communities," theme parks are becoming common components of a growing dynamics of the commercialization, "McDonaldization," and spectacularization of most aspects of everyday life (Ritzer 1999). This process thus occurs within the broader scenario of a transformation of the role of leisure in society, which could be summarized by: a high valorization of free time as a central component of contemporary life in developed societies; the leadership of a small number of large entertainment corporations on a worldwide scale with enormous financial capacity and technological possibilities; the incorporation of issues related to leisure into all facets of life; customer acceptance of high levels of theatrical authenticity (see AUTHENTICITY) in entertainment products; a demand for high levels of comfort, safety (including the reinforcement of sanitary procedures during and after the pandemic), security, and environmental aesthetics; as well as the growing importance of media, information, and digital fun (Anton Clavé 2007; Castronova 2007). Additionally, a fundamental change in values helps to explain the theme park and entertainment industry's current development: "a change away from social-collective and work-based virtues like diligence, discipline, and performance to a more hedonistic attitude" (Schmid 2009, 7).

As a consequence, theme parks—including indoor parks, nature- and culture-based facilities, and water parks—have been growing all over the world. These facilities are part of what Ritzer (1999) defined as "cathedrals of consumption," where "consumer religion" is practiced—a "religion" with a complex liturgy that incorporates play, fun, leisure, travel, learning, information, and the strengthening of middle-class values (Barber 1996, 128; see WORLDVIEWS) through a theme. Indeed, by definition, the theme park's basic characteristic is the theme, a concept that unifies disparate elements (see THEMING) and, according to Younger (2016, 5), "creates experiential benefits which in turn brings financial benefits." Theming thus reflects the way in which leisure and entertainment is consumed and produced in our society from the global to the local, from the dynamics led by a few transmedia operators such as Disney and Universal to the growing number of local players elsewhere in the world.

This chapter seeks to disentangle the system of the global theme park industry by outlining the different types of theme parks, the importance of the main theme park companies and brands in the evolution of the industry, and the role of the many diverse players that participate in the delivery of the theme park experience.

2 Business Typologies and Corporate Developments

The global theme park industry is diversifying toward multiple areas of amusement, leisure, and consumption. New and different types of theme parks (as well as other out-of-home entertainment facilities which are not discussed in this chapter) are currently emerging. Identifying, measuring, and analyzing theme parks is not an easy task, and the development of a basic typology is useful to understand the complexity of the industry and its development trends.

2.1 Theme Park Types and Characteristics

Wanhill (2008) has drawn up a broad classification of theme parks, noting that many of them are actually hybrids. He notes that water parks introduce thrill rides to compete with "dry" parks whereas "dry" parks add water parks adjacent to them; and that in the Middle East, Korea, and South East Asia many "leisure parks" are designed as indoor attractions combined with shopping, whereas in Europe and the U.S. large shopping complexes and malls add entertainment to their portfolio of activities. While categories are thus blurring, Wanhill (2008, 63) distinguishes destination parks, regional parks, traditional parks, and water parks. Based on their size, but especially on their ability to attract tourists, Anton Clavé (2007, 29) also differentiates between destination parks, regional parks, urban parks, and niche parks. Similarly, Younger (2016, 15) states that "themed attractions operate at a variety of scales, each one evolving business models that cater to their market size." According to Younger, these scales are transient, local, regional, and global.

Based on these approaches, as well as with data and information from The Park Database (2020a), Table 1 seeks to give a comprehensive classification of the diverse types of theme parks according to their dimension and the relation to their potential markets, paying particular attention to the types of resources used to design them. The table lists six types of theme parks (some of which could also exist independent of their theming) that can be regularly found in the global theme park industry. It is important to note that this classification includes commonly acknowledged theme parks, but also amusement and indoor parks with themed attractions as well as nature- and culture-based entertainment parks and themed waterparks. Conversely, the classification does not include standalone themed attractions such as indoor skydives, observatory wheels, VR attractions, escape rooms, water plays, or flying theatres, even though these form part of the conceptual analysis done by The Park Database (2020a). Neither does the table include Family Entertainment Centers, art and culture or interactive/discovery museums, gardens, factory tours, participatory parks (adventure parks and agricultural parks), and other themed venues such as themed village/retail entertainment centers and themed resort-based attractions (these have, however, been included in Anton Clavé's (2022) wider classification of themed visitor attractions).

Table 1 Types of theme parks. *Source* Authors' elaboration with information from The Park Database (2020a) and Anton Clavé (2022) and with insights from Wanhill (2008) and Younger (2016)

Category	Type	Characteristics	Examples	Size	Business model
Major mega theme parks		Mega theme parks usually integrated in entertainment and tourism; highly branded themed resorts with intellectual property driven attractions. High cost (at least US$ 1 billion). International market catchment area and positioning. Pay One Price business model. Low competition. Public subsidies/partnerships. Multiple parks in single resorts are designed to draw and entertain visitors over multiple days. Destination Parks	Parks in Disney Resorts, Parks in Universal Resorts, Parks in Chimelong Resorts	25–150 ha	Ticket, F&B, retail, hotels

(continued)

Table 1 (continued)

Category	Type	Characteristics	Examples	Size	Business model
Outdoor attractions theme parks	Attractions theme parks	Parks with most rides and attractions located outdoors and grouped around a theme. Sometimes developed or branded by entertainment and/or theme park corporations/chains. Regional or metropolitan/urban catchment areas. Increasing diversification with hotels and complementary activities located close to the park in a complex entertainment resort. Largest regional parks have increasing tourism attractiveness. Cost of several hundreds of millions of US$. Pay One Price business model. Medium to low competition. Sometimes hybrid with water parks	Europa-Park theme park, Efteling theme park, LEGOLAND parks, PortAventura park	25–50 ha	Ticket-dominant, F&B, retail, increasingly hotels

(continued)

Table 1 (continued)

Category	Type	Characteristics	Examples	Size	Business model
	Dark ride theme parks	Most rides and attractions are located in indoor buildings scattered throughout the grounds in themed warehouse-like shells. Shows, media-based attractions, and experiences	Futuroscope theme park, Fantawild Oriental Heritage parks	10–50 ha	Ticket-dominant
	Amusement parks with themed rides and attractions	Antecedents in carnivals and funfairs. Some individual theming. Mix of rides, games, and attractions. Some seasonal. Low Pay One Price. Some Pay-Per-Play, concessionaires. Small tourist markets. High to medium competition	Six Flags parks, Cedar Fair parks	10–50 ha	Ticket-dominant
Indoor attractions theme parks	Indoor theme parks	Includes rides, shows, exhibits, and attractions. Appropriate for locations experiencing inclement climates. Environmental control. Cost of constructing the shell	Lotte World	0.3–10 ha	Ticket-dominant

(continued)

Table 1 (continued)

Category	Type	Characteristics	Examples	Size	Business model
	Indoor ice, ski, and snow facilities	Indoor ice-skating rinks or skiing domes with artificial snow. Used for training. Year-round attraction	Ski Dubai	0.4–4 ha	Ticket-dominant
	Indoor water, surf and pool venues	Indoor water environment facilities providing artificial controlled water related experiences, sports, and activities	Kelly Slater Surf Ranch	0.4–2 ha	Ticket dominant
	Indoor themed role play	Indoor environments that provide immersive experiences of "playing grown-up." Hands-on crafts, games, and play activities based on different occupations	KidZania parks	0.4–1 ha	Ticket, sponsors, non-profit (grants/donations)
Nature-based parks	Wildlife parks	Exhibit and show-based animalia, zoos, and safaris. Standalone or included in larger parks. Education-oriented	Disney's Animal Kingdom, Chimelong Safari Park	10–125 ha	Ticket, non-profit (grants/donations)

(continued)

Table 1 (continued)

Category	Type	Characteristics	Examples	Size	Business model
	Marine animal parks	Themed around marine animal shows and exhibits. Marine mammals are the main attraction (dolphins, orcas, belugas, or seals). Between aquaria (emphasis on education and conservation) and for-profit attractions. Social criticism	SeaWorld parks, Ocean Park, Siam Paak	10–30 ha	Ticket-dominant
	Marine interaction parks	Hands-on encounters with marine wildlife. High ticket price and low volume of visitors	Discovery Cove, Dolphin Island	<3 ha	Ticket-dominant
Culture-based parks	Ethnic parks	Focus on regional cultures. Shows, exhibits, and hands-on experiences rather than rides. Some are non-profit (grants, donations, subsidies, and operating budget contributions from corporations, organizations, municipalities, and government)	Yunnan Ethnic Village	2–20 ha	Ticket, non-profit (grants/donations)

(continued)

Table 1 (continued)

Category	Type	Characteristics	Examples	Size	Business model
	Cultural parks	Focus on the showcasing of assets, characteristics, or developments of cultural heritage in terms of economic activity, social values, architecture, professional knowledge, local identity, environmental engagement. Mainly based on shows, exhibits, and hands-on experiences. Can include reproductions of cultural landscapes, preserved historical or industrial buildings, or co-creative heritage experiences. In some cases close to the concept of ethnic parks but including a larger array of cultural manifestations	Xcaret, Landschaftspark Duisburg-Nord, Château de Guêdelon	20–50 ha	Ticket, non-profit (grants/donations)

(continued)

Table 1 (continued)

Category	Type	Characteristics	Examples	Size	Business model
	Religious parks	Focus on religions or religious issues. Shows, exhibits, and hands-on experiences rather than rides. Some are non-profit (grants, donations, subsidies, and operating budget contributions from corporations, organizations, municipalities, and government)	Holy Land, Tierra Santa, Big Buddha	2–20 ha	Ticket, non-profit (grants/donations)
	History-based parks	Parks whose theme is a particular historical event or local historical episodes. Includes rides, shows, exhibits, and attractions	Isla Mágica	20–50 ha	Ticket-dominant
	Show-based parks	The defining feature is a landmark show or series of shows. Themed venues for shows. May be considered a specific version of other culture-based parks mostly featuring shows and exhibits	Puy du Fou, Song Dynasty Town	10–30 ha	Ticket-dominant
	Minilands	Exhibit based miniature simulacra featuring buildings and scenes of the world or a particular region of the world	OCT Windows of the World, Mini-Europe	10–50 ha	Ticket-dominant

(continued)

Table 1 (continued)

Category	Type	Characteristics	Examples	Size	Business model
Themed waterparks		Themed rides, slides, and pools. Usually seasonal, partly or wholly. Some indoor to offset seasonality. Engaging, kinetic experiences. In the U.S., smaller parks are municipally operated as public pools	Chimelong, Blizzard Beach	4–30 ha	Ticket, non-profit (grants/donations)

As highlighted in Table 1, there are significant differences between the various types of theme parks, depending on the resources deployed to attract visitors; the type of attractions included in the park; the characteristics of the facility; its ownership; the size of the amenity; its business model; and its target market (local, regional, international). Unfortunately, what we know about the characteristics of the different types of parks is limited. This is because existing statistics about attendance at both world scale and park level (most importantly, the TEA/AECOM *Theme Index and Museum Index: The Global Attractions Attendance Report*; see e.g. Rubin 2020 or 2021) include only the most attended venues among the major mega theme parks and outdoor attractions theme parks, while but a few of the other types of theme parks are listed. Only water parks have separate attendance statistics on a global scale, but accuracy is still a challenge. In fact, the existing statistical definitions and industry classification boundaries often overlap, and many gaps exist. Alternative databases such as The Park Database (https://www.theparkdb.com/) can provide information on individual parks that can be useful to characterize the industry (see METHODS).

2.2 Industry Development Trends and Players Worldwide

Until 2019, the last year before the COVID-19 pandemic, the global attraction industry was growing all over the world. In total, the global theme park industry counted over 1,137 million visits in 2019 (IAAPA and WGA 2021). Combining the two indicators of attendance and per capita spending estimated by IAAPA and WGA (2021) that are reported in Table 2, total spending in the worldwide theme and amusement park industry was US$ 37.2 billion in 2014 and US$ 50.9 billion in 2019. Results declined to US$ 21.2 billion in 2020 because of the pandemic. The overall growth until 2019 has shown different characteristics and has occurred at different speeds in different regions. It has been driven both by the opening of new parks, especially in regions starting from a less developed base, and by operators reinvesting in new rides and attractions, hotels, and special events and celebrations in the existing parks (Rubin 2020).

Table 2 reports existing estimated comparable data on theme and amusement parks at a worldwide scale, as provided by IAAPA (IAAPA and WGA 2021). Note that all the data refers to venues that feature rides as the primary attraction and thus include both theme and amusement parks. Expenditures at water parks or indoor, nature- and culture-based parks that do not feature rides are not included. While thus only reflecting a part of the industry, the information in the table can be used as a proxy for the global dynamics and the general trends of the entire theme park industry. Spending information is estimated from admission fees and, if applicable, food, souvenirs, and other related purchases made at the parks.

In Asia/Pacific, the main contributors to the growth of the industry until 2019 were: new global and regional brand parks such as Shanghai Disneyland or Fantawild Oriental Heritage; enhancements to and new themed areas within existing parks in Japan; the growing number of visitors from the People's Republic of China to Hong

Table 2 Recent global and regional recent evolution of the main indicators for the theme and amusement park industry, 2014, 2019, and 2020 estimates. *Source* Authors' elaboration with data from IAAPA and WGA (2021)

	Attendance (Millions)			Per capita spending (US$)		
	2014	2019	2020	2014	2019	2020
Asia/Pacific	390.0	504.9	242.9	30.53	33.91	34.69
Europe	161.1	176.9	77.3	33.39	38.69	47.01
North America	371.9	417.5	134.5	52.58	63.06	69.14
Latin America	30.7	30.6	15.2	8.57	10.78	11.18
Middle East & Africa	2.9	7.5	3.4	32.07	37.60	37.69
GLOBAL	956.6	1137.4	473.3	38.88	44.75	44.78

Kong parks; and the opening of parks in Vietnam and India. Nevertheless, in 2019 per capita spending was still significantly below the global average, even though the region had a booming economy and a growing middle class, especially in the People's Republic of China (Ren 2013; see ANTECEDENTS). Indonesia, South Korea, and Singapore are also on the rise, and new high-profile brand parks or expansions have opened in 2021, most notably Universal Beijing Resort (People's Republic of China) and Super Nintendo World at Universal Studios in Osaka (Japan).

In Europe, the theme and amusement park industry benefited from faster-growing disposable income during the 2014–2019 period. In general, the drivers that contributed to the increase of visitations and per capita spending in Europe were the opening of additional attractions such as Wicker Man in Alton Towers (Alton, UK); new water parks like Rulantica at Europa-Park; major enhancements in existing parks; new hotels, such as those recently opened at Europa-Park or PortAventura; and special events, celebrations, and new attractions, such as "Peppa Pig Land" at Gardaland. Before COVID-19, successful season pass packages also contributed to growth, especially in the Netherlands and France. Some new parks also opened during this period, such as Gulliver's Valley (Rotherham, UK) with 26 rides and attractions, and Puy du Fou (Toledo, Spain), which like its French counterpart features historical performances. The opening of Dream Island, an indoor park in Moscow (Russia), in 2020 highlights the potential of Eastern European markets.

North America is the largest theme and amusement park region in the world, with a spending of US$ 26.3 billion in 2019, which was 51.7% of the whole industry. Per capita spending is the principal driver of growth for this market. This can be explained by the role of average per capita spending at destination parks, which in 2019 was US$ 119.7 compared to less than US$ 39 at regional parks (IAAPA and WGA, 2021). During the years before the pandemic, destination parks opened strong new brands and transmedia attractions such as "Pandora: The World of Avatar" at Disney's Animal Kingdom, "Toy Story Land" at Disney's Hollywood Studios, Fast and Furious: Supercharged at Universal Studios Florida, "The Wizarding World of Harry Potter" at Universal Studios Hollywood and Florida, "Star Wars: Galaxy's Edge" at both Disneyland and Disney's Hollywood Studios, Jurassic World—The

Ride at Universal Studios Hollywood, "Sesame Street Land" at SeaWorld Orlando, and Hagrid's Magical Creatures Motorbike Adventure at Universal Studios Orlando (see MEDIA). Results during this period also benefited from the transition of SeaWorld Orlando from an animal show into a destination park that also offers thrill rides such as Infinity Falls. This was the company's proactive response to criticisms of its animal-based attractions (see WORLDVIEWS). Both destination and regional parks also benefited from themed events such as food festivals, as well as Christmas, Easter, and Halloween events (see TIME). New attractions and enhancements at regional parks, as well as a trend to install hotels around regional parks, also helped the growth in performance of the North American theme and amusement park industry.

During the 2014–2019 period, the theme park industry in the Middle East/Africa region expanded from a very small base with the opening of new parks such as IMG Worlds of Adventure, the three Dubai Parks and Resorts parks—Motiongate, Bollywood, and Legoland—and Warner Bros Abu Dhabi on Yas Island, joining Yas Waterworld and Ferrari World Abu Dhabi. Likewise, Saudi Arabia is emerging as a potential theme and amusement park market, while Qatar and Bahrain also have the potential for growth. In Latin America, Mexico and Brazil are the dominant markets (IAAPA and WGA 2021). These two countries account for 60% of the region's total spending in 2019.

All of this dramatically changed in 2020 because of COVID-19. The pandemic situated the theme park industry into a general context of downsized mobility and long- and medium-haul travel restrictions (Rubin 2021). In 2020 almost all theme parks across the world closed for several months and, in most cases, reopened with strict operational regulations and limited capacity. Available evidence shows that, in general and also depending on the restriction policies in each country and the moment of the year of the pandemic peak, theme parks that largely relied on resident markets were less vulnerable (Anton Clavé et al. 2022). This was the case, for example, with the People's Republic of China's OCT parks, which reported an average attendance decline of only 14%, Fantawild parks, which dropped an average of 20%, or Efteling, which was hit by COVID-19 later in the season and declined by only 45% (Rubin 2021). As a result, IAAPA and WGA (2021) estimated that in 2020, the first year of the pandemic, global theme park attendance fell by 58.4%, with North America (67.8%) and Latin America (50.3%) registering the highest and the lowest decline rates, respectively.

By contrast, according to IAAPA and WGA (2021) estimates, per capita spending in theme parks rose during 2020. This was due to the higher spending propensity of theme park enthusiasts, who were the first to come back to the venues after they reopened; the higher per capita spending when the density of visitors is lower; and the fact that annual pass holders, who generally spend less per visit, visited the parks less frequently. Growth spending has been variable in the different regions of the world. Estimates range from a minimum of 0.2% in the Middle East to a maximum of 9.6% in North America. Even though per capita spending was thus significantly higher than in 2019, due to lower attendance numbers the North American region nevertheless experienced a significant decline in its share of the global market to 43.9% in 2020, representing only US$ 9.3 billion.

Table 3 lists the top ten theme park operating corporations in the world in 2019, the year before the pandemic. It also includes attendance figures for each corporation's most visited park in 2019 and 2020. This data has been selected to illustrate the differential impact of COVID-19 on the attendance of parks depending on the business model of the companies, the location of particular venues, and the diffusion patterns of the pandemic. Total attendance data for 2019 includes the total number of visits to their theme parks but also to all the facilities they have in their portfolio of amenities, including amusement parks, water parks, zoos, and other out-of-home entertainment facilities.

The companies with the largest theme parks are Disney and Universal with twelve and four parks, respectively, among the top 25 most visited parks in the world in 2019. Aside from these two, among the top ten park operating companies only the Chinese OCT Group and Chimelong Group own individual parks that ranked among the top 25. However, in the case of OCT, even its most visited venue in 2019 is only a medium-sized park (Happy Valley Beijing, with over 5.1 million visitors). The rest of the companies likewise operate only small- and medium-sized theme parks and other amusement and entertainment facilities. This is the case, for example, of Merlin Entertainment. While the company is ranked second in terms of its total attendance and operates nine Legoland parks plus six other theme parks, a water park, and many other entertainment attractions, it does not operate any of the top parks in the world. In fact, Gardaland (Castelnuovo del Garda, Italy), Merlin Entertainment's most visited park in 2019 (2.9 million visitors), only ranked eighth among the top European parks.

Only Disney and Universal currently attract most of their attendance in parks that are among the worldwide top 25 (97.3% of Disney parks' attendance is attracted by the company's twelve parks, which all rank among the top 25; 87.7% of the Universal attendance is attracted by the company's four parks, all ranking among the top 25). According to The Park Database (2020b), only these two companies have the financial capacity to develop such massive parks, and only they can generate over US$ 1 billion of revenue from a park's annual operation. As stated by Weidenfeld et al.,

> the enormous investment costs involved in starting up mega theme parks [...] deters competition, both because of the scale of the resources involved and their ability to protect their core product or identity from imitation by competitors, by the uses of patents, trademarks and other measures. (2016, 65)

Disney and Universal also have another important common characteristic: "their theme parks are a location-based delivery system for the rest of their media business with television, movie, and other media properties that allow cross-pollinating [...] business models between their theme parks and the rest of their assets" (The Park Database 2020b; see MEDIA).

The list of the main theme park operators in 2019 also includes companies with a highly diversified portfolio of entertainment amenities and strong global positioning including venues in different regions of the world: the previously mentioned Merlin Entertainment (operating 130 attractions in 2019, from theme parks to urban standalone visitor attractions in 25 countries all over the world) and Parques Reunidos

Table 3 Worldwide top ten theme park operating corporations, 2019. *Source* Authors' elaboration with data from Rubin (2020)

Company	Total attendance 2019 (in millions) [A]	Parks within the top 25	Total attendance of parks within the top 25 2019 [B]	% [B/A]*100	Most visited theme park 2019	Attendance of the most visited theme park of the company 2019 (in millions)	Attendance of the 2019 most visited theme park of the company in 2020 (in millions)	% Change 2020/2019
Disney	155,991,000	12	151,859,000	97.3	Magic Kingdom at WDW	20,963,000	6,941,000	−66.9
Merlin	67,000,000	–	–	–	Gardaland	2,920,000	1,350,000	−53.8
OCT Parks	53,970,000	1	5,160,000	9.6	Happy Valley Beijing	5,160,000	3,950,000	−23.4
Universal	51,243,000	4	44,944,000	87.7	Universal Studios Japan	14,500,000	4,901,000	−66.2
Fantawild	50,393,000	–	–	–	Zhengzhou Fantawild Adventure	3,840,000	3,421,000	−10.9
Chimelong	37,018,000	2	16,641,000	44.9	Chimelong Ocean Kingdom	11,763,000	4,797,000	−59.1
Six Flags	32,811,000	–	–	–	Six Flags Magic Mountain	3,610,000	686,000	−81.0

(continued)

Table 3 (continued)

Company	Total attendance 2019 (in millions) [A]	Parks within the top 25	Total attendance of parks within the top 25 2019 [B]	% [B/A]*100	Most visited theme park 2019	Attendance of the most visited theme park of the company 2019 (in millions)	Attendance of the 2019 most visited theme park of the company in 2020 (in millions)	% Change 2020/2019
Cedar Fair	27,938,000	–	–	–	Knott's Berry Farm	4,238,000	811,000	−80.9
SeaWorld	22,624,000	–	–	–	SeaWorld Orlando	4,640,000	1,598,000	−65.6
Parques Reunidos	22,195,000	–	–	–	Warner Park Madrid	2,232,000	450,000	−79.8
TOTAL	521,183,000	18	218,505,000	41.9	–	–	–	–

(operating about 60 attractions including amusement parks, aquariums, zoos, water parks, and other out-of-home entertainment amenities in Europe, the United States, Middle East, and Australia in 2019). Another type of company is represented by Six Flags, an American amusement park company currently expanding worldwide with parks and projects in Canada, Mexico, Saudi Arabia, and the People's Republic of China. Finally, among the main theme park operators there are regionally based corporations with connections to other entertainment-related and unrelated economic activities such as zoos, circus, water parks, nature conservation, real estate, and attraction design and development such as the OCT Group and Chimelong (operating only in the People's Republic of China), SeaWorld (currently operating only in the United States), Cedar Fair (operating only in the United States and Canada), and Fantawild (operating mainly in the People's Republic of China).

Even though the available data only allows us to analyze the performance of the largest parks, two significant points stand out here: firstly, most of the existing them parks in the world (and likely most future park developments) fall into the category of investments of less than US$ 100 million. Secondly, aside from the classical destination major mega theme parks and the outdoor and indoor attractions theme parks targeting a regional market, other types of parks have a high potential for growth and, in many cases, higher profit margins than those obtained by the classical assets of the industry. As noted by The Parks Database (2020b), mid-sized indoor attractions such as some of those operated by Merlin Entertainment, indoor role play parks such as KidZania, and themed water parks are quite profitable. In any case, as stated by Younger (2016), what should be kept in mind when analyzing the global theme park industry is that it is largely a prototype industry, where each project can be completely different from that which precedes it.

Table 3 also illustrates that COVID-19 has had a major impact on theme parks' capacity to attract visitors. Using the most visited parks of the top ten operating corporations in 2019 as a sample to explore the impact of the pandemic on park attendance in 2020, the available data indicates that both destination and tourism-oriented theme parks as well as regional and metropolitan theme parks have suffered a massive loss of visitors. Only domestic parks in the People's Republic of China, such as Zhengzhou Fantawild Adventure and Happy Valley Beijing, report declines below the 25% line. The figures of such a large park as Chimelong Ocean Kingdom (People's Republic of China) are closer to those of other major mega destination parks in the table, like Magic Kingdom (Orlando, Florida) and Universal Studios Japan. Finally, due to the opening and capacity restrictions (mainly during the peak season), medium-sized American and European parks also report important decreases in attendance, with Six Flags Magic Mountain (Los Angeles, California) and Knott's Berry Farm (Buena Park, California) declining by more than 80% in 2020 compared to 2019. Nevertheless, even during the pandemic theme park operators have been actively investing in new products, attractions, and experiences. Two trends can be observed: the use of digital technology, not only to create experiences but also to adapt operations (e.g. advanced ticketing and reservations, incorporation of Artificial Intelligence, big data analytics to personalize the guest experience); and the rethinking of business strategies towards sustainability (see ECONOMIC STRATEGY).

3 Theme Park Clusters and Networks

Two important questions concerning the spatial economic role of the theme park industry emerge: firstly, theme parks transform local/regional societies and allocate dominant entertainment consumption practices, imagination, symbols, and utopias. Secondly, the dynamics they start can, under certain circumstances, catalyze production processes that might shift the focus of value creation at a regional level from the visitor economy to industrial and other service sectors.

Regarding the first point, it is important to highlight that before the pandemic, 20 out of the top 40 theme parks in the world were located in only three regions of the world, namely Central Florida (Orlando–Tampa), the Greater Pearl River Delta (Shenzhen–Zhuhai–Guangzhou–Hong Kong–Macau), and Southern California (Los Angeles–San Diego). Asia has four other privileged locations with more than two top parks (Japan, the Shanghai–Jiangsu–Zhejiang region, Northeastern China, and South Korea) plus Singapore. The top parks in Western Europe are distributed across five different countries (France, Netherlands, Germany, Denmark, and Spain), whereas two out of the top 40 global theme parks are located in Midwestern North America. All these regions also include other mid- and smaller-sized parks as well as a variety of other out-of-home entertainment facilities.

Importantly, the three top global theme park agglomerations at the beginning of the 2020s—Central Florida, the Greater Pearl River Delta, and Southern California—act as systems where the future of the theme park industry as one of the leading industries that shape tastes and cultures in our contemporary world is actively negotiated, created, and changed. These three regions constitute central nodes from where theme park culture as a form of cultural and economic globalization is disseminated. This is because, among other factors, the co-location of theme parks promotes change through competition and induces productivity increases, risky innovation, and competitiveness in a context with distinctive features including substitutability, perishability, and the blurring of production and consumption (Weidenfeld et al. 2016).

While their role as theme park clusters may change (especially because of the dynamics of industry evolution in the emerging theme park regions in Asia and especially in the People's Republic of China, and because of the location of large transnational theme park corporations in the Shanghai and Beijing areas), Central Florida, the Greater Pearl River Delta, and Southern California are currently the leading regions of the theme park industry. Following Goodman, Goodman, and Redclift (2010, xi), they are regions where the sense of theme parks as consumption spaces is tested and proved, where the meaning of entertainment is taught and learned, where benchmarks and challenges are established, and where consumers define the standards of their responses to the experiences delivered by the industry. These three top theme park regions are also places where tourism and entertainment are, despite their importance, not the only activities; on the contrary, it is added value services, urban dynamics, and high-tech industry that are dominant. Therefore, these regions host an important services and products industry that supplies the theme park industry. Hence, they are clusters that define hegemony, create legitimacy, and irradiate experience.

This point is related to the second issue introduced at the beginning of this section. As there is a large number of industries that work as suppliers for the amusement park industry, the locations of these industries can highlight the dominant role of certain regions of the world in terms of the definition of global theme park culture, practices, and expectations. Table 4 thus allows us to explore to what extent theme parks catalyze, in their capacity as characteristically consumption-based spaces, the creation of productive activities linked to their value chain in the three top global theme park agglomerations previously identified (Central Florida, the Greater Pearl River Delta, and Southern California).

For the purposes of the table, the IAAPA's classification of non-theme park members has been adapted and summarized, resulting in the following set of supplier categories: planning and design; site development and management (construction, engineering, facility and grounds, people moving equipment, theming, and scenery); rides (ride planning, design, and management, rides); shows and productions; multimedia attractions; games and play equipment (games and devices, participatory play equipment); general management (business services, human resources, marketing and branding, insurance, customer services, communication systems, printing, computer systems, web services); and operations (cash handling and

Table 4 Estimated worldwide share of specialized products and services supplied to the theme park industry located in the top three theme park destinations per category. Data only includes firms that are members of IAPPA and TEA or participated in their events, 2018. *Source* Authors' elaboration with data from IAAPA and TEA

	% of firms supplying every type of good/service over the total number of world firms supplying the same type of good/service by region		
Products and services supplied by specialized firms per categories	Central Florida	Great Pearl River Delta	Southern California
Planning and design	9.98	3.20	19.21
Site development and management	7.64	4.23	12.05
Games and play equipment	3.40	6.94	4.73
Multimedia attractions	5.74	2.73	15.30
Rides	8.06	3.78	5.29
Shows and productions	9.43	2.26	22.26
Operations	7.85	4.47	12.81
General management	7.97	4.83	10.66
Other/no data	5.23	11.57	8.54
All products/services	7.27	4.85	11.75

payment, clothing, food and beverages, merchandise and gifts, security and safety, theatrical equipment and supplies, water-related equipment and supplies).

The table, built upon data from members and participants in professional trade shows organized by the IAAPA and members of the Themed Entertainment Association (TEA) in 2018, reflects the share of specialized products and services provided by explicitly theme park-oriented suppliers located in the three top theme park regions of the world. Of course, services or industrial activities required by theme parks can be hired elsewhere in the world, supplier firms can sometimes be geographically located far from their own surroundings, and there are other firms not registered in the IAAPA and TEA databases that also supply products and services to the theme parks operators. But the point here is to observe how the entertainment visitor economy creates industrial and service-oriented value in places where entertainment is consumed. Moreover, as will be elaborated below, the differences between the three regions also highlight how each region's entertainment path dependency, industrial specialization, level of economic development, and other related institutional factors influence the typology of suppliers that they have. Finally, entertainment specialization in each region also has to do with the specific dimension and intensity of other entertainment industries located there, including movie, digital, or game production (see MEDIA).

Hence, Table 4 is useful to understand that, beyond consumption, the location of various firms supplying goods and services to the theme park industry can create industrial clusters based on entertainment, which may also be corroborated by analogous studies of other regions of the world. The results confirm that, at least in these three top regions, the attention to the consumption-based nature of the space created by the entertainment industry should not make us forget the strong production flows tied to it.

As shown in Table 4, the three largest theme park regions of the world also have an important role as providers of services, products, knowledge, and expertise, as they host 11.75% of the specialized globally supplied services and products by firms listed in the IAAPA and TEA membership and events databases for 2018 in the case of Southern California, 7.27% in the case of Central Florida, and 4.85% in the case of the Great Pearl River Delta. Furthermore, Southern California hosts a significant share of shows and production (22.26%), planning and design (19.21%), and multimedia attractions (15.30%) world suppliers and, by comparison, a significant number of rides producers are located in Central Florida (8.06%). Finally, 6.94% of the world's suppliers of games and participatory equipment have a firm located in the Great Pearl River Delta region. The table also shows that aside from some specialization in each region, all three of them can provide a broader spectrum of solutions to theme park facilities located there. Thus, from this perspective, the theme park industry creates linkages with other industrial sectors and positions each of the three regions and their firms for the international competition.

Thus, in these three cases, the concept of the theme park agglomeration or entertainment global destination must add to its *consumption*-oriented conceptualization some of the characteristics of *production* clusters, such as (1) a pool of labor (see LABOR), (2) with flexible division, (3) shared inputs, infrastructure, and services (particularly transport), and (4) a diffusion of knowledge and technology (Weidenfeld et al. 2016). Furthermore, due to their location in these specialized clusters, suppliers acquire capacities that they can export to other regions at the local and international scales. Thus, high connectivity, industry networks, and the exchange of both formal and informal knowledge all play an important role and help them to define their business model.

It should also be noted here that, as explained by Younger (2016), there are two main forms of suppliers: partner companies and independent companies. Partner companies are owned by a theme park operator or are part of the same corporate group, such as Disneyland Imagineering, Universal Creative, Merlin Studios, and Efteling Imagineering. This is also the case, to name one further example, with the family of partner companies of the Mack Group, which includes Mack Rides, VR Coaster, MackMedia, Mack Solutions, and the Europa-Park theme park and resort in Germany. Through these corporations, the Mack group develops innovative media content (Coastiality VR rides, interactive attractions, flying theatre films, 4D films; see MEDIA) but also creates story worlds, builds rides and coasters, provides theming solutions, and develops innovations not only for its own facility but also for other park operators (the SeaWorld parks, for example, which feature a range of Mack rides).

With a different business model, in the People's Republic of China the Shenzhen-based Fantawild group offers and exports a range of services, from theme-park master planning to attractions and project design. In addition to managing their own different theme park brands (Fantawild Adventure, Fantawild Dreamland, Fantawild Water Park, and Fantawild Oriental Heritage; see Baker 2022) with more than 20 parks in the People's Republic of China, the Fantawild business model includes solutions related to innovation and design, research and development, content support, equipment, construction, and operation. The company also develops business concepts and has other business areas including animation, special-effects films, film production, live performances, consumer products, and equipment manufacture. Partner companies facilitate processes for parks as they allow them to always have a resource at their disposal.

Independent companies, from planners and designers like BRC Imagination Arts or Thinkwell to ride designers and builders like Intamin or Vekoma, have no structural ties to a theme park and are instead hired in by operators for a specific project depending on their technical abilities and the demand requirements. As discussed above, these companies can supply services in more than one specialized field and can have a broader market than the theme park industry itself. This is the case, for instance, with the Canadian company Forrec, which not only offers solutions to the theme park industry but also develops resorts and plans mixed-use (living, working, shopping, and entertainment) experiential urban areas (see Grice 2017). Focusing on the creation of immersive environments, companies like Mycotto, based in Pasadena, specialize in creating compelling experiences from the entertainment strategy to the experience design.

In addition to inter-company linkages within the theme park industry value chain, it is important to point out that the theme park and related attractions sector has well-established trade groups which include not only parks but also most of the providers that do business with the entertainment operators, from designers to managers and producers. The most influential organization is the International Association of Amusement Parks and Attractions (IAAPA), which is also the largest. With about 6,000 members, the association's global headquarters are in Orlando, and it has regional offices in Brussels, Hong Kong, Shanghai, Mexico City, and Alexandria, Virginia. The IAAPA includes professionals from amusement parks, theme parks, museums, water parks, wildlife attractions, and the entire list of location-based entertainment facilities included at the beginning of this chapter. The association seeks to promote safe operations, global development, professional growth, and the commercial success of its members. The emphasis is on providing reliable data, training, statistical analysis, and branding, with a growing focus on sustainability, corporate social responsibility, and digitization. A second relevant and influential trade group, with about 1,600 members, is the Themed Entertainment Association (TEA), an international worldwide nonprofit association that represents creators, developers, designers, consultants, suppliers, owners/operators, and producers of themed experiences, whose mission is to provide channels and opportunities to network, educate, and assist the entertainment business as well as personal careers within the industry. Both associations gather companies from the entire theme park value chain that, working together, help set the standards for the industry, improve processes, innovate, and guarantee quality.

References

Anton Clavé, Salvador. 2007. *The Global Theme Park Industry.* Wallingford: CABI.

Anton Clavé, Salvador, Joan Borràs, Jonathan Ayebakuro Orama, and Maria Trinitat Rovira-Soto. 2022. The Changing Role of Tourism-Oriented Theme Parks as Everyday Entertainment Venues during COVID-19. In *Tourism Dynamics in Everyday Places: Before and after Tourism*, ed. Aurélie Condevaux, Maria Gravari-Barbas, and Sandra Guinand, 245–261. London: Routledge.

Anton Clavé, Salvador. 2022. Themed Visitor Attractions. In Encyclopedia of Tourism Management and Marketing, ed. Dimitrios Buhalis. Cheltenham: Edward Elgar. https://www.elgaronline.com/view/book/9781800377486/b-9781800377486.themed.visitor.attractionss.xml. Accessed 28 Aug 2022.

Baker, Carissa. 2022. A Chinese "High-Tech Theme Park Full of Stories": Exploring Fantawild Oriental Heritage. Cultural History 11 (2): 199–218.

Barber, Benjamin R. 1996. *Jihad vs McWorld.* New York: Ballantine.

Castronova, Edward. 2007. *Exodus to the Virtual World: How Online Fun is Changing Reality.* New York: Palgrave Macmillan.

Goodman, Michael K., David Goodman, and Michael Redclift. 2010. Situating Consumption, Space and Place. In *Consuming Space: Placing Consumption in Perspective*, ed. Michael K. Goodman, David Goodman, and Michael Redclift, 3–40. Abingdon: Routledge.

Grice, Gordon. 2017. Temporality and Storytelling in the Design of Theme Parks and Immersive Environments. In *Time and Temporality in Theme Parks*, ed. Filippo Carlà-Uhink, Florian Freitag, Sabrina Mittermeier, and Ariane Schwarz, 241–257. Hanover: Wehrhahn.

IAAPA and WGA. 2021. *IAAPA Global Theme and Amusement Park Outlook: 2020–2024*. Orlando: International Attractions and Amusement Parks Association and Wilkofsky Gruen Associates.
Ren, Hai. 2013. *The Middle Class in Neoliberal China: Governing Risk, Life Building, and Themed Spaces*. Abingdon: Routledge.
Ritzer, George. 1999. *Enchanting a Disenchanted World: Revolutionizing the Means of Consumption*. Thousand Oaks: Pine Forge.
Rubin, Judith, ed. 2020. *2019 Theme Index and Museum Index: The Global Attractions Attendance Report*. Burbank: Themed Entertainment Association and AECOM.
Rubin, Judith, ed. 2021. *2020 Theme Index and Museum Index: The Global Attractions Attendance Report*. Burbank: Themed Entertainment Association and AECOM.
Schmid, Heiko. 2009. *Economy of Fascination: Dubai and Las Vegas as Themed Urban Landscapes*. Berlin: Gebrüder Borntraeger.
The Park Database. 2020a. The Definitive Guide to Attraction Concepts. https://www.theparkdb.com/blog/conceptguide/. Accessed 28 Aug 2022.
The Park Database. 2020b. The Business of Theme Parks (Part I): How Much Money Do They Make? http://www.theparkdb.com/blog/the-business-of-theme-parks-part-i-how-much-money-do-they-make/. Accessed 28 Aug 2022.
Wanhill, Stephen. 2008. Economic Aspects of Developing Theme Parks. In *Managing Visitor Attractions*, ed. Alan Fyall, Brian Garrod, Anna Leask, and Stephen Wanhill, 59–79. Second ed. Oxford: Butterworth-Heinemann.
Weidenfeld, Adi, Richard W. Butler, and Alan M. Williams. 2016. *Visitor Attractions and Events: Locations and Linkages*. Abingdon: Routledge.
Younger, David. 2016. *Theme Park Design & the Art of Themed Entertainment*. N.P.: Inklingwood.

Labor: Working Conditions, Employment Trends, and the Job Market in the Theme Park Industry

Abstract This chapter discusses labor practices prevalent in theme parks across the globe. The chapter first introduces the theme park industry's general internal principles with regard to employment, from formal to informal types of labor organization. This part specifically focuses on industry-specific forms of organization, such as distinctions between "onstage" and "backstage" work or highly specialized departments; contexts of labor and working conditions, especially recruitment practices, worker demographics, and broad employment shifts in the service industry ("McDonaldization" and "Disneyization"); human resources, corporate culture, and training, as notably expressed in industry-wide practices known as emotional and performative labor (sometimes resulting in the typecasting of performers); and informal hierarchies, as well as staff adoption or rejection of behavioral guidelines and other elements of corporate culture. The chapter further deals with the parks' labor practices in the broader context of the industry, which is notably shaped by labor organizations, especially regarding hours and wages protection or discriminatory hiring practices ("typecasting"); by the parks' legal framework (e.g. industry-specific visas, artists' contracts, etc.); and by migration patterns, some of which are national, and others global. Finally, the chapter discusses different types of "audience labor," a concept developed to acknowledge the visitors' own contributions to the theme park experience.

1 Introduction

By the industry leaders' own admission, workers and their labor constitute a key component of the success of the theme park guest experience (Kinni 2003; Lipp 2013); at the same time, the industry's labor standards have come to symbolize the poor conditions of service economy employees everywhere. More recently, the topic of theme park labor has gained new prominence as the working conditions of Disney

This work is contributed by Thibaut Clément, Scott A. Lukas, Salvador Anton Clavé, Florian Freitag, Astrid Böger, Filippo Carlà-Uhink, Sabrina Mittermeier, Céline Molter, Crispin Paine, Ariane Schwarz, Jean-François Staszak, Jan-Erik Steinkrüger, Torsten Widmann. The corresponding author is Thibaut Clément, Faculté des Lettres, Sorbonne Université, Paris, France.

F. Freitag et al., *Key Concepts in Theme Park Studies*,
https://doi.org/10.1007/978-3-031-11132-7_9

park employees have repeatedly made headlines, alerting the public to the realities painstakingly concealed from view by management. This chapter discusses labor practices prevalent in theme parks across the globe. The first section, "Theme Parks as Labor Environments," deals with the theme park industry's broad internal principles with regards to employment, from formal to informal types of labor organization, and focuses on:

- industry-specific forms of organization, such as distinctions between "onstage" and "backstage" work or highly specialized departments;
- contexts of labor and working conditions, especially recruitment practices, worker demographics, and broad employment shifts in the service industry ("McDonaldization" and "Disneyization");
- human resources, corporate culture, and training, as notably expressed in industry-wide practices known as emotional and performative labor (sometimes resulting in the "typecasting" of performers);
- informal hierarchies, as well as staff adoption or rejection of behavioral guidelines and other elements of corporate culture.

The second section, "Theme Parks, Labor, and Society," deals with the parks' labor practices in the broader context of the industry, which is notably shaped by:

- labor organizations, especially regarding hours and wages protection or discriminatory hiring practices ("typecasting");
- the parks' legal framework (e.g. industry-specific visas, artists' contracts, etc.);
- national and global migration patterns.

In order to highlight how even theme park patrons are, to some extent, put to work before, during, and even after their visits, scholars have sometimes drawn on the concept of "audience labor." The last section of this chapter, entitled "Audience Labor," will thus introduce different levels on which this operates in theme parks.

2 Theme Parks as Labor Environments: Internal Organizational Principles

2.1 Contexts of Labor

2.1.1 Forms of Labor and Official Organizational Structures

The vast majority of theme park workers consists of frontline workers in low-skilled, low-paid operational positions. Most of these jobs entail repetitive tasks in often seasonal or part-time positions, resulting in a high employee turnover (up to 50%) and a higher-than-average proportion of young (35 years of age or less) employees with no college education (Anton Clavé 2007, 402). A smaller share of employees works in backstage positions—most of them in equally low-paid, low-skilled positions (for example as dishwashers, landscape gardeners, or maintenance workers), with only a minority of managers working permanent back-office jobs (for example in marketing or human resources). At Walt Disney World (Orlando, Florida), two-thirds of the resort's 65,000 employees are estimated to work "onstage," and the remaining third "backstage" (Johnson 2011, 919).

As the industry's largest employer (with a roster of 130,000), Disney has long been recognized for its career advancement opportunities and has gained a reputation for internal mobility. For example, a number of second-generation "Imagineers" (Disney's designation for its park designers) started their careers in the late 1970s as ride operators. Disneyland Paris likewise boasts that 80% of its managers and senior managers are the product of internal promotion, with employee loyalty now averaging nine years of employment with Disney (EuroDisney SCA 2017, 25).

In addition to full/part time or seasonal/permanent positions, the theme park sector is characterized by a broad variety of employment types, including:

- Apprenticeships and internships, as notably represented by the Disney College Program (which some union leaders have described as a "migrant college-worker program"; see Billman and Sheffield 2007) and, for international students, the Disney Cultural Exchange Program or Academic Exchange program (see also below).
- Volunteer work, as in religious theme parks like the Akshardham Cultural Complex in Delhi, India (Brosius 2012, 161–248; on volunteer work in religious theme parks, see Paine 2019, 71–72), or the 4,000 volunteers who take part in the "Cinéscenie" pageant at Puy du Fou (Les Epesses, France).
- Subcontracting, in particular represented by lessees. Disney is particularly noted for its reliance on outside partners, with a number of early Disneyland (Anaheim, California) restaurants or shops leased to outside businesses, and all of the pavilions at Epcot (Orlando, Florida) originally sponsored and financed by major corporations—businesses whose responsibility it was to staff their locations (Clément 2018). Though most leases and sponsorships have expired, the situation prevails at Walt Disney World's non-Disney hotels, some of which are managed by Marriott, Hilton, or Starwood.

Owing to the far-ranging nature of their operations, theme parks present fairly complex organizational structures. Younger (2016, 20) has identified nine divisions typically represented in theme park corporations (Business development, Design, Finance, HR, Legal, Marketing, Operations, Sales, Technology) and just as many operational departments (Attractions, Custodial, Entertainment, Food & Beverages, Guest Services, Hotels, Maintenance, Merchandise, Security). Theme park workers are likewise organized into such hierarchical structures as, from bottom to top: Host/Attendant, Operator, Lead, Team Manager, Department Manager, Operations Manager, Park Manager, and, finally, Resort Manager. Such typologies are of course prone to variation: the corporate structure of PortAventura (Vila-seca and Salou, Spain), for instance, dedicates entire divisions to such operational concerns as Shows, Hotels, Technical Services (or Maintenance, in Younger's categorization), alongside the more expected categories of Finance, Human Resources, or Development (i.e. Design; The PortAventura Group 2018, 23).

2.1.2 Performative Labor and Broader Changes in the Service Sector

No matter which specific position in which particular department they may occupy, most theme park workers, and especially those who work in front of visitors, are always also hired as performers. In his 1991 article "The Smile Factory: Work at Disneyland," John van Maanen—one of several former Disney employees turned theme park scholar—has maintained that the theme park industry relies as much on "the symbolic resources put into place by history and park design" as it does "on an animated workforce that is more or less eager to greet guests, pack the trams, push the buttons, deliver the food, dump the garbage, clean the streets, and, in general, marshal the will to meet and perhaps exceed customer expectations" (59). According to van Maanen, then, working at a theme park involves not only performing specific tasks at the park's rides, shops, restaurants, and service areas, but also performing these tasks with a specific attitude that is not rude, careless, insincere, sleepy, or bored, but friendly, helpful, sincere, and attentive.

More than a decade after the publication of van Maanen's article, this basic idea was captured by Alan Bryman in the terms "performative labor" and "emotional labor." In *The Disneyization of Sociey* (2004), Bryman describes a system, exemplified by Disney theme parks, that according to him has come to dominate more and more sectors of society. Alongside theming, hybrid consumption, and merchandising, one of the cornerstones of this system is performative labor: "There is a growing trend for work, particularly in the service industries," Bryman writes, "to be construed as a performance, much like in the theater" (103). This is perhaps best illustrated by the Walt Disney Company's management vocabulary, which refers to workers as "cast members" who wear "costumes," operate "on stage" or "backstage," and whose "performances" are evaluated against "good show/bad show" criteria. A central element of performative labor is emotional labor, which refers "to employment situations in which workers as part of their work roles need to convey emotions" (104).

As van Maanen had already noted, however, the specific organization of work at theme parks leaves employees little room to perform or convey emotions: as he points out, most of the tasks performed by frontline employees "require little interaction with customers and are physically designed to practically insure that is the case. The contact that does occur typically is fleeting and swift, a matter usually of only a few seconds" (1991, 60). There is, then, very little room for any kind of interaction between employees and visitors, let alone for an elaborate performance. What little room there is needs to be filled as efficiently as possible, by a quick, simple act that instantly and unambiguously communicates friendliness, helpfulness, joy, and attentiveness: namely, the smile. According to Bryman, the smile indeed epitomizes emotional labor. Along with such other minimalist signs of a positive emotional display as eye contact, it "transfers to the service transaction a generally upbeat set

Fig. 1 At Dismaland (Weston-super-Mare, UK), Banksy's 2015 art installation parodying theme parks, employees were specifically instructed not to smile (see also PARATEXTS AND RECEPTION). *Photograph* Steve Taylor ARPS/Alamy Stock Photo

of impressions that are meant to improve the aura surrounding the service" (2004, 105; see Fig. 1).

Smiling and eye contact are key terms included in many theme park training employee manuals: Younger, for instance, quotes from Disney's Seven Guest Service Guidelines, the first of which simply reads: "Be happy .... Make eye contact and smile" (2016, 277). Much more interesting is the following excerpt from a Disneyland employee handbook quoted by Raz, which notes: "At Disneyland we get tired, but never bored, and even if it's a rough day, we appear happy. You've got to have an honest smile. It's got to come from within" (1999, 114). Hence, theme parks do not just require performative or emotional labor, they also need what sociologist Arlie Hochschild has referred to as "deep acting" (see Hochschild 1983). In contrast to surface acting, which merely requires workers to go through the motions of displaying the correct emotional form, deep acting asks emotional laborers to really feel the emotions that they are supposed to exhibit. By requiring deep acting, (Disney) theme parks seek to counter the widespread critique according to which employees merely mimic the robotic movements and facial expressions of their audio-animatronic colleagues. As early as 1968, Richard Schickel disdained the Disney University for "training employees in the modern American art forms [...] of the frozen smile and the canned answer delivered with enough spontaneity to make it seem unprogrammed" (318; see also below).

Though now adopted industry-wide, Disney's conception of performative labor and deep acting should obscure neither the company's debt to its predecessors nor broader shifts in management that Disney parks merely exemplify, rather than instigate. For example, in a formulation that long predates Disney's own service standards, Fred Thompson, co-founder and -owner of Coney Island's Luna Park, wrote in 1908:

> Courtesy on the part of the employee is as necessary as decency on the part of the visitor. If I hear of one of my employees resenting an insult offered by a visitor, I dismiss him. I tell him that so long as he wears my uniform he is representing me, and that I am the only person who can be insulted inside the gates. (Thompson 2015, 106)

Moreover, in its values-driven management and other key dimensions, Disney's corporate philosophy also offers parallels with Walmart—yet another behemoth symbolic of managerial transformations in the service economy. Much like Walt Disney then and now, Sam Walton serves as something of a "cult figure" whose "persona embodied Wal-Mart's reputation as an earthy and virtuous enterprise" (Lichtenstein 2009, 84) and whose folksy, paternalistic management practices were likewise resolutely anti-union. Walmart trades in the same kind of symbolic leveling and imaginary social landscapes as the Disney corporation, labeling all its employees "associates" (much like Disney workers are branded as "cast members") or likewise operating as a first-name basis environment. And much like Disney, Walmart's small-town, Midwestern values are central to its perception by both the public and its workers, generating the "wholehearted dedication [and the] devoted, empathetic psychologically humane engagement" also typically expected from Disney employees (Lichtenstein 2009, 110).

Disney and Walmart thus illustrate broader changes in the service sector after World War II, processes famously dubbed as "McDonaldization"—after the company that best embodies the dimensions of efficiency, calculability, predictability, and control characteristic of today's working environment (Ritzer 2008). McDonald's, Disney, and Walmart—as well as "McDonaldization" and "Disneyization"—overlap, particularly in their approach to management: these companies have come to exemplify corporate control over employee behavior through surveillance.

2.1.3 Surveillance

Because of the on-site, performative, and affective nature of labor in the theme park industry, the surveillance of workers is broad and multi-faceted (Lukas 1999, 2007). Many theme parks utilize members of their company training staff or outside assessors to conduct location audits of employees. For example, audits may be performed by plain clothes members of management who conduct employee reviews while attempting to blend in as park visitors. Employee audits focus on a variety of performance areas that are common to other service industry workplaces: safety, friendliness, attitude, and knowledge of park information (locations, hours, times, costs), among other areas. Such surveillance may also extend to the theme park corporation's policing of current or former employees in online and social media contexts. Many theme parks also encourage employees to monitor each other (and reward good "performances"; see Davis 1997, 91–92) and/or offer visitors opportunities to identify particularly good or bad employees.

At Disney theme parks, onstage surveillance of employee behavior is conducted by "area supervisors," whose monitoring techniques are more or less covert, with some favoring the quasi-panoptic view from "eagle's nests" hidden throughout the park, and others posing as tourists ghost-riding attractions or even using cameras to document deviations (van Maanen 1991, 68–69; Koenig 2002, 73). Disney's supervision has even been found to extend to employees' "offstage life" (including, rumor has it, their sex lives or drug habits), with the result that the "feeling of being watched is a rather prevalent complaint among Disneyland people" (van Maanen and Kunda 1989, 69). Offstage surveillance takes the form of police checks on current or potential employees, or the alleged monitoring of online meeting places of employees. As some cast members warn on forums for fellow park employees: "if they can identify you from your post(s), they can take steps that you will no longer have to worry about what you say on line. The Company wants to micromanage the content available on line, especially from its own cast" (qtd. in Clément 2016, 144).

2.1.4 Other Working Conditions

Other factors that contribute to the relatively high turnover rates in some theme parks (in the 1990s, only 45% of employees at Six Flags AstroWorld (Houston, Texas) remained in their company positions from the beginning until the end of the season) include long work hours, extreme weather, the psychological and social demands of service interaction, challenging relationships between management and workers, and low pay and benefits. Indeed, while the public perception may be that work in the theme park industry is "fun," working in theme parks—as in many service industry forms—can be extremely challenging (Hochschild 1983; Lukas 1999, 2007).

The issue of theme park employee pay has been a particular concern, with numerous studies pointing to low wages and a lack of health and welfare benefits (Dreier and Flaming 2018; Quinnell 2018). In addition, employee housing has emerged as an important topic of critical interest: for example, the considerable reliance of the Walt Disney Company on interns and foreign workers in most of its parks (except for its Asian locations) has made temporary housing key to its operations, and this also applies to many other theme parks in which some staff come from abroad (mostly because of typecasting; see below), as in the case of the European and African performers and dancers in E-Da Theme Park (Kaohsiung, Taiwan). As per the terms of their visas, interns at the Walt Disney World International Program are thus required to be housed on-site in dorm-like apartment complexes serviced by shuttles to the resorts' various work locations and local merchants. Apartments accommodate up to eight participants, and the US$114–205/week rent is automatically deducted from the participants' paycheck (DisneyCareers 2018). On-site housing is also open to Domestic College Program interns, although they are free to opt out. Disney's need for on-site housing shows no sign of abating: in December 2018, a partnership was announced with American Campus Communities, a leader in the student accommodation business, to build a 10,400-bed project to be completed by May 2023 (Walt Disney World N.D.). While nowhere near comparable in size with that of Walt Disney World, the internship program at Disneyland in California likewise entails accommodation being available for 260 participants at the company's Carnegie Complex, two miles away from the Disneyland Resort and also serviced by a shuttle service to work locations (Disney Internships and Programs N.D.): each apartment can accommodate two to five participants and is available at a rate of US$140/week.

Critically lacking, though, is affordable housing for the park's permanent workers, whose low pay is generally insufficient to afford homes in the parks' vicinity. At US$25.46, the hourly wage needed to afford a median-priced one-bedroom apartment in Anaheim far exceeds Disneyland's worker minimum hourly wage of US$11 (Velasco 2017). And with a growth rate twice the national average (5.6%), average rents in the Orlando area have likewise quickly outpaced the purchasing power of Walt Disney World workers, keeping many out of the rental market and forcing some to live in motels (Sainato 2018). At Disneyland Paris, the company provides housing for both temporary and permanent workers: 2,000 apartments with shared bedrooms for up to four people are available at a moderate rent (approximately €290/month) for non-local temporary workers; permanent contracts are eligible for temporary residence (up to two years) in one-person studios located in six nearby housing complexes (Lecompte 2012; Disneyland Paris 2018).

2.2 *Casting and Training*

2.2.1 General Worker Demographics

Unfortunately, research on worker demographics is extremely scarce. Where information is available, it shows wide variations from park to park, though top management positions still remain almost entirely male-dominated. At French leisure park and mountain resort operator Compagnie des Alpes, women make up 57% of all employees working in non-permanent positions but only 43% of the company's managers (Compagnie des Alpes 2016, 18). At 66%, PortAventura's workforce is likewise overwhelmingly female—except, again, in top management roles: women particularly dominate the 36–45-year-old as well as 25 and younger age groups, with 406 and 276 women workers out of a total of 1922 employees, respectively (The PortAventura Group 2018, 55).

2.2.2 (Type-)Casting

If theme park labor has been theoretically conceptualized as well as discursively framed as performative labor (see above), the question arises as to what kind of role employees play on the theme park stage. Younger (2016, 277) distinguishes between "costumed characters" who are fully integrated into the theme world of the park (i.e. performers) and "employees in more accessible positions" who use such themed elements as theme-appropriate uniforms or greetings in their performances but for whom safety and courtesy nevertheless rank higher than the show (i.e. performative/emotional laborers). Regardless of the themed area in which they work, the latter primarily play the generic part of the ever smiling, ever courteous, and helpful theme park employee. Younger cites the example of employees operating the Pirates of the Caribbean ride who "are explicitly not termed pirates in employee training documents as the designers are wary that such instructions might produce performances inconsistent with the friendly atmosphere of the park" (2016, 278).

However, much like "real" performers (e.g. Caucasians who perform as costumed face characters and dancers in Disney's Asian resorts, as they correspond more closely to the medial image derived from Disney movies), performative laborers have frequently been subject to "typecasting": a theme park training manual entitled *On Stage, Please!* and distributed to employees at Disneyland Paris in 2000/01, for example, insists that cast members "also have to 'look the part' to appear in a show" (N.N. N.D., 9). While the manual only gives examples from the realm of performers ("It would be difficult to imagine one of Snow White's dwarves with the physique of a rugby player"), the history of theme parks is replete with examples of positions in attractions and/or operations that have been cast with a keen eye to gender, race, and/or ethnicity. A July 1995 article in the *Orange County Register*, for instance, reports that after 40 years of gender-specific hiring, Disneyland finally introduced a "'unisex' policy" for ride operators that allowed women to work on rides like the Storybook Land Canal Boats or the Jungle Cruise, which had previously been operated only by men (see Fisher 1995).

Typecasting according to race and ethnicity is at least as old and widespread (originally, employees at Disneyland's Indian War Canoes were both male and of Native descent; see Strodder 2008, 216), but has also proven more resilient. Perhaps the most frequently cited contemporary example of racial and ethnic typecasting in theme parks is Walt Disney World, which has used its International Program and, from 1990 onwards, the Q-1 "Disney" visa to staff the individual national pavilions at Epcot's World Showcase or the Asian and African "lands" at Disney's Animal Kingdom with workers from the respective countries and continents (see Koenig 2007, 319; see also below). Technically, however, the Disney International Program is based on citizenship (and "culture") rather than race and ethnicity. In a postcolonial and increasingly globalized world, these two criteria do not necessarily correlate, although they (still) do in the minds of some visitors: Hermanson writes of a "native of South Africa working at [Disney's Animal Kingdom] [who] occasionally encountered guests who refused to believe he was African. Despite his accent and insistence, they could not overlook his white skin. He was perceived as fake because he did not conform to the image of Africa" (2005, 221).

True racial/ethnic typecasting does exist, however. Phantasialand (Brühl, Germany), for instance, staffs its "Deep in Africa" section primarily with dark-skinned workers (regardless of their actual nationality; see Steinkrüger 2013, 273–274); and in Chinese "ethnic" parks such as China Folk Culture Villages (Shenzhen,

China), representatives of ethnic minorities within the People's Republic people the life-size reconstructions of various ethnic villages. In fact, casting at China Folk Culture Villages is intersectional, as the company hires not just individuals who are "immediately recognizable as a member of a particular ethnic group," but also women rather than men in order "to satisfy the majority male visitors" (Ren 1998, 73). At the neighboring Window of the World park (Shenzhen, China), these very same ethnic minorities represent other racial and ethnic groups from outside China: in the park's African section, for instance, visitors may enjoy performances by members of the Wa people from southwestern China, "chosen, no doubt, for the color of their skin," while Mongolians and members of the Miao people are cast as Native Americans and dancers from Yunnan perform as Maori (see Ren 1998, 94; Stanley 2002, 286).

2.2.3 Training and Education

In order to prepare newly cast employees for their various tasks, theme parks frequently operate internal training divisions, with the Disney University, created in 1962, constituting the oldest and most famous example. In his autobiography, the University's first director, Van France, explains the motivations for the university's founding:

> [W]e needed something new, something that would impose responsibility and self-discipline on all of our key people. [...] We were also fighting what I call "amusement park thinking." Since I had heard it straight from Walt, I preached that Disneyland was NOT an amusement park. So what was it then? We came up with the concept that it was a WORLD SPECTACULAR SHOW, played on a large stage with the Southern California Sky as a giant backdrop. (1991, 74)

Such foundational principles have come to define the Disney Corporation's training and management policies. In addition, the Disney University has helped establish now industry-wide service standards, known, in order of importance, as safety, courtesy, show, and efficiency (originally capacity). These four keys are introduced during Disney Traditions (the corporation's orientation day and first dip into its 40-h apprenticeship program), alongside Disney's overall "service theme" or corporate statement: "We create happiness."

The Disney University has also been adopted—as well as adapted—for Disney's non-domestic resorts in Europe and Asia. Employee training at Tokyo Disneyland and Hong Kong Disneyland has been analyzed in detail by Raz (1999, 73–101) and Choi (2007, 317–326), respectively. Both have noted the introduction of significant, culture-specific changes to the established training program of the U.S.-based parks. Comparing the Orlando Disney park and the Walt Disney World College Program's version of Disney Traditions with the Hong Kong Disneyland version, for instance, Choi notes a "relative neglect of workers' safety and rights" (319) in the Hong Kong version.

With the creation of the Disney Institute in 1996, the Disney Corporation has become the foremost promoter of its management philosophy, training outside employees to its parks' and hotels' service standards. In 2000, the Disney Institute began providing consulting services and professional development courses for clients such as Volvo, Siemens, and the NFL. Training seminars typically take place at the Orlando resort and purport to allow clients to

> take a look at our time-tested customer experience best practices as you explore our business insights firsthand, in our parks and resorts "living laboratory." This is a powerful professional development opportunity, to learn how we think and how you could adapt our principles to your own organization. (Disney Institute N.D.)

Most theme parks have extensive training programs that mirror those of other corporate and industry settings. Like Six Flags AstroWorld, many parks conduct a general Human Resources employee onboarding training that introduces employees to company policies, working benefits, and practical matters like wardrobe, security, and pay. After the general HR training comes a more specific training that is focused on the specific employee department of work—Rides, Grounds Quality, Attractions, Retail, Security, Parking Lot, etc. In the 1990s, the Six Flags AstroWorld training department included a number of Training Coordinators who were responsible for the specific curriculum provided to all Rides and Grounds Quality employees. The initial training focused on safety and OSHA (Occupational Safety and Health Administration) principles, visitor-focused interaction philosophies, employee policies, and AstroWorld operational procedures. The training curriculum was developed collaboratively with management and training coordinators. Some of AstroWorld's training approaches were adapted from other theme park industry programs, such as Disney, including their emphasis on "Guest First" philosophies for employees. Many training classes included role playing and forms of dramaturgy in which new employees were asked to role play situations that included, "a guest is lost in the park, how would you help her out?" "A guest is upset that a ride is closed for the day, how would you react to this guest?" These scenarios were intended to stress a guest-first approach to visitor interaction as well as performative and improvisational skills (Lukas 1999, 2007).

Following this initial operations training, employees experienced an even more specific location training in which they learned about the on-the-job duties of the ride to which they were assigned. Such training was conducted by a ride's lead or area supervisor who would go through all of the specific duties associated with the work location and then complete an extensive checklist related to ride duties, safety, and guest interaction. For legal purposes, all rides outlines were kept on record at the AstroWorld Training Center. After this training was complete, the employee would return to the Training Center to take an exam to determine if they understood all of the duties required to become a CRA (Certified Rides Attendant). If the employee wished to become a CRO (Certified Rides Operator), additional on-the-job training, examinations, and other assessments were required. Additional training, such as visitor-focused classes and re-trainings, were also routinely scheduled throughout a given season or work at AstroWorld. In some cases, outside training groups were brought in to conduct rides lead and supervisor trainings as needed (Lukas 1999, 2007).

Due to the harsh working conditions, theme park employees occasionally try to "get back at" visitors or their employer by deliberately violating the rules of the workplace. At Disney parks, resistance to or deviation from standard operational procedures and recommended standards take various forms. These range from the subtle or surreptitious (for example, the overzealous application of safety procedures by tightly squeezing seatbelts on unruly visitors, or breaking from the pre-approved script; see van Maanen 1991, 71; Clément 2016, 207; Choi 2007, 329–332) to the public and overtly confrontational: this might entail workers alerting the press and going public with their complaints toward management (Choi 2007, 326; Clément 2016, 210) or even staging strikes—as Disneyland Paris workers do with some regularity. All such events directly contravene "good show" imperatives to never discuss employee working conditions in public (see INCLUSION AND EXCLUSION). By contrast, workers may also on occasion demonstrate the internalization of behavior guidelines. This is especially evident in the park's social hierarchy—the result of workers' identification with their roles and characters, as notably reflected in their costumes which "provide instant communication about the social merits or demerits of the wearer within the little world of Disneyland workers" (van Maanen 1991, 62). Employees at high-capacity attractions may also compete for the highest ride count (Koenig 1995, 83).

2.3 *Informal Organization and Worker Interactions*

Like other service industry places of work, the work structure of theme parks is often characterized by rich and expressive social interaction among employees. A common practice in theme parks is to schedule days and times in which employees can enjoy the park rides and attractions when the park is closed to visitors. In the 1990s at Six Flags AstroWorld, these parties were known as "section parties" and were often designed to increase employee morale during particularly challenging times of employment. In the 2000s at Disneyland Paris, the entire team of each unit (ride, shop, restaurant, etc.) would be invited to dinner at one of the on-site restaurants once a year.

Within some theme park workplaces, there are informal hierarchies among workers that mirror the formal hierarchies evident in the organization. At Six Flags AstroWorld in the 1990s, for example, such hierarchies were often based on employee perceptions of the prestige or relative value of one park job compared to another (Lukas 2007). The department in which such hierarchies were most prominent was Rides. The formal organizational structure of most rides included an Area Supervisor, a Rides Lead, a CRO (Certified Rides Operator), and a CRA (Certified Rides Attendant). Training periods for a CRA were notably less than those for a CRO, due to the greater responsibilities that were involved in operating a ride (see above). Any given ride location had additional subdivisions of these positions, including

MCO (Main Control Operator), Unloader, Loader, Grouper, Height Checker, and Gate Person. Staffing of such positions on a given ride depended on the complexity of a ride—typically, a roller coaster required more greeter staffing than a troika—and the availability of staffing on a given day of operation.

Informal senses of prestige were not rooted in the specific duties at a given ride, but were instead tied to the type of ride, or even its theme (Lukas 2007). Bugs Bunny Land, which was a children and family play area, was often interpreted as a less-than-serious place to work in Rides. The railroad (610 Limited), conversely, was seen as a prestigious Rides location, in no small part due to the fact that the park's train was an actual and fully operational locomotive. Spinning rides or troikas, like the Gunslinger, were often seen as being of less work value due to the lesser degree of complexity in the ride's operation. The most prestigious rides were roller coasters, like the Texas Cyclone and Batman the Ride. In popular culture, the roller coaster has achieved a mythical status (Cartmell 1986), with ride enthusiast groups often based solely on these forms of rides; thus, it is not surprising that such perceptions of ride value were maintained in the AstroWorld workplace. An additional sense of value was noted in some employee perceptions of a ride's theme being connected to their own interest in being assigned to that location. When Batman the Escape opened in 1993, there was a great deal of employee interest in working at the ride due to its theme being connected to the popular films and comics, the extensive theming of the ride queue house, and employee outfits that mimicked some of the fashion design from the Batman films. Employees who worked on these rides felt that they were part of a transmedia story that also had relevance outside the world of the theme park (see MEDIA).

Evidence suggests that such informal hierarchies exist elsewhere, based on the perceived skillsets and autonomy required for any given job. At Disney theme parks, such hierarchies have been described as ranging from "the upper-class prestigious Disneyland Ambassadors and Tour Guides" and, just below them, ride operators performing "skill work" (i.e. live narrations or operating costly vehicles), with other ride operators, "proletarian sweepers," and the "sub-prole or peasant status Food and Concession workers" occupying the lower three echelons of the park's class system (van Maanen 1991, 61–62).

3 Theme Parks, Labor, and Society: Parks Within Their Broader Environment

3.1 Impact of Theme Parks on the Labor Market and Local Economies

The impact of theme parks on local labor markets has long been held as positive—to the point that in France many economically depressed municipalities or regional governments have considered them a potential solution to chronic unemployment

(Mercante 2015). Yet all too often, the seasonal nature of many theme park operations means that the vast majority of positions are only temporary: for example, Parc Astérix (Plailly, France), long the country's second most visited theme park, only boasts 200 permanent employees and 1,000 temporary workers per year (most on minimum wage and some working only weekends or during summer months), with a 30–50% return rate (Courrier Picard 2017). Looking to better meet the needs of the industry and stimulate sluggish labor markets, joint programs and agreements have been developed in France and Spain between parks, public authorities, and institutions of learning: the French and Catalan employment services have thus established training programs with Parc Astérix and PortAventura Park for the training of job seekers (Courrier Picard 2017; The PortAventura Group 2018, 52). Similar agreements exist between Disneyland Paris and the nearby Université Paris-Est Marne-la-Vallée, as well as between PortAventura Park and 25 universities and schools for the training of students (EuroDisney SCA 2017; The PortAventura Group 2018, 53).

The parks' impact on the labor market is for the most part indirect, notably by contributing to local tourist economies and creating indirect employment. In 2017, PortAventura thus spent US$122 million on 1,048 suppliers (81% of whom were Spanish), as well as generating employment for 958 contractor-employed staff working in the park's resort facilities (The PortAventura Group 2018, 49, 110). It is likewise estimated that, in 2013, the Puy du Fou generated up to 4,700 jobs (including the park's 1,375 employees), as well as injecting €193 million into the local tourist economy, in addition to the park's €64 million revenue (Litzler et al. 2014).

Owing to its sheer scale and impact on labor markets, the Disney Corporation has invited more scrutiny than any other actor in the theme park business. In part as a result of contractual obligations with the French government, Disneyland Paris is estimated to sustain 56,000 jobs (including the park's own 15,000 workers). It has likewise contributed an average of €2.7 billion to the French economy, generating €80 billion worth of tourist revenues (i.e. 6% of all tourist revenues in France) and buying over €14 billion worth of supplies over 25 years (EuroDisney SCA 2017, 12–14). Yet elsewhere, the Walt Disney Company's contribution to the local economies and labor markets of Orlando and Anaheim is very much in dispute. At Walt Disney World, the institution of a two-tier pay system in 1998 (with new hires receiving 12.2% lower pay) resulted in the loss of 178 jobs and US$23.4 million in goods and service production for the year 2006 (Nissen et al. 2007, 2). Likewise, average hourly wages at Disneyland in California dropped 15% in real dollars from 2000 to 2017, in one of the United States' most expensive metro areas. As a result, a 2018 survey conducted by a group of Disneyland unions, the Coalition of Resort Labor Unions, found that, of its 5,000 respondents, three-fourths did not earn enough money to cover their living expenses, leaving two-thirds insecure and one in ten homeless in the last two years (Merrit 2018).

3.2 *Artists' Contracts*

In France, Germany, and other countries with publicly subsidized arts sectors, theme parks have often relied on contracts that are specific to the performing and entertainment arts. In France, for example, positions for performers, theater technicians, costume designers, makeup artists, and hair stylists are often based on industry-specific contracts for artists working on a job-by-job basis and paid by the day: such contracts allow show-business professionals to receive unemployment benefits between gigs, provided they have worked an annual minimum of 507 h (or 43 gigs). With 2,000 hires and a total of 13,820 days of employment offered under this arrangement, Disneyland Paris is France's largest provider of such contracts (along with the Paris National Opera), with contracts typically spanning from one day to several months (Disneyland Paris 2016; La Croix 2010). At Parc Astérix, 250 staff members (or approximately one-fourth of the park's seasonal workforce) are employed under this contract (N.N. 2017). While theme parks have much to gain from the flexibility of such contracts—notably by allowing them to easily match their workforce with demand—employees also enjoy the benefits of paid rehearsals and expanded networking possibilities, as Disneyland Paris touts on its website (Disneyland Paris 2016). Just as importantly, parks help aspiring actors meet the minimum working hours required for unemployment benefits, with some claiming that theme parks represent 60% of their earnings but only 30% of their working hours (La Croix 2010). And while theme parks contribute to the vitality of the performing arts sector by providing actors with additional employment opportunities in-between gigs, it might also be argued that such contracts amount to a public subsidy to the theme park sector. The Disney corporation has recently come under fire for the unilateral termination of such contracts during the COVID-19 pandemic and the forced closure of its French operations, thus compromising the artists' minimum hours requirements and, therefore, their unemployment compensation (N.N. 2020).

3.3 *Labor Organizations and Collective Bargaining*

Because of the parks' staff-heavy nature (as with much of service industry), unions have often played a key role in shaping the labor environment of park workers—a role highlighted whenever labor disputes, which operators work so hard to keep backstage, shatter the parks' audience-friendly image. Of course, wide variations exist in union roles and union representation, depending on location and local labor regulations. At Disneyland Paris, seven unions are represented, with the comparatively moderate CFTC (Confédération Française des Travailleurs Chrétiens) enjoying the widest support. At the opening of the park in 1992, the park's unions almost immediately objected to American policies alleged to violate French labor laws and customs, famously and repeatedly locked horns with management, and, aided by disappointing economic results, successfully forced a change of directorate, with

French Philippe Bourguignon replacing Robert Fitzpatrick as EuroDisney's CEO in 1993. The unions' favored modes of action have notably included strikes: in 1998, 300 face characters demanding higher pay staged a 17-day strike—only three days longer than that of the park photographers in 2006, for similar motives (60% of employees at Disneyland Paris earn the French minimum wage; see Lecompte 2012). Unions have also resorted to litigation in order to force Disney to align internal, US-imported practices with French labor law (which Disney, in its negotiations with the French government, initially asked to be exempt from; see Packman and Casmir 1999, 481). In a high-profile 1994 case, the "Disney look," which strictly regulates acceptable hairstyles, make-up, or tattoos for employees, and which was variously branded an "attack on individual liberty" by the Communist-led Confédération Nationale du Travail or a violation of "human dignity" by the CFTC (see N.N. 1991), was thus judged discriminatory, allowing unions to burnish their credentials by taking on the American goliath. Other judicial rulings have since forced the company to comply with the collective bargaining agreement in force in the parks and recreation sector (1998), to regard time spent traveling from backstage to onstage areas as worktime (2006), and to negotiate the terms for the application of the 35-h workweek (adopted nationwide in 2000) and the conditions for possible exemptions (2007; see N.N. 2007).

Industrial action is typically less developed in the United States. For example, Walt Disney World, Disney's largest operation, is located in the right-to-work state of Florida. As a result, workers do not have to join unions to gain the benefits of union memberships, thus stunting organized labor's clout and power of attraction. In addition, working conditions at Disneyland and other Disney parks have long been reported as superior, including "quite satisfactory" benefits and salaries for part-time employees (Spinelli 1995, 11) or, starting in 1995, health coverage for live-in partners of gay and lesbian employees (while widely held to signal change in corporate America, in reality Disney was among the last Hollywood studios to do so; see N.N. 1995; Woodyard and Lee 1995; see also INCLUSION AND EXCLUSION).

However, eroding pay standards in both the company's Florida and California locations have recently led unions to take center stage and to dramatically increase their public profile, demanding and obtaining raises in starting wages: in a major union victory, hourly wages have been set to increase to US$15 by 2019 at Disneyland and by 2021 at Walt Disney World—up from US$10 in Florida (above the state's US$8.25 minimum) and US$11 in California (equal to the state's minimum; see Martin 2018). Expected to contribute an extra US$1 billion to Central Florida's economy, the raise at Walt Disney World followed a nine-month bargaining process (including protests near the parks' entrances) led by the Service Trades Council Union, representing six unions and 38,000 workers or half of all Disney World employees (Caron 2018). At Disneyland, the raise applied to 9,700 workers (approximately one-third of all Disneyland employees) and anticipated California's minimum wage increase by three years.

With respect to Asia, Ren (1998) and Raz (1999) have stressed continuities between working conditions in theme parks and at other local companies, e.g. how the labor contract system adopted in Shenzhen's Splendid China Parks (e.g.

China Folk Culture Villages) reflects that of other companies operating in Chinese Special Economic Zones (Ren 1998, 72–73) and how Tokyo Disneyland's grooming standards and general organizational culture correspond to those of other Japanese corporations and institutions (Raz 1999, 85–86; 107). The respondents from Hong Kong Disneyland in Choi (2007), by contrast, complained about "physically toilsome work, certain inconsiderate management practices, undesirable visitors, and conflicting beliefs concerning social fairness, meritocracy, rule by contract, rule by law, and free choice in a market economy," but nevertheless failed to join the Hong Kong Disneyland Cast Members' Union (HKDCMU) due to their "ambivalence toward HKDL management, labor unions' reputation for confrontation, and workers' fear that labor-union membership would negatively affect their careers" (328–329). Finally, the religious theme parks in Asia researched by Paine (2019) distinguish themselves by attracting volunteers and basing their labor relations on religious precepts, e.g. the Akshardham parks in India, where, in accordance with the teachings of Swaminayaran Hinduism, "employees must be paid properly and treated fairly; every worker must fulfill his tasks with careful deliberation" (60).

3.4 Migration Patterns: Domestic and International Dynamics

The theme park industry is marked by migration patterns at both the domestic and international levels. This is due to the at times seasonal nature of their operations, as well as the number of positions open to non-local, low-skilled workers from less dynamic work markets, e.g. Puerto Rico in the case of U.S. parks, or Southern Europe in the case of Northern European locations. In the case of Disneyland Paris, for example, the European Union's unique opportunities for worker mobility thus mean that non-French workers represent 18% of park's total workforce—approximately one-third of whom hail from Italy. Cross-European recruitment is facilitated by the organization of "casting tours" across 13 European towns, which advertise the 8,000 positions open for employment in the park (EuroDisney SCA 2017, 25, 27).

At Walt Disney World, migration may be domestic or international, as well as temporary or permanent (or at least long term). Disney's proactive role in instigating work migration is notably illustrated by continued efforts to attract Puerto Rican workers to its Florida resort: looking to benefit from the island's sluggish economy and attractive workforce, the Disney Corporation has actively recruited recent graduates and students from the University of Puerto Rico, offering free airfare and US$1,500 in relocation fees (Baribeau 2011). Disney's efforts do not stop at American borders and extend internationally with its Disney International Program, which includes the Cultural Exchange Program and the Disney Academic Exchange Program. Hired for twelve months (except for Brazil), cultural representatives are expected to "represent their cultures, traditional [sic!], and history of the entire country" in the pavilions at Epcot's "World Showcase" or at some of Animal Kingdom's African or Asian sections (DisneyCareers N.D.).

In a demonstration of Disney's lobbying power, a visa specifically tailored to the needs of Walt Disney World's Cultural Representatives program—the Q-1 Visa—was adopted in 1990. In fact, according to the 1990 Immigration Act, the visa was meant for participants in "an international cultural exchange program approved [...] for the purpose of providing practical training, employment, and the sharing of the history, culture, and traditions of the country of the alien's nationality" (Johnson 2011, 926). With good reason, the visa soon came to be known as the "Disney visa"—as demonstrated by the fact that in 2007, 54% of all 2,412 Q visas granted that year were issued to participants in Disney's program (Johnson 2011, 935). Another way to attract international workers is through J visas, reserved for international student exchanges and now used by half of Disney World's international workers (Johnson 2011, 937). Under the guise of its Disney Academic Exchange program, the Walt Disney Company works with higher-education institutions, which sponsor applicants and allow them to engage in on-field "academic training" at Disney World. While participants are considered full-time students (and as such must take course work, either from their home university, their American sponsor, or Disney's own classes), many positions bear little to no relevance to the participant's academic training, making the program a clear case of violation of "legislative intent" (Johnson 2011, 949).

The use of temporary workers—especially international ones—comes with added benefits for the Disney corporation: not only are they paid less than workers with seniority, but foreign employees are not subject to FICA taxes that apply to domestic workers. The company likewise earns extra money on the rent collected for their on-property accommodation (see earlier) while enjoying a "uniquely dependable" workforce—one whose stay in the U.S. is entirely dependent on Disney.

In contrast to Walt Disney World, Chinese "ethnic" theme parks such as China Folk Culture Villages (Shenzhen) exclusively hire domestic workers as cultural representatives, whether they represent ethnic minorities from within China or particular racial and ethnic groups from outside the country (see above). Other Asian theme parks also tend to hire domestically—with, for instance, Tokyo Disneyland preferring ethnic Japanese over other Asians for frontline positions, according to Raz's informants (1999, 87). Nevertheless, east–west migration takes place in the realm of performance, as due to typecasting most characters at the Disney resorts in Tokyo, Hong Kong, and Shanghai, for example, are usually played by Caucasians (see above).

4 Audience Labor

At least according to some scholars, it is not only the employees who are put to work in the theme park, but also the visitors. Drawing on ideas concerning the blurring of the boundary between labor and leisure in general (Rosa 2013), and the tensions and ambiguities that underlie theme parks as both places of fun and places of employment in particular, researchers have identified at least four different forms or types of "audience labor," not all of which take place during the limited time of the theme park visit: (1) the time and effort visitors have to invest into planning and organizing their visit; (2) visitors' performances as a key component of the success of the theme park's "show" and experience; (3) the commoditization of theme park audiences, who are sold to corporate sponsors; and (4) the audience's interpretive activity as part of the media's production process.

Various concepts have sought to capture the idea that theme park visitors typically spend time, effort, and sometimes even money (e.g. to purchase a guidebook) to plan and organize their visit: among them are Aldo Legnaro's "Erlebnisarbeit" ("experience work"; Legnaro 2000, 293), H. Jürgen Kagelmann's "Spaß-Arbeit" ("fun work"; Kagelmann 2004, 175), and Rebecca Williams's "anticipatory labor" and "plandom" (Williams 2020, 67–99). However, whereas Kagelmann interprets this form of audience labor as a rational attempt to get the maximum amount of fun (measured, for instance, in the number of attractions one has been able to visit) from a limited amount of time in the park, Williams stresses the additional fun, particularly from fans and regular visitors, that may be drawn from planning their next visit: "This form of 'plandom fandom' can actually offer an integral part of the theme park experience beyond simply organizing one's trip. For many fans, there is a clear sense of pleasure to be gained from the act of planning" (2020, 69; see also VISITORS).

Visitor performances during the visit are critical to the success of the theme park experience: much like park workers, not only are visitors and their emotions literally put on stage (for example in thrill rides; see ATTRACTIONS), but some element of play-acting (or at least a willingness to play along) is also expected from them (see VISITORS). Indeed, as Walt Disney once remarked while touring his soon-to-open Disneyland park, "people are [the park's] biggest attraction": "[y]ou fill this place with people, and you'll really have a show" (qtd. in Thomas 1994, 14). In other words, it might be argued that

> visitors' very presence [...] provides extra show material and consequently helps management extract additional surplus value from its park. [This turns] visitors into unpaid and often unacknowledged workers, whose successful emotional enlistment is critical to the success of the park's "show." (Clément 2015, 154)

Another form of audience labor during the theme park visit takes the form of visitor exposure to corporate messages from outside companies. Dallas Smythe has argued that as media audiences are sold to corporations, so the audiences are put to work: "Because audience power is produced, sold, purchased and consumed, it commands a price and is a commodity. [...] Your audience members contribute your unpaid work time and in exchange you receive the program material and the explicit advertisements" (1981, 233). In theme parks, exposure to corporate messages and advertisements often—though not exclusively—takes the form of attractions that are sponsored by major corporations. For example, when Disneyland opened, corporate exhibitions made up most of the park's otherwise mostly empty "Tomorrowland" section. As one critic has noted, the fact that such "corporate exhibitions were originally branded as 'free' (when other attractions required admission tickets, before the adoption of all-inclusive entry tickets in 1982) paradoxically points to their transactional nature, allowing the audience's attention time to serve as payment for their educational or entertainment value" (Clément 2018). Likewise, the presence of intellectual property promoting large media franchises, such as Six Flags' rights to the DC Comics universe, helps reinforce the brand equity of their original producers.

Finally, the notion that media consumption qualifies as a meaning-making and, therefore, a production activity in its own right, was first identified by French historian and philosopher Michel de Certeau (De Certeau 1984). It was later further elaborated on by media scholars John Fiske and Henry Jenkins (Fiske 1992; Jenkins 1992), whose work on fans highlights audience efforts to influence the production process of media products and "participate in the construction of the original text" (Fiske 1992, 40). In Disney theme parks, visitors' interpretive activity during and beyond the theme park visit has helped expand the park's appeal to ever-wider audiences: this form of underground labor is occasionally evidenced by fans or subgroups, who inject the park with meanings that speak to their specific interests, for example by making the family-centric environment of theme parks a platform for gay rights (see Clément 2016, 160; see also INCLUSION AND EXCLUSION).

5 Conclusion

Broadly conceived, the notion of "labor" in the context of theme parks encompasses the contributions of *all* theme park actors to the theme park experience—employees as well as visitors and fans. Among the former, the designers in particular have received considerable prominence with the general public through paratexts that offer behind-the-scenes glimpses at the making of theme parks (see, for instance, the various coffee-table books published by Disney about the company's "Imagineers"; see PARATEXTS AND RECEPTION). By contrast, theme park researchers have instead focused on frontline employees and, more recently, theme park visitors and fans. The paratextual promotion of designers on the one hand and the academic interest in the on-stage side of theme park labor on the other may reflect theme parks' attempts to generate a "revered gaze" ("a response marked as much by recognition of the labor and effort involved in creating the spectacle as in the spectacle itself," Griffiths 2008, 286; see also Boorstin 1961, 38) as well as scholars' methodological difficulties in accessing backstage areas (see METHODS), respectively. Yet as much as it may be rendered invisible, rhetorically disguised as performance, or identified as pure fun, labor in theme parks occurs both on stage and backstage, among those who get paid for it and those who pay for it. Our broad approach to theme park labor thus not only reveals the broad variety of laborers that are necessary to "produce" the theme park experience, but also illustrates the theme park's inherent tension or ambiguity as both a place of leisure and a place of work.

References

Anton Clavé, Salvador. 2007. *The Global Theme Park Industry*. Wallingford: CABI.

Baribeau, Simone. 2011. Puerto Ricans in Central Florida's Tourism Hub are Driving Hispanic Growth. *Bloomberg News*, March 18. http://www.bloomberg.com/news/2011—03—18/puerto-ricans-in-florida-s-tourism-hub-drive-hispanic-growth.html. Accessed 12 Dec 2019.

Billman, Jeffrey C., and Deanna Sheffield. 2007. Of Mouse and Man. *Orlando Weekly*, December 27. https://www.orlandoweekly.com/orlando/of-mouse-and-man/Content?oid=2257267. Accessed 3 March 2021.

Boorstin, Daniel J. 1961. *The Image: Or, What Happened to the American Dream*. New York: Atheneum.

Brosius, Christiane. 2012. *India's Middle Class: New Forms Of Urban Leisure, Consumption And Prosperity*. London: Routledge.

Bryman, Alan. 2004. *The Disneyization of Society*. London: Sage.

Caron, Christina. 2018. Walt Disney World Workers Reach Deal for $15 Minimum Wage by 2021. *The New York Times*, August 25. https://www.nytimes.com/2018/08/25/business/disney-world-minimum-wage-union.html. Accessed 28 Aug 2022.

Cartmell, Robert. 1986. *The Incredible Scream Machine: A History of the Roller Coaster*. Bowling Green: Bowling Green State University Popular Press.

Choi, Kimburley. 2007. *Remade in Hong Kong: How Hong Kong People Use Hong Kong Disneyland*. Diss., Lingnan University.

Clément, Thibaut. 2015. "Whistle While You Work": Work, Emotions, and Contests of Authority at the Happiest Place on Earth. In *ICT and Work: The United States at the Origin of the Dissemination of Digital Capitalism*, 145–165. London: Palgrave Macmillan.

Clément, Thibaut. 2016. *Plus vrais que nature: Les parcs Disney, ou l'usage de la fiction de l'espace et le paysage*. Paris: Presses de la Sorbonne Nouvelle.

Clément, Thibaut. 2018. "They All Trust Mickey Mouse": Showcasing American Capitalism in Disney Theme Parks. *InMedia: The French Journal of Media Studies* 7.1. http://journals.openedition.org/inmedia/1021. Accessed 28 Aug 2022.

Compagnie des Alpes. 2016. *Annual Report 2015*. https://www.compagniedesalpes.com/sites/default/files/brochures/2016/ra_cda_gb_2015_web_vdef.pdf. Accessed 12 Dec 2019.

Davis, Susan G. 1997. *Spectacular Nature: Corporate Culture and the Sea World Experience*. Berkeley: University of California Press.

De Certeau, Michel. 1984 [1980]. *The Practice of Everyday Life*. Trans. Steven Randell. Berkeley: University of California Press.

DisneyCareers. 2018. Housing Fees and Policies: Disney International Programs. *DisneyCareers*. https://jobs.disneycareers.com/housing-information. Accessed 28 Aug 2022.

DisneyCareers. N.D. Cultural Representative Program. *DisneyCareers*. https://jobs.disneycareers.com/cultural-representative-program. Accessed 28 Aug 2022.

Disney Institute. N.D. Case Studies, Client Impact, Disney Institute. *Disney Institute*. https://disneyinstitute.com/client-impact/case-studies/. Accessed 28 Aug 2022.

Disney Internships & Programs. N.D. Carnegie Plaza Provides California-Based Participants the Perfect Home Away from Home. *Disney Internships & Programs Blog*. https://disneyprogramsblog.com/carnegie-plaza-disneyland-resort-housing/. Accessed 12 Dec 2019.

Disneyland Paris. 2016. Intermittents du spectacle: Bienvenue! *Disneyland Paris Carrière*, September 6. http://careers.disneylandparis.com/fr/intermittents-du-spectacle-bienvenue. Accessed 8 March 2021.

Disneyland Paris. 2018. What Housing Options are Available for Cast Members with Permanent (CDI), Apprenticeship or Professionalisation Contracts? *Disneyland Paris Careers*, October 5. http://careers.disneylandparis.com/en/faq/what-housing-options-are-available-cast-members-permanent-cdi-apprenticeship-or. Accessed 8 March 2021.

Dreier, Peter, and Daniel Flaming. 2018. Disneyland's Workers are Undervalued, Disrespected and Underpaid. *The Los Angeles Times*, February 28. https://www.latimes.com/opinion/op-ed/la-oe-dreier-flaming-disneyland-employee-survey-20180228-story.html. Accessed 8 March 2021.

EuroDisney SCA. 2017. Disneyland Paris, 25 ans de contribution économique et sociale. *EuroDisney SCA*, February 27. http://disneylandparis-news.com/wp-content/uploads/2017/03/Disneyland-Paris-25-ans-de-contribution-économique-et-sociale.pdf. Accessed 12 Dec 2019.

Fisher, Marla Jo. 1995. "Unisex" Policy at Disney Theme Parks: Disneyland's Yearlong Program Means Women in Submarines and Men on Storyland Boats as Hosts. *Orange County Register*, July 12. N.P.

Fiske, John. 1992. The Cultural Economy of Fandom. In *The Adoring Audience: Fan Culture and Popular Media*, ed. Lisa A. Lewis, 30–49. London: Routledge.

France, Van Arsdale. 1991. *Window on Main Street: 35 Years of Creating Happiness at Disneyland Park*. Nashua: Stabur.

Griffiths, Alison. 2008. *Shivers Down Your Spine: Cinema, Museums, and the Immersive View*. New York: Columbia University Press.

Hermanson, Scott. 2005. Truer than life: Disney's Animal Kingdom. In *Rethinking Disney: Private Control, Public Dimensions*, ed. Mike Budd and Max H. Kirsch, 199–227. Middletown: Wesleyan University Press.

Hochschild, Arlie. 1983. *The Managed Heart: Commercialization of Human Feeling*. Berkeley: University of California Press.

Jenkins, Henry. 1992. *Textual Poachers: Television Fans and Participatory Culture*. New York: Routledge.

Johnson, Kit. 2011. The Wonderful World of Disney Visas. *Florida Law Review* 63: 915–958.

Kagelmann, H. Jürgen. 2004. Themenparks. In *ErlebnisWelten: Zum Erlebnisboom in der Postmoderne*, ed. H. Jürgen Kagelmann, Reinhard Bachleitner, and Max Rieder, 160–180. Munich: Profil.

Kinni, Ted. 2003. *Be Our Guest: Perfecting the Art of Customer Service*. New York: Disney Editions.

Koenig, David. 1995. *Mouse Tales: A Behind-the-Ears Look at Disneyland*. Irvine: Bonaventure.

Koenig, David. 2002. *More Mouse Tales: A Closer Peek Backstage at Disneyland*. Irvine: Bonaventure.

Koenig, David. 2007. *Realityland: True-Life Adventures at Walt Disney World*. Irvine: Bonaventure.

La Croix. 2010. Disneyland Paris, vivier de jeunes comédiens en mal d'emploi. *La Croix*, August 11. https://www.la-croix.com/Culture/Actualite/Disneyland-Paris-vivier-de-jeunes-comediens-en-mal-d-emploi-_NG_-2010-08-11-555479. Accessed 28 Aug 2022.

Lecompte, Francis. 2012. Disneyland Paris: Dans les coulisses d'une usine à rêve. *Capital*, March 27. https://www.capital.fr/entreprises-marches/disneyland-paris-dans-les-coulisses-d-une-usine-a-reves-707758. Accessed 28 Aug 2022.

Legnaro, Aldo. 2000. Subjektivität im Zeitalter ihrer simulativen Reproduzierbarkeit: Das Beispiel des Disney-Kontinents. In *Gouvernementalität der Gegenwart*, ed. Ulrich Bröckling, Susanne Krasman, and Thomas Lemke, 286–314. Frankfurt: Suhrkamp.

Lichtenstein, Nelson. 2009. *The Retail Revolution: How Wal-Mart Remade American Business, Transformed the Global Economy, and Put Politics In Every Store*. New York: Metropolitan.

Lipp, Doug. 2013. *Disney U: How Disney University Develops the World's Most Engaged, Loyal, and Customer-Centric Employees*, 2nd ed. New York: McGraw-Hill Professional.

Litzler, Jean-Bernard, and Service Infographie. 2014. Spectacles vivants: menaces sur le recours aux bénévoles. *Le Figaro*, February 6. https://www.lefigaro.fr/emploi/2014/02/06/09005-20140206ARTFIG00225-spectacles-vivants-menaces-sur-le-recours-aux-benevoles.php. Accessed 28 Aug 2022.

Lukas, Scott A. 1999. An American Theme Park: Working and Riding Out Fear in the Late Twentieth Century. In *Late Editions 6, Paranoia within Reason: A Casebook on Conspiracy as Explanation*, ed. George E. Marcus, 405–428. Chicago: University of Chicago Press.

Lukas, Scott A. 2007. How the Theme Park Gets Its Power: Lived Theming, Social Control, and the Themed Worker Self. In *The Themed Space: Locating Culture, Nation, and Self*, ed. Scott A. Lukas, 183–206. Lanham: Lexington.

Martin, Hugo. 2018. Disneyland Resort Workers Approve Contract That Raises The Minimum Hourly Wage to $15 by Next Year. *Los Angeles Times*, July 27. https://www.latimes.com/business/la-fi-disney-contract-vote-20180726-story.html. Accessed 28 Aug 2022.

Mercante, Agathe. 2015. Les collectivités veulent toutes leur parc d'attractions. *Les Echos,* September 30. https://www.lesechos.fr/2015/09/les-collectivites-veulent-toutes-leur-parc-dattractions-254741. Accessed 29 Aug 2022.

Merrit, Kennedy. 2018. Some Disneyland Employees Struggle to Pay For Food, Shelter, Survey Finds. *NPR*. https://www.npr.org/sections/thetwo-way/2018/02/28/589456403/some-disneyland-employees-struggle-to-pay-for-food-shelter-survey-finds. Accessed 8 March 2021.

N.N. N.D. *On Stage, Please! Values, Rules and Requirements for the Entertainment Professions.* Marne-la-Vallée: Disneyland Paris.

N.N. 1991. A Disney Dress Code Chafes in the Land of Haute Couture. *The New York Times*, December 25. https://www.nytimes.com/1991/12/25/business/a-disney-dress-code-chafes-in-the-land-of-haute-couture.html. Accessed 28 Aug 2022.

N.N. 1995. Disney Co. Will Offer Benefits to Gay Partners. *The New York Times*, October 8. https://www.nytimes.com/1995/10/08/us/disney-co-will-offer-benefits-to-gay-partners.html. Accessed 28 Aug 2022.

N.N. 2007. Eurodisney fête ses quinze ans. *Le Nouvel Observateur*, April 12. https://www.nouvelobs.com/economie/20070412.OBS1643/eurodisney-fete-ses-quinze-ans.html. Accessed 3 March 2021.

N.N. 2017. Le Parc Astérix, un chaudron aux emplois saisonniers. *Courrier Picard*, January 16. https://www.courrier-picard.fr/art/4905/article/2017-01-15/le-parc-asterix-un-chaudron-aux-emplois-saisonniers. Accessed 28 Aug 2022.

N.N. 2020. Disneyland Paris met fin aux contrats de ses intermittents. *Le Figaro*, April 4. https://www.lefigaro.fr/flash-eco/disneyland-paris-met-fin-aux-contrats-de-ses-intermittents-20200404. Accessed 28 Aug 2022.

Nissen, Bruce, Eric Schutz, and Yue Zhang. 2007. *Walt Disney World's Hidden Costs: The Impact of Disney's Wage Structure on the Greater Orlando Area.* Miami: Research Institute on Social and Economic Policy, Center for Labor Research and Studies, Florida International University. http://www.risep-fiu.org/2007/03/walt-disney-world%E2%80%99s-hidden-costs-the-impact-of-disney%E2%80%99s-wage-structure-on-the-greater-orlando-area/. Accessed 12 Dec 2019.

Packman, Hollie Muir, and Fred L. Casmir. 1999. Learning from the Euro Disney Experience. *International Communication Gazette* 61 (6): 473–489.

Paine, Crispin. 2019. *Gods and Rollercoasters: Religion in Theme Parks Worldwide.* London: Bloomsbury.

Quinnell, Kenneth. 2018. Worker Death Shines Spotlight on Disney's Poverty Policies. *AFL-CIO*, March 27. https://aflcio.org/2018/3/27/worker-death-shines-spotlight-disneys-poverty-policies. Accessed 28 Aug 2022.

Raz, Aviad. 1999. *Riding the Black Ship: Japan and Tokyo Disneyland*. Cambridge: Harvard University Press.

Ren, Hai. 1998. *Economies of Culture: Theme Parks, Museums and Capital Accumulation in China.* Diss., The University of Washington.

Ritzer, George. 2008. *The McDonaldization of Society*. Thousand Oaks: Pine Forge.

Rosa, Hartmut. 2013. *Social Acceleration: A New Theory of Modernity.* Trans. Jonathan Trejo-Mathys. New York: Columbia University Press.

Sainato, Michael. 2018. When Working at Disney World Means Being Trapped in Poverty. *Vice*, June 1. https://www.vice.com/en_us/article/xwma3q/when-working-in-disney-world-means-being-stuck-living-in-a-cheap-motel. Accessed 28 Aug 2022.

Schickel, Richard. 1968. *The Disney Version: The Life, Times, Art and Commerce of Walt Disney.* New York: Simon and Schuster.

Smythe, Dallas W. 1981. *Dependency Road: Communications, Capitalism, Consciousness, and Canada*. Norwood: Ablex.

Spinelli, Maria-Lydia. 1995. What's in a Theme Park? Exploring the Frontiers of Euro Disney. *Archive of European Integration*. http://aei.pitt.edu/7021/. Accessed 28 Aug 2022.

Stanley, Nick. 2002. Chinese Theme Parks and National Identity. In *Theme Park Landscapes: Antecedents and Variations*, ed. Terence Young and Robert Riley, 269–289. Washington, D.C.: Dumbarton Oaks Research Library and Collection.

Steinkrüger, Jan-Erik. 2013. *Thematisierte Welten: Über Darstellungspraxen in Zoologischen Gärten und Vergnügungsparks*. Bielefeld: Transcript.

Strodder, Chris. 2008. *The Disneyland Encyclopedia*. Salona Beach: Santa Monica.

The PortAventura Group. 2018. *PortAventura 2017 Corporate Social Responsibility Report*. https://www.investindustrial.com/dam/Investindustrial/Social-responsibility/Portoflio-Companies-CSR-Reports/PortAventura-2017-Corporate-Social-Responsibility-Report/PortAventura%202017%20Corporate%20Social%20Responsibility%20Report.pdf. Accessed 28 Aug 2022.

Thomas, Bob. 1994. *Walt Disney: An American Original*. New York: Hyperion.

Thompson, Frederic. 2015 [1908]. Amusing the Million. In *A Coney Island Reader: Through Dizzy Gates of Illusion*, ed. Louis Parascandola and John Parascandola, 103–108. New York: Columbia University Press.

Van Maanen, John. 1991. The Smile Factory: Working at Disneyland. In *Reframing Organizational Culture*, ed. Peter J. Frost, Larry F. Moore, Meryl Reis Louis, Craig C. Lundberg, and Joanne Martin, 58–77. Newbury Park: Sage.

Van Maanen, John, and Gideon Kunda. 1989. Real Feelings: Emotional Expressions and Organization Culture. *Research in Organizational Behavior* 11: 43–102.

Velasco, Paulina. 2017. Cinderella Is Homeless, Ariel "Can't Afford to Live on Land": Disney under Fire for Pay. *The Guardian*, July 17. https://www.theguardian.com/us-news/2017/jul/17/disneyland-low-wages-anaheim-orange-county-homelessness. Accessed 28 Aug 2022.

Walt Disney World. N.D. Walt Disney World Begins Construction of New Housing Complex for College Program. *Walt Disney World News*. https://aboutwaltdisneyworldresort.com/releases/walt-disney-world-begins-construction-of-new-housing-complex-for-college-program/. Accessed 12 Dec 2019.

Williams, Rebecca. 2020. *Theme Park Fandom: Spatial Transmedia, Materiality and Participatory Cultures*. Amsterdam: Amsterdam University Press.

Woodyard, Chris, and Don Lee. 1995. Disney to Extend Health Benefits to Gay Partners. *Los Angeles Times*, October 7. https://www.latimes.com/archives/la-xpm-1995-10-07-mn-54276-story.html. Accessed 28 Aug 2022.

Younger, David. 2016. *Theme Park Design & the Art of Themed Entertainment*. N.P.: Inklingwood.

Media and Mediality in Theme Parks

Abstract This chapter focuses on two different aspects of theme parks as media phenomena: their plurimediality, a characteristic they inherited from virtually all of their antecedents; and their relevance to contemporary chains of remediation or convergence culture, a direct result of recent structural shifts in the theme park industry away from individual operator-ownership to the integration of an increasing number of parks worldwide into multinational transmedia conglomerates. On the one hand—just like the pleasure gardens, amusement parks, world's fairs, etc. that came before them—theme parks combine multiple media and art forms that have conventionally and historically been considered distinct, including architecture, painting, landscaping, music, language, or performance, in order to create multisensory environments. Theme parks have thus been referred to as "hybrid," "composite," or "metamedia." On the other hand, with respect to their choice of themes, the parks have increasingly focused on transmedia franchises, particularly those from the movie industry (e.g. *Star Wars*, *Harry Potter*), and/or have themselves spawned franchises (e.g. *Pirates of the Caribbean* or *Boonie Bears*). This development has occurred parallel to the rise of self-reflective or autotheming, i.e. the creation of more and more themed spaces in which the theme park reflects on itself and its history as a medium.

1 Introduction

This chapter locates the theme park within multiple inter-, pluri-, and transmedial networks, both diachronically and synchronically. Following Bolter and Grusin, who in *Remediation* (1999) stress the "hypermediacy of [...] sophisticated theme parks like Disneyland and Disney World" (170), it discusses (1) the origins of the theme park within a specific media ecology, (2) the plurimediality of the theme park, (3)

This work is contributed by Astrid Böger, Florian Freitag, Sabrina Mittermeier, Salvador Anton Clavé, Filippo Carlà-Uhink, Thibaut Clément, Scott A. Lukas, Céline Molter, Crispin Paine, Ariane Schwarz, Jean-François Staszak, Jan-Erik Steinkrüger, Torsten Widmann. The corresponding author is Astrid Böger, Institut für Anglistik und Amerikanistik, Universität Hamburg, Hamburg, Germany.

F. Freitag et al., *Key Concepts in Theme Park Studies*,
https://doi.org/10.1007/978-3-031-11132-7_10

remediations within the theme park, and (4) medial convergences within the theme park industry.

2 Media Ecologies and the Development of the Theme Park

Theme parks are rooted in a broad variety of two- and three-dimensional immersive media, including gardens, zoos, panoramas, world's fairs, and exhibitions (see ANTECEDENTS). Yet it is perhaps no coincidence that in its contemporary form, the theme park emerged not only when it did—namely, during the mid-1950s with their quest for suburban landscapes of "homogeneity, containment, and predictability" (Avila 2004, 6) on the one hand and their appreciation for movement, (auto)mobility and "togetherness" (Marling 1991, 174–177; Schwartz 2020) on the other—but also where it did, namely, in Southern California and in close proximity to Hollywood, then the center of American movie and television production. Today, the top three global theme park agglomerations—Central Florida, the Greater Pearl River Delta (China), and Southern California—are the regions where visitors find a high concentration of major theme parks and other place-based attractions. This is also where numerous services and product industries that supply the theme park industry have clustered, companies that provide solutions to issues of planning and design, site development and management, rides, shows and productions, etc., all tailor-made for the theme park medium (see INDUSTRY). Such a sophisticated supply chain was not (yet) available in 1950s Southern California—or anywhere else, for that matter. It was the movie and television industry that provided the emerging theme park industry with much of the talent, expertise, and infrastructure necessary to conceptualize, design, and even market the first parks.

No wonder, then, that visiting Disneyland felt (and perhaps still feels), as Thomas Hine has pointed out, "like walking into a movie" (1986, 151): as numerous critics have noted (see e.g. Borrie 1999, 74; Finch 2011, 456), the designers who worked on the park were all recruited by Disney from the staff of his own movie studio. They "did not think as architects; they were filmmakers. And what they designed was not a group of buildings or a park but an experience. They thought in very literally cinematic terms as they designed the place as a movie that could be walked into" (Hine 1986, 151). For instance, Herbert Ryman and Peter Ellenshaw had worked as an illustrator and a matte/special effects designer, respectively, for various movie studios before joining Disney and the first team of "Imagineers" in charge of the design of the park. Long before the 1955 opening of Disneyland, Ryman and Ellenshaw created large overhead views of the park that not only determined its layout, its themes, and its overall mood, but that also played key roles in marketing the project to investors and the general public (see PARATEXTS AND RECEPTION). Accustomed to working on movies, Ryman and Ellenshaw would include techniques borrowed from the construction of movie sets (e.g. "forced perspective") and the pacing of cinematic storytelling into the design of these maps and, in the case of Ryman, into his later work on the park itself (see Neuman 2008, 91–92; Freitag 2021a, 155–160; see also the discussion of "Main Street, U.S.A." below).

The designers' background in cinema and television also had an impact on their conception of the audience. The first visitors to Disneyland may not have been entirely sure whether the park constituted an adaptation of Disney's eponymous weekly TV show or vice versa—or whether the two were "all part of the same," as Walt Disney himself noted in the first episode of *Disneyland* (aired on 27 October 1954; see Florey and Jackson 04:09–04:10). After all, even the show's individual segments bore the exact same names as the park's various sections (see PARATEXTS AND RECEPTION). And perhaps it didn't matter either, as park visitors were expected to behave exactly like television viewers: sit back, relax, perhaps change the channel/explore a different land, and enjoy themselves (see Marling 1991, 205; see VISITORS). Disney itself, as well as early commentators (e.g. King 1981), regularly highlighted the overall "newness" of the theme park medium (see ANTECEDENTS), but it is important to also stress the continuities with established media in general and film and television in particular.

Creating and operating a theme park involves much more than deciding on a layout and designing themed landscapes, of course (see THEMING). For a long time, Disney and other modern park owners and operators had to rely on outside suppliers and contractors that had little experience in either the movie/television industry or in theme parks or similar leisure projects. For instance, prior to being selected as the prime contractor for the construction of Disneyland, Los Angeles-based McNeil Construction Co. had been principally involved in housing and hotel projects, corporate buildings, and industrial plants. And as late as the early 1990s, the construction crews contracted to work on "Mickey's Toontown" (also at Disneyland) were supposedly unable to understand the nature of the project: "When they installed a door, they would ask, 'do you want to hang it straight?' And [the designers'] response would be, 'Of course not. Crooked is more toony'" (The Imagineers 1996, 144). The humorous self-stylization of the park designers in this quote as quirky outsiders within their profession masks the fact that today, an entire range of companies readily offer their services and expertise in theme park planning, construction, and operation. However, it is noteworthy that those individuals responsible for the structure, look, and feel of the first theme parks all came from the movie and television industry and left their cinematic imprint on the park's landscapes. Nonetheless, while from a historical perspective the theme park in its contemporary form is rooted in the specific media ecology of 1950s Hollywood, the parks have also—and from their very beginnings—constituted plurimedial environments.

3 Plurimediality

Theme parks seek to immerse visitors into multisensory environments by combining kinetics with a variety of different art forms, semiotic systems, or means of communication, including architecture, landscaping, painting, sculpture, music, theater, and (analog and digital) film. From a medial perspective, the theme park can thus be described as a "hybrid," "composite," or "meta-medium" (see Rajewsky 2002, 203; Wolf 2007, 253; and Geppert 2010, 3, respectively)—that is, as a medium in its own right that combines and fuses various other media that have been conventionally and historically regarded as distinct (see Freitag 2017, 706). In general, the "multimediality" (Wolf 2007, 253–255; Rippl 2015, 12) or "plurimediality" (Pfister 1977, 24–29) of the theme park medium has received surprisingly scant critical attention. Rather, scholars have focused on the forms and functions of individual media within the theme park's complex medial or semiotic mix—notably those of architecture (see e.g. Marling 1991), music (see e.g. Carson 2004; Camp 2017), film (see e.g. Freitag 2017), performance (see e.g. Kokai and Robson 2019), and language (see e.g. Freitag 2021b), while the contributions of other parts have been rather neglected. Nevertheless, it is precisely the combination, or fusion, of all these individual media that characterizes the theme park's plurimedial status.

A good example of plurimediality in practice is Disneyland's "Main Street, U.S.A.," which belongs to the first themed park environments, having originated at Disneyland in Anaheim in 1955, and which serves as the main entrance to all the so-called "castle parks" operated by the Walt Disney Company around the world. The architecture recalls, as Francaviglia has shown in detail (1996, 145–163), the Victorian, Second Empire, and Queen Anne commercial styles of the main streets of late nineteenth and early twentieth century American small towns. The general arrangement of the buildings (around a rectangular square and alongside a central corridor) as well as the landscaping (the subdivision of the ground into "walkways" and "street"; street furniture; well-kept flower beds) further suggest public urbanity, while smaller decorations (historical U.S. flags; shop and other signs in English) help to establish the Americanness of "Main Street, U.S.A." At the same time, the flags (specifically the number of stars on them), the fonts used on the signs as well as their highly ornate visual design, but also the alleged years of construction painted on building facades contribute to establishing the temporal setting of the space. Painting (murals and illustrated attraction posters) and sculptures (a so-called "Cigar Store Indian" in front of a no longer existing tobacco shop) complete the visual plurimediality of "Main Street, U.S.A."

Music, in the form of the lively ragtime soundtrack that pervades all the exterior and interior spaces (Carson 2004, 229), not only adds to the sense of time, but also gives a certain rhythm to the place. The musical soundtrack is supplemented by recorded language—supposedly the dialogues between the various "inhabitants" of "Main Street, U.S.A.," including a less than skillful dentist and his unfortunate patients—as well as live sounds, notably the quirky sounds from the Main Street vehicles (the honking sounds of old-fashioned horns and the clip-clop of horseshoes). Performance—itself a hybrid, composite, or meta-medium that may comprise (stage) architecture, costumes, spoken language or singing, music, and other media—features in various forms on "Main Street, U.S.A." In addition to regular live performances by marching bands, choirs, and other musical ensembles as well as the appearances of Disney characters (which all rely on a minimum of props), there are the daily and nightly parades with their dancers and elaborate floats (which have occasionally also included filmic projections and other special effects) as well as the nighttime spectaculars, which combine music, spoken words, fireworks, projections, lighting, and other effects. Another frequently overlooked type of performance is the performative labor of the employees in the shops and restaurants of "Main Street, U.S.A." (as well as elsewhere in the park), which involves period-appropriate costumes and often tightly scripted lines of dialogue and which is complemented by the performative "fun work" or "experience work" of park visitors (see LABOR).

Film, another medium that forms an integral part of "Main Street, U.S.A."'s plurimediality while being of a multimedial nature itself, may at first sight be restricted to the Main Street Cinema, which has been continuously showing *Steamboat Willie* (1928) along with other classic Disney shorts since the 1955 opening of the park. Yet cinematic techniques have also influenced the layout and design of "Main Street, U.S.A.," as well as the park's entrance area, in a much more fundamental manner. Critics have long argued that Disneyland as a whole translates the temporal logic of film (or TV) into the spatial logic of the theme park (see e.g. King 1981, 127; Fjellman 1992, 257; Marling 1991, 205), but it is this area in particular that references cinematic conventions and practices: the first thing Disneyland visitors encounter when they arrive from the parking lot are rows of ticket booths, which resemble the small, isolated box offices that are considered an iconographic feature of early cinema architecture. Like the vestibules surrounding these box offices, the area around the ticket booths serves as a display area for posters that advertise individual rides and shows within the park. Having passed through the turnstiles, visitors enter one of the two tunnels underneath the tracks of the Disneyland Railroad to enter the park, with the brief dimming of daylight recalling the dimming of the lights at movie theaters. Having passed the tunnels and entering "Main Street, U.S.A.," visitors are confronted with a view of Sleeping Beauty Castle, framed to their left and right by the buildings of "Main Street, U.S.A." What they see is, in fact, a closed composition that resembles a classic establishing shot of a movie. As visitors walk down "Main Street, U.S.A." towards the castle, the buildings slowly move out of their field of vision. Just as in the establishing shot of, for instance, *Snow White and the Seven Dwarfs* (1937), the camera zooms in on the castle and the trees that frame the image slowly glide out of frame. Hence, using a "sequentializing concept of architectural experience" (Lonsway 2009, 44), the park entrance and "Main Street, U.S.A." formally imitate the process of entering a movie theater and the beginning of a movie, in what will be referred to below as an intermedial system reference.

Finally, at the end of "Main Street, U.S.A.," visitors can find a board that uses digital screens to inform them of the current waiting times for individual attractions, as well as the park's entertainment schedule; a somewhat more elaborate version of this board can be found in roughly the same location at Disneyland Paris in France (see TIME). Indeed, the concept of "Main Street, U.S.A.," along with its plurimediality, has been adopted, and to some extent adapted to, almost all Disney-licensed theme parks around the world, as well as to other theme parks (see, for instance, "Berlin" at Phantasialand in Brühl, Germany, "Via Antiqua" at Parc Astérix in Plailly, France, or the central avenue at Fantawild Dreamland in Xiamen, China). In all of these "main streets"—as well as in other themed areas in theme parks worldwide (see e.g. Gottwald and Turner-Rahman 2019)—one finds similar or even more complex plurimedial arrangements, often with one specific medium that occupies a somewhat dominant position (as in the case of film and Disneyland's "Main Street, U.S.A."). Indeed, some parks, while drawing from and fusing a large variety of media in order to create a plurimedial, multisensory experience, nonetheless stress one specific art form—or an individual artifact realized in one specific medium—within individual rides, restaurants, shops, or shows. For aesthetic, commercial, and/or historical reasons, theme parks reproduce selected elements from other medial artifacts, fully adapt other artifacts, or even systematically refer to and draw on other media in various forms of remediation.

4 Remediations

The theme park's plurimedial or medially hybrid nature allows it to selectively reproduce specific elements of artifacts realized in a different medium. These partial reproductions may be part of a fully-fledged adaptation of the artifact in question (see below) but may also take the form of what has been termed in intermediality theory as an implicit "individual reference" (Wolf 2007, 254–255): reproducing a part of or quoting from a medial artifact without explicitly mentioning the source, the theme park points to this artifact in order to thematically exploit visitors' associations with it. Movies constitute particularly popular and interesting sources of intermedial references, as the following two examples will illustrate.

To complement its then-new water coaster Atlantica Supersplash, Europa-Park (Rust, Germany) opened a nautically themed attraction shop in 2006, variously referred to as Atlantica Souvenirs or Atlantica Shop. While the coaster features a "Portuguese conquistador" theme, the shop has also offered generic pirate-themed merchandise, no doubt in an attempt to cash in on the pirate craze launched by the successful *Pirates of the Caribbean* film series (2003–). To further strengthen the pirate subtheme, the designers added the theme of *The Curse of the Black Pearl* (2003), the first film of the series, to the shop's soundtrack (excerpts from the movie's soundtrack had apparently also been used in the coaster's waiting area but were later replaced). Somewhat paradoxically, Pieces of Eight, the attraction store of the Disneyland Pirates of the Caribbean dark ride that originally inspired the film series (see Schweizer and Pearce 2016), uses the ride's original theme as well as other pirate-themed songs rather than the movie's soundtrack as area music. Europa-Park's own pirate-themed dark ride, Piraten in Batavia (1987; a lightly rethemed copy of Pirates of the Caribbean), never featured a merchandise location.

Phantom Manor at Parc Disneyland (Paris) similarly uses a partial reproduction of a specific movie to communicate its theme to visitors. Unlike its counterparts in Anaheim, Orlando, Tokyo, and Hong Kong, the Paris version of the Haunted Mansion ride is not housed in a seemingly well-maintained, otherwise inconspicuous building, but in a dilapidated mansion that strongly resembles the Bates mansion from Alfred Hitchcock's *Psycho* (1960). Except for the shared horror theme, the ride itself has no connections whatsoever with the movie, but the designers apparently felt that the name change from Haunted Mansion to Phantom Manor (with "phantom"/ "fantôme"/ "Phantom" meaning "ghost" in English, French, and German) was not enough to communicate the ride's potentially frightening nature to a multilingual European audience—hence the intermedial reference to *Psycho* and its iconic set (see Lainsbury 2000, 61).

Partial reproductions—remediations of specific medial artifacts into the medium of the theme park—are also frequently employed in theme park adaptations. In *Fairground Attractions* (2012), Deborah Philips identifies a number of genres that have shown a "stubborn persistence" (6) across theme parks worldwide, including the fairy tale, the Western, and science fiction. Philips traces their successive adaptation from folk history and oral tradition to (illustrated) chapbooks, dioramas and panoramas, theater plays, cinema, and finally theme parks (2012, 4; see Anton Clavé 2007, 35; Steinecke 2009, 8; and Younger 2016, 50–55 for largely congruent lists of popular genres or themes in theme parks; see THEMING). In addition to adaptations of such "theme archetypes" (Younger 2016, 50), theme parks also feature adaptations of individual medial artifacts such as specific movies, novels, etc., which may or may not belong to some overarching genre—and they have done so from their very beginnings, as will be shown below using the case of Disneyland's "Fantasyland" dark rides (see below).

Generally, ride adaptations of movies (for example) have relied on and partially reproduced the visuals and soundtracks of their sources, whereas plot lines have been subject to frequent and liberal alterations. A particularly interesting case in point is Splash Mountain, a log flume ride at Disneyland that was originally based on (portions of) Disney's 1946 live-action/animated musical film *Song of the South*, itself an adaptation of Joel Chandler Harris's 1886 story collection *Uncle Remus: His Songs and His Sayings*. In *Disney's Most Notorious Film* (2012), Jason Sperb describes Splash Mountain as an example of "strategic remediation," a process in which media companies only use carefully selected parts of a given text for further distribution across media platforms, with the result that "intellectual property diffuses across the dispersed texts of media convergence culture" (21). Indeed, Splash Mountain's designers not only completely dropped the live-action frame narrative of the movie (which had been greatly expanded from the literary original, where it consisted of just one short paragraph), they also changed what had come to be regarded as the most racist element of the animated core segments, namely the "Tar Baby" storyline (see Mauro 1997; Sperb 2012, 163–164). Thus, purging the movie plot of its most controversial elements, Disney sought to keep *Song of the South* palatable—and profitable —in the post-Civil Rights movement era (see Bringardner 2019, 116). Yet, in June 2020, in what seems to be a direct reaction to the Black Lives Matter demonstrations and ongoing discussions about structural racism in the U.S. and abroad, Disney announced it would re-theme the ride to the animated, Louisiana-set *The Princess and the Frog* (2009), yet another movie adapted from literature.

Until the turn of the twenty-first century, theme park adaptations, whether based on movies or other medial artifacts (see Younger 2016, 60–61 for an overview of sources), most often took the form of individual elements (dark and other rides, restaurants, shops, shows, etc.). Early land- or park-sized adaptations, such as "Camp Snoopy" at Knott's Berry Farm (California; 1983), based on Charles M. Schulz's *Peanuts* comic strips (1950–2000), or Parc Astérix (France; 1989), based on René Goscinny and Albert Uderzo's *Astérix* comic book series (1959–), constituted the exceptions rather than the rule. Since the opening of Universal's Islands of Adventure (Florida) in 1999, which debuted with no fewer than three IP-based lands ("Jurassic Park," "Marvel Super Hero Island," and "Seuss Landing"), however, entire lands based on individual works and franchises have become more and more frequent (see Younger 2016, 80 for a comprehensive list up to 2016). While allowing visitors to explore the world of a specific medial artifact in much more depth than e.g. a single ride (as well as allowing companies to profit from synergy effects; see below), franchise-based lands potentially pose problems of expandability and longevity (see Younger 2016, 80–81) and thus tend to be reserved only for extremely popular IPs.

As Bolter and Grusin have pointed out, intermedial references and adaptations are not the only ways in which theme parks "remediate other media" (1999, 171). Like any other medium, the theme park also pays homage to and rivals earlier and newer media by "appropriating and refashioning [their] representational practices" (Bolter 2005, 14; see also Bolter and Grusin 1999, 19) through what Werner Wolf may call "system references" (2007, 254). For example, in addition to directly incorporating filmic (projection) technology into its plurimedial mix (see above), the theme park also evokes or imitates the conventions and practices of cinema through other means. Lukas has pointed out how ride pacing evokes the distinctly filmic technique of jump cuts: "the newest dark rides use the unexpected—a quick turn of the ride and a sudden jolt of a monster, scene or other sight—to heighten the sensory experience of the ride. Similar to jump cuts in film, ride pacing creates constant visual and kinetic situations" (2008, 26). Building on Lukas, Freitag (2016b) has used the examples of Pinocchio's Daring Journey at Disneyland and Journey to the Center of the Earth at Tokyo DisneySea (Tokyo) to argue that dark rides imitate not only jump cuts, but cuts in general. In the former case, sharp turns on the ride's tracks and hydraulic doors in-between show scenes prevent riders from seeing the next scene until the very last moment, thus employing visual barriers that imitate filmic cuts. In contrast, Journey to the Center of the Earth imitates cuts through the use of so-called Enhanced Motion Vehicles—ride vehicles that, among other characteristics, slow down and accelerate at specific points during the ride. When the ride vehicles suddenly speed up after the pivotal encounter with the lava monster, for example, the ride "cuts" from a close-up of the monster to a frantic point-of-view movie ride.

First-generation shooter rides such as Buzz Lightyear's Space Ranger Spin (Magic Kingdom, Florida; 1998) or El Laberinto del Minotauro (Terra Mítica, Spain; 2000), in which riders use guns or similar devices to shoot at targets hidden in the sets to gain points, have in turn imitated the logic of video games. Indeed, in contrast to later incarnations (e.g. Toy Story Midway Mania! at Disney's Hollywood Studios, Florida, 2008; or Maus au Chocolat at Phantasialand, 2011), which replaced physical sets with digital ones and thus actually incorporated the screen interface of video games, these early shooter rides merely evoked or referenced the video game medium by translating its digitally created on-screen worlds into the three-dimensional sets of "classic" dark rides. The physical movement of the ride vehicles imitates the virtual movement of the player through the game world, whereas in the case of shooter rides with digital targets there is, as David Younger has pointed out, "no intrinsic reason why the ride vehicles need to move between screens (rather than the screens cycling through the games)" (2016, 422)—except to give riders a short break and to add a kinetic element to the ride.

Intermedial system references also occur on a land- and park-wide scale, with the entrance to Disneyland and "Main Street, U.S.A." likely providing the earliest example (see above). Examples of other parks that employ intermedial system references on a larger, even park-wide scale include the Legoland parks in Europe, North America, and Asia with their "cubist" aesthetics and references to the famous toy building blocks, Dollywood (Tennessee) with its references to music, or Efteling (Kaatsheuvel, the Netherlands) with its references to children's books illustrations (particularly those by Anton Pieck, the park's first designer). As in the case of Efteling, several other parks have started to use system references to explore their own genealogy. Indeed, the turn of the twenty-first century not only saw the rise of IP-based lands (see above), but also of "autothemed" lands (see Freitag 2016a), such as "The Boardwalk" at Knott's Berry Farm (California) or "Paradise Pier" at Disney California Adventure (California), which are themed to amusement parks and piers, trolley parks, fun fairs, and roadside attractions, and thus to some of the theme park's direct antecedents (see ANTECEDENTS and THEMING). One of the main reasons for the popularity of this type of autotheming may indeed be the fact that it provides designers with a comparatively cost-efficient frame to bring together a number of generic, off-the-shelf carnival rides (see Younger 2016, 55). However, as illustrated by lands such as "Lost Kennywood" at Kennywood (Pennsylvania; see Samuelson and Yegoiants 2001, 138) and parks such as Luna Park on Coney Island (New York), theme park remediations of early amusement and trolley parks also offer fascinating, if somewhat nostalgic glimpses into the history of both the very places where they are located and the medium in general.

Remediation practices are employed by all parks, whether they are run independently or form part of large entertainment companies such as Disney or Universal. The latter in particular, however, frequently use theme parks as but one of several medial platforms—alongside movies, TV shows, video games, books, websites, consumer products, etc.—in order to distribute their content, a practice known as transmedia storytelling.

5 Transmediality and Convergence Culture

The Walt Disney Company pioneered the theme park in its contemporary form, and with it perfected the concept of synergy, the economic practice at the heart of transmedia storytelling. As Waysdorf and Reijnders argue

> What the park offers is the chance to physically interact with characters and narratives that are well-loved, made all the more powerful for the long history of Disney media, its existing cultural meanings and the interaction of different Disney texts with each other. The Disney model of design and cross-marketing is the archetype on which contemporary theme parks are based. (2018, 176)

Sleeping Beauty Castle, the original Disneyland's park icon, is the bona fide example of this practice. When the park opened in 1955, Disney's animated version of the classic fairy tale had not yet been released—and would not be for another four years. The castle itself also never actually resembled the design of the finalized film, yet its mere name and presence would act as a powerful marketing tool of what was to come, prominently appearing in the opening sequence of the *Disneyland* TV show and, from 1995 onwards, serving as the logo for the entire company. In 1957, Disney also installed a walk-through exhibit in the castle, a further marketing effort (Mittermeier 2021, 45).

Synergy was generally strongest in "Fantasyland" on opening day, with its dark rides directly based on Disney's other popular animated films: Snow White's Adventures, Mr. Toad's Wild Ride, and Peter Pan Flight constituted adaptations of Disney's animated movies *Snow White and the Seven Dwarfs* (1937), *The Adventures of Ichabod and Mr. Toad* (1949), and *Peter Pan* (1953), respectively. In contrast to the movies' omniscient narrative perspective, however, these rides cast visitors in the roles of the main protagonists to re-tell the movie's plots in either their original or slightly altered versions (see Rahn 2000; Strodder 2012, 329). More importantly, although the movies themselves constituted adaptations (of the Grimm fairy tale, Kenneth Grahame's 1908 children's novel *The Wind in the Willows*, and J.M. Barrie's 1904 play *Peter Pan; Or, the Boy Who Wouldn't Grow Up*, respectively), the rides were based directly on the films rather than their original literary sources, as indicated by the fact that they borrowed from and partially reproduced the movies' visuals for the sets and their soundtracks for the audio.

Subsequent versions of these rides (as well as later adaptations of other artifacts) generally dropped the first-person narrative perspective—as designer Ken Anderson has noted, "nobody got it. Nobody actually figured out that they were Snow White" (qtd. in Rahn 2000, 22)—but continued to rely on the film adaptations rather than the literary originals. This even applies to Pooh's Hunny Hunt (2000), a trackless dark ride at Tokyo Disneyland's "Fantasyland" whose façade depicts a giant, open-faced illustrated book that invites visitors to step inside (in fact, through a hole in the "page," visitors can spot people in the waiting line walking through the "book"). Decorated with huge printed and illustrated book pages, the waiting line continues the metaphor of visitors being drawn into the book. Yet what may appear to be a nod to the literary origins of the character of Winnie-the-Pooh, which first appeared in the eponymous 1926 story collection by author A.A. Milne and illustrator E.H. Shepard, in reality constitutes a faithful adaptation of the 1977 film *The Many Adventures of Winnie the Pooh*, whose live-action opening sequence depicts a nursery with a Winnie-the-Pooh-titled book that suddenly opens itself and whose illustrations magically come to animated life (see Fig. 1).

Fig. 1 A.A. Milne's books merely serve as decorative items at the Pooh Corner shop at Tokyo Disneyland (Japan). *Photograph* Florian Freitag

Generally, then, most of these dark ride adaptations fail to provide riders with a narrative that is understandable if one does not know the material they are based on, and solely rely on the visual imagery of the "Disney version," meaning that even knowledge of the original fairy tales of Snow White or Pinocchio would likely not be of much help. As Aronstein has argued, these attractions "are effective only because the [...] narratives on which they are based are omnipresent; without the intertextual interpretive context provided by the films, these rides would be meaningless—a series of disconnected images" (2012, 67). This at first may seem like a risky practice, as it appears to exclude those not familiar with Disney's products, but in many ways this is what the company capitalizes on, as it increases the chance of park visitors seeking out their movies. A recent example of this logic is the incorporation of a Captain Marvel character meet and greet on the Disney Cruise Line's Marvel Day at Sea, which began two months before the release of the eponymous film in March 2019. Banking on the popularity of the Marvel Cinematic Universe, the character meet proved a big draw for guests and likely sparked interest in the film, even for those less invested in the franchise.

Since 1955, synergy has extended to all of Disneyland and other Disney and non-Disney theme parks alike and has thus become a defining factor of the medium. In fact, Disney is acutely aware that building a theme park is still a surefire way to penetrate a new market, as exemplified by Shanghai Disneyland, which opened the almost untapped potential of the massively growing Chinese middle-class to their conglomerate (Mittermeier 2021, 171). Since the People's Republic of China only allows twenty international films to be released per year, and many works over the years have thus not reached a Chinese audience, the theme park makes it possible for Disney to introduce a number of its IPs all at once, ranging from 1930s Mickey Mouse cartoons all the way through to Iron Man.

Meanwhile, Chinese theme park companies are emulating Disney's strategies. For instance, Fantawild (Fangte), one of the most successful and aggressively expanding theme park operators in the People's Republic of China alongside OCT and Chimelong, has developed its own animation and consumer products departments, with the *Boonie Bears* and *Chicken Stew* series among their most popular brands. Fantawild's successful expansion into animation will allow the company to introduce exclusive, localized content into their theme parks via synergy: in the future, according to the 2017 *Global Attractions Attendance Report*, Fantawild will seek to "develop new parks that feature their own IP and characters from movies, television and cartoons produced by the Fantawild Animation company and well known in the domestic Chinese markets" (Rubin 2018, 41). Somewhat paradoxically, then, Fantawild has adopted a strategy practiced by Western companies such as Disney precisely to introduce more localized, Chinese content into their parks. European parks have followed suit, if on a somewhat more modest scale. Both Europa-Park and Efteling, for instance, have started (co-)producing their own movies—*Happy Family* (2017) and *Fabula* (2019), respectively—which are prominently featured in the parks and which, in the case of the former, also made it to movie theaters. More and more, even parks that are not part of a transmedial entertainment corporation are using transmedial storytelling practices.

The aforementioned Marvel Cinematic Universe's gradual incorporation into Disney's theme parks provides another example of how transmedia storytelling is becoming more and more intricate and international. Walt Disney Imagineering has begun to tie in ride stories across theme parks around the world, effectively extending the serialized storytelling of the franchise to its attractions. With Marvel-based lands, the so-called "Avengers Campuses," opening in Disney California Adventure, Hong Kong Disneyland, Epcot, Tokyo Disneyland, and Walt Disney Studios in Paris, the company is planning to weave ties between the rides. This not only provides an incentive to visit all their theme parks worldwide, but also deepens immersion for those fans who pay attention to detail—a strategy previously tested with the original theme park backstory of S.E.A. (The Society of Explorers and Adventurers) that connects such rides as Tokyo DisneySea's version of the Tower of Terror and Hong Kong Disneyland's Mystic Manor (Mittermeier 2021, 152). As Månsson has argued, "one way for producers to use media convergence is intertextuality: media products' interrelationship through either hidden or open reference" (2011, 1637); Disney uses this practice to engage their fans even more in their story worlds. Such intertextuality between attractions and across parks strengthens narrative coherence and thus immersion, as it relies on visitors' active participation to decipher these cross-references (Mittermeier 2020, 32; see THEMING). This participatory element is integral to the theme park experience and to what Henry Jenkins has famously defined as "convergence culture" (2006, 3). Theme parks, as previously argued, thus emerge as prime examples of this phenomenon (see also Williams 2020).

"The Wizarding World of Harry Potter" is another classic example of transmedia storytelling in theme parks, even though it benefits differently from synergy, since Universal/Comcast merely holds theme park licensing rights rather than the ownership of its movies (which are produced by Warner Brothers Studios, while the book rights lie with different publishers depending on the country). The theme park lands at Universal Resort Orlando, Hollywood, and Osaka, Japan, are direct adaptations/remediations of the movies rather than the books, copying the art design down to the smallest details. This applies to rides and the theming of show buildings, restaurants, and shops, as well as to merchandise such as the popular Chocolate Frogs. Universal has also incorporated an interactive wand feature that makes it possible for visitors to de facto cosplay as wizards and witches, complete with Ollivander's where the wand chooses you, allowing for even more immersion within the transmedia story world (Mittermeier 2020, 31; see Fig. 2; see ATTRACTIONS).

Licensing is a common practice among both smaller IP rights holders as well as independently run theme parks as it allows them to benefit from synergy, despite not being directly involved in the theme park business or being integrated into a larger transmedia entertainment company such as Disney, respectively. In 1990, for instance, family-owned Europa-Park obtained the theme park rights for Vicky the

Fig. 2 If the wand has not chosen you, you may still choose a wand at Ollivander's at Universal Studios Hollywood (California). *Photograph* Salvador Anton Clavé

Viking (a popular 1970s animated children's series known in Germany as *Wickie und die starken Männer*) and transformed a previously unthemed playground into "Wickieland." Twenty years later, the characters from two other 1970s children's programs, *Maya the Bee* and *Heidi*, started to appear in the shops of independently run Terra Mítica. The temporary nature of such licensing agreements may prevent parks from investing too heavily in these IPs, however. Indeed, Europa-Park's "Wickieland" eventually received a more generic Viking theme before switching to an Irish theme, while Terra Mítica started to rely on its own mascots again, and Wickie, Maya, and Heidi moved to the parks of the Plopsa group (e.g. Holiday Park in Haßloch, Germany, and Plopsaland in De Panne, Belgium), whose parent company Studio 100 bought the rights to the characters in 2008 (see THEMING).

6 Conclusion

As the previous two sections have shown, the theme park in its current form—whether part of multinational transmedia conglomerates or not—has been firmly integrated into chains of remediations and transmedia story worlds. And yet the theme park is more than "simply another depiction" (Waysdorf and Reijnders 2018, 180) in these transmedia experiences. As Victoria Godwin and Rebecca Williams have recently argued, theme parks occupy an exceptional position within transmedia storytelling due to their materiality and spatiality. Godwin, for instance, notes that as "material interfaces simultaneously engaging multiple senses to immerse visitors in a variety of story worlds," theme parks "allow *haptic dominance* and the discovery of minutiae that could pass unnoticed or unrecorded onscreen" (2017, para 1.10 and 3.6; emphasis original). Williams, in turn, refers to theme parks as "'located transmedia'"—"a form of spatial transmedia experience that is resolutely rooted in a specific place" and that offers a "form of embodiment" to even animated and virtual characters (2020, 51, 139). In short, theme parks offer a "home" to (trans)media story worlds and characters—sometimes quite literally so, as in the case of "Mickey's Toontown" at Disneyland in California or Moominworld in Naantali (Finland) where visitors can not only see, but also touch and enter the homes of Mickey Mouse and his friends or the Moomin characters, and even interact with them. This is ultimately due to theme parks' plurimediality, which also includes three-dimensional and embodied media such as architecture, landscaping, and performances.

At the same time, theme parks occupy an exceptional position within transmedia story worlds as visitors never experience the theme park medium in a direct, unmediated manner. Before, after, and even during the visit, theme park paratexts such as advertisements, maps, apps, postcards, or coffee table books and other souvenirs intervene between visitors and the theme park landscape, providing the former with more or less tight "scripts" on how to use and experience the latter (see PARATEXTS AND RECEPTION). As especially destination parks such as the Walt Disney World Resort or Universal Resort Orlando rely more and more on advance reservation systems for rides (e.g. Disney's FastPass) or restaurants (e.g. the Disney Dining

Plan), visitor interaction with these paratexts becomes increasingly mandatory even before the visit—not just for fans (see e.g. Williams 2020, 67–99), but also for regular visitors. Hence, the visual and auditory, but also the spatial and haptic experience of theme parks has always been previously mediated and continues to be mediated even during the act of consumption itself, to a much larger extent than in the case of movies or TV series (see Gray 2010). While the origins of theme parks in specific medial ecologies, their plurimediality or hybrid, composite, and meta-mediality, as well as their participation in remediation and transmediality have already received considerable attention, their own mediatedness through theme park paratexts—their paramediality, so to speak—still needs to be examined in greater detail.

References

Anton Clavé, Salvador. 2007. *The Global Theme Park Industry*. Wallingford: CABI.

Aronstein, Susan. 2012. Pilgrimage and Medieval Narrative Structures in Disney's Park. In *The Disney Middle Ages: A Fairy-Tale and Fantasy Past*, ed. Tison Pugh and Susan Aronstein, 57–76. New York: Palgrave Macmillan.

Avila, Eric. 2004. *Popular Culture in the Age of White Flight: Fear and Fantasy in Suburban Los Angeles*. Berkeley: University of California Press.

Bolter, Jay David. 2005. Transference and Transparency: Digital Technology and the Remediation of Cinema. *Intermédialités* 6: 13–26.

Bolter, Jay David, and Richard Grusin. 1999. *Remediation: Understanding New Media*. Cambridge: The MIT Press.

Borrie, William T. 1999. Disneyland and Disney World: Constructing the Environment, Designing the Visitor Experience. *Loisir et Société/Society and Leisure* 22 (1): 71–82.

Bringardner, Chase A. 2019. Disney-fying Dixie: Queering the 'Laughing Place' at Splash Mountain. In *Performance and the Disney Theme Park Experience*, ed. Jennifer A. Kokai and Tom Robson, 107–125. Cham: Palgrave Macmillan.

Camp, Gregory. 2017. Mickey Mouse Muzak: Shaping Experience Musically at Walt Disney World. *Journal of the Society for American Music* 11 (1): 53–69.

Carson, Charles. 2004. "Whole New Worlds": Music and the Disney Theme Park Experience. *Ethnomusicology Forum* 13 (2): 228–235.

Finch, Christopher. 2011. *The Art of Walt Disney from Mickey Mouse to the Magic Kingdoms*. New ed. New York: Harry N. Abrams.

Fjellman, Stephen M. 1992. *Vinyl Leaves: Walt Disney World and America*. Boulder: Westview.

Florey, Robert, and Wilfred Jackson, dir. 1954. The Disneyland Story. *Disneyland* 1 (1), October 27. Burbank: Walt Disney Productions.

Francaviglia, Richard V. 1996. *Main Street Revisited: Time, Space, and Image Building in Small-Town America*. Iowa City: University of Iowa Press.

Freitag, Florian. 2016a. Autotheming: Themed Spaces in Self-Dialogue. In *A Reader in Themed and Immersive Spaces*, ed. Scott A. Lukas, 141–149. Pittsburgh: ETC.

Freitag, Florian. 2016b. Movies, Rides, Immersion. In *A Reader in Themed and Immersive Spaces*, ed. Scott A. Lukas, 125–130. Pittsburgh: ETC.

Freitag, Florian. 2017. Like Walking into a Movie: Intermedial Relations between Disney Theme Parks and Movies. *The Journal of Popular Culture* 50 (4): 704–722.

Freitag, Florian. 2021a. *Popular New Orleans: The Crescent City in Periodicals, Theme Parks, and Opera, 1875–2015*. New York: Routledge.

Freitag, Florian. 2021b. This Way or That? Par ici ou par là? Language in the Theme Park. *Visions in Leisure and Business* 23 (1). https://doi.org/10.25035/visions.23.01.07. Accessed 28 Aug 2022.

Geppert, Alexander C.T. 2010. *Fleeting Cities: Imperial Expositions in Fin-de-Siècle Europe*. New York: Palgrave Macmillan.
Godwin, Victoria L. 2017. Theme Park as Interface to the Wizarding (Story) World of Harry Potter. *Transformative Works and Cultures* 25. https://doi.org/10.3983/twc.2017.01078. Accessed 28 Aug 2022.
Gottwald, Dave, and Greg Turner-Rahman. 2019. The End of Architecture: Theme Parks, Video Games, and the Built Environment in Cinematic Mode. *International Journal of the Constructed Environment* 10 (2): 41–60.
Gray, Jonathan. 2010. *Show Sold Separately: Promos, Spoilers, and Other Media Paratexts*. New York: New York University Press.
Hine, Thomas. 1986. *Populuxe*. New York: Alfred A. Knopf.
Imagineers, The. 1996. *Imagineering: A Behind the Dreams Look at Making the Magic Real*. New York: Hyperion.
Jenkins, Henry. 2006. *Convergence Culture: Where Old and New Media Collide*. New York: New York University Press.
King, Margaret J. 1981. Disneyland and Walt Disney World: Traditional Values in Futuristic Form. *The Journal of Popular Culture* 15 (1): 116–140.
Kokai, Jennifer A., and Tom Robson, eds. 2019. *Performance and the Disney Theme Park Experience: The Tourist as Actor*. Cham: Palgrave Macmillan.
Lainsbury, Andrew. 2000. *Once Upon an American Dream: The Story of Euro Disneyland*. Lawrence: University Press of Kansas.
Lonsway, Brian. 2009. *Making Leisure Work: Architecture and the Experience Economy*. New York: Routledge.
Lukas, Scott A. 2008. *Theme Park*. London: Reaktion.
Månsson, Maria. 2011. Mediatized Tourism. *Annals of Tourism Research* 38: 1634–1652.
Marling, Karal Ann. 1991. Disneyland, 1995: Just Take the Santa Ana Freeway to the American Dream. *American Art* 5 (1–2): 168–207.
Mauro, Jason Isaac. 1997. Disney's Splash Mountain: Death Anxiety, the Tar Baby, and Rituals of Violence. *Children's Literature Association Quarterly* 22 (3): 113–117.
Mittermeier, Sabrina. 2020. Theme Parks: Where Media and Tourism Converge. In *The Routledge Companion to Media and Tourism*, ed. Maria Månsson, Annæ Buchmann, Cecilia Cassinger, and Lena Eskilsson, 27–34. London: Routledge.
Mittermeier, Sabrina. 2021. *A Cultural History of the Disneyland Theme Parks: Middle Class Kingdoms*. Chicago: Intellect.
Neuman, Robert. 2008. Disneyland's Main Street, USA, and Its Sources in Hollywood, USA. *Journal of American Culture* 31 (1): 83–97.
Pfister, Manfred. 1977. *Das Drama: Theorie und Analyse*. Munich: Fink.
Philips, Deborah. 2012. *Fairground Attractions: A Genealogy of the Pleasure Ground*. London: Bloomsbury.
Rahn, Suzanne. 2000. Snow White's Dark Ride: Narrative Strategies at Disneyland. *Bookbird: A Journal of International Children's Literature* 38 (1): 19–24.
Rajewsky, Irina O. 2002. *Intermedialität*. Tübingen: Francke.
Rippl, Gabriele, ed. 2015. *Handbook of Intermediality: Literature—Image—Sound—Music*. Berlin: de Gruyter.
Rubin, Judith, ed. 2018. *2017 Theme Index and Museum Index: The Global Attractions Attendance Report*. Burbank: Themed Entertainment Association and AECOM.
Samuelson, Dale, and Wendy Yegoiants. 2001. *The American Amusement Park*. St. Paul: MBI.
Schwartz, Vanessa. 2020. *Jet Age Aesthetic: The Glamour of Media in Motion*. New Haven: Yale University Press.
Schweizer, Bobby, and Celia Pearce. 2016. Remediation on the High Seas: A Pirates of the Caribbean Odyssey. In *A Reader in Themed and Immersive Spaces*, ed. Scott A. Lukas, 95–106. Pittsburgh: ETC.

Sperb, Jason. 2012. *Disney's Most Notorious Film: Race, Convergence, and the Hidden Histories of Song of the South.* Austin: University of Texas Press.

Steinecke, Albrecht. 2009. *Themenwelten im Tourismus: Marktstrukturen—Marketing—Management—Trends.* Munich: Oldenbourg.

Strodder, Chris. 2012. *The Disneyland Encyclopedia*, 2nd ed. Salona Beach: Santa Monica.

Waysdorf, Abby, and Stijn Reijnders. 2018. Immersion, Authenticity and the Theme Park as Social Space: Experiencing the Wizarding World of Harry Potter. *International Journal of Cultural Studies* 21 (2): 173–188.

Williams, Rebecca. 2020. *Theme Park Fandom: Spatial Transmedia, Materiality and Participatory Cultures.* Amsterdam: University of Amsterdam Press.

Wolf, Werner. 2007. Intermediality. In *Routledge Encyclopedia of Narrative Theory*, ed. David Herman, Manfred Jahn, and Marie-Laure Ryan, 252–256. London: Routledge.

Younger, David. 2016. *Theme Park Design & the Art of Themed Entertainment*. N.P.: Inklingwood.

Methods: Facing the Challenges of Studying Theme Parks

Abstract This chapter focuses on the variety of methods used by scholars and the challenges they face when researching, studying, and analyzing particular theme parks or the industry as a whole. The earliest studies on theme parks and their antecedents were mostly reflections of a journalistic nature (for example, Maxim Gorky's diatribes against Coney Island). With the development of the theme park industry in the 1950s and 1960s, a shift towards new and diverse research agendas can be noted. This chapter will introduce the purposes and contexts of theme park research within academia and the different methodological approaches that scholars have applied. It points to the challenges, constraints, and confounds of each type of research, including the availability of comparable data, diachronic research, NDAs (non-disclosure agreements), human subjects review, access to informants, ethnographic studies, content analysis, and relevant theme park data. There will be an emphasis on how these challenges may be overcome in terms of advancing quality-focused and successful theme park research in the future.

1 Introduction

How do you study a phenomenon that simultaneously constitutes a popular cultural artifact, a multi-billion-dollar business, a medium of stories and worldviews, a man-made (linguistic) landscape, a complex performance, a text, a topic or discourse in other texts, an object of fandom, a place of leisure and education, a space in which people interact with the space and with each other, and a place that changes over time in response to numerous factors? Quite obviously, there is not—there cannot be—one single method that allows us to capture and analyze all facets of theme parks and understand them in their totality. As a field of research, theme park studies is inherently transdisciplinary and forces students and scholars to engage with other disciplines and the methods the latter have developed—if only in a "derivative" way,

This work is contributed by Florian Freitag, Salvador Anton Clavé, Filippo Carlà-Uhink, Thibaut Clément, Scott A. Lukas, Sabrina Mittermeier, Céline Molter, Crispin Paine, Ariane Schwarz, Jean-François Staszak, Astrid Böger, Jan-Erik Steinkrüger, Torsten Widmann. The corresponding author is Florian Freitag, Institut für Anglophone Studien, Universität Duisburg-Essen, Essen, Germany.

F. Freitag et al., *Key Concepts in Theme Park Studies*,
https://doi.org/10.1007/978-3-031-11132-7_11

by reading the works of theme park scholars from a variety of disciplines. However, as in the case of the present chapter—and the volume of which it forms a part—theme park studies also often assume a more "real" form of transdisciplinary collaboration, or a blend of "virtual" and "actual" work across disciplinary boundaries, with research projects developed and carried out by groups rather than by individual scholars. Hence, while there is no individual method that is tied to or associated with theme park studies and that an aspiring theme park scholar must be familiar with, transdisciplinary collaboration functions as a "meta-method," and the willingness to transcend one's own disciplinary home and to learn from others may be considered the sine qua non—the entry ticket, so to speak—of this field.

Moreover, one of the reasons why academic theme park research has been conceptually, practically, and, most importantly for this chapter, methodologically limited until now is that knowledge about theme parks has traditionally circulated outside of academia and its channels of knowledge dissemination. In fact, while scholars examine questions that could be of immediate interest to professionals in the industry domain, most of the knowledge amassed by the industry is basically in the hands of the operating companies and of the design and planning consultants (Anton Clavé 2007, xiv). However, the industry is continuously re-thinking its business strategies in order to adapt to the evolution of technologies, to face changing consumer needs, to decide about investments in new products and experiences, and to manage the increasing global challenges related to sustainability, inclusivity, and prosperity. As a result, a stronger involvement of universities and research centers, as well as new partnerships between the industry and academia in the context of specific research initiatives, may lead towards a process of adoption, transformation, refinement, development, dissemination, and even creation of research methods. An example—among many others and in this case with the participation of one of the co-authors of this book—is the joint effort of PortAventura World (Vila-seca and Salou, Spain), the University Rovira i Virgili (Vila-seca, Spain), and the Science and Technology Park for Tourism and Leisure of Catalonia, since 2019 integrated into Eurecat Technology Centre of Catalonia (Barcelona), to measure changes in visitors' electrodermal activity (i.e. changes in their sweat gland activity as a physical manifestation of emotional responses) during a visit to the theme park (see Orellana et al. 2016; see also SPACE).

Therefore, this chapter not only draws on the practical experience of the authors to discuss some of the challenges that researchers may face when they apply specific methods such as archival work or field research to the theme park, it also reflects upon the radically transdisciplinary and collaborative nature of theme park studies, which requires strategic decisions and creative solutions especially in the realm of publishing. It also advocates for an increasing amount of collaboration between operators, consultants, industry associations, governments, universities, and researchers. The chapter starts, however, by addressing head-on the "stigma" that has often surrounded this particular research topic—the more or less openly expressed bemusement that fellow academics, friends, and strangers sometimes offer in response to the statement "I study theme parks." The authors of this chapter have all been there—to the archives that have sometimes refused access, to the parks that have sometimes refused cooperation, to the editors that have sometimes refused publication, but also to the conferences and workshops where the topics of their papers have sometimes raised some eyebrows. Anticipating and countering such reactions, we believe, also belongs to the theme park researcher's toolkit—the methodology of theme park studies.

2 Queer Academia

As a field for scholarly research, theme parks are a bold choice. "I have tenure, and I'm going to Disney World!" Stephen M. Fjellman supposedly thought when he decided to write *Vinyl Leaves: Walt Disney World and America* (1992, xv), suggesting that it was only *after* he had secured a tenured position at an academic institution that he would dare engage in theme park studies. In fact, especially in prefaces and acknowledgements, theme park scholarship offers numerous anecdotes that talk about the various forms of demotivating and career-endangering resistance from colleagues that (aspiring) students of theme parks have been confronted with. For one thing, fellow critics may object to the topic itself, considering it unsuitable for "serious" research: as early as 1981, for instance, Margaret J. King complained about critics' unwillingness to "grant any symbolic and evocative function to these amusement park features as one would for works of art or literature" (117). Almost 40 years later, and the "boom in 'Disney Studies'" from the 1990s onwards notwithstanding, Janet Wasko still found reason to note that in "some academic circles, the study of Disney in particular, and popular culture in general, has been perceived as an irrelevant, frivolous, 'Mickey Mouse' occupation" (2020, 5).

Moreover, even when not objecting to the topic as such, fellow scholars may expect theme park studies to go in one specific direction: Susan Willis, for example, co-author of *Inside the Mouse: Work and Play at Disney World* (1995), reports that during a conference panel on Walt Disney World, one listener commented: "'Why are you so critical? Wasn't anything fun?'" (1). At the other end of the ideological spectrum, in her work on theme park fandom Rebecca Williams has met with "outright refusals to acknowledge that we can move beyond political economy-rooted approaches grounded in critique" (2020, 248–249). Problems multiply if, as Williams does, one (openly) self-identifies as not only a student, but also a fan of theme parks—i.e. an aca-fan—as one may alternately "be seen as a traitor to the Cultural Studies cause, guilty of falling blindly in-step in support of a global corporation" (Williams 2020, 249), or as wrongly prioritizing some aspects of and responses to theme parks over others (see Wasko 2020, 238).

In response to these and other forms of resistance to (certain types of) theme park studies, some theme park scholars have offered detailed accounts of the various difficulties they have encountered in the context of their work on theme parks (see e.g. Lukas 2016a, 159–160 or Freitag 2021, 140–141)—perhaps in an attempt to illustrate that far from simply being "frivolous" or "fun," this work also involves "complicated production and articulation" (Banet-Weiser 1999, 4). Others have decided to anticipate and forestall any speculation by openly addressing their multiple allegiances as scholars and fans of theme parks, with the following passage from Fjellman's *Vinyl Leaves* providing a famous example:

> I love it [Disney World]! I could live there. I love its infinitude, its theater, its dadaism. I love its food, its craft, its simulations. It gets me to think, to remember, and to make up new fantasies. I appreciate its civility and its safety. I crave its contradictions. I like walking in its streets. I am writing this book because it has allowed me to spend lots of time there—and lots of money. If ever there were self-funded research, this is it. (Fjellman 1992, 17)

Written in a defiant, assertive and at the same time liberatory, celebratory tone, this passage has something of a "coming out"—indeed, other researchers have quoted it to show their solidarity (see e.g. Steinkrüger 2013, 29). It is perhaps no coincidence that one of the co-authors of this book has suggested calling this section "queer academia." As "queer academics"—as critics who are not afraid of, accept the challenge to, and even enjoy working on the craft, the kitsch, and the contradictions of theme parks—theme park scholars need to acknowledge and know that (a part of) academia may have a "problem with pleasure" (Willis 1995). Yet they also need to know that this has not kept others out there from putting out excellent contributions to the scholarship about theme parks and from building their academic careers upon it—or from having fun while doing so, either.

3 Failed Projects

Doing theme park research does not necessarily involve going to a theme park. For one thing, the theme park (or themed area, or individual attraction) under discussion may have never existed in the first place. Analyzing themed spaces that were spoken of, planned, but never built may seem paradoxical. However, failed projects are extremely relevant and illuminating as they reveal much about the workings of the industry, the public perception of theme parks, their impact on specific locations, their role in entertainment, education, and even in politics. To be sure, it is often hard to find material for a more substantial analysis, especially if the never realized projects are several years or even decades old. Yet particularly in the case of Disney, fans have often taken on the labor of unearthing some of these materials, posting released concept art or announcements, and essentially archiving them online, whether on dedicated websites or on social media platforms such as Facebook, Instagram, or Tumblr. Such fan archives frequently fill in the gaps left when company archives refuse access (see below).

Perhaps the most well-known among these never realized theme park projects is Disney's America. Announced in November 1993, the park was supposed to be built in Haymarket, Virginia, and to introduce visitors to several periods and key events of American history, but eventually fell victim to the Culture Wars (see Mittermeier 2016; Carlà-Uhink 2020, 14–15). As it caused widespread controversy, there is a wealth of source material on it—both scholarly treatments by historians and researchers from other disciplines as well as countless media reports. Indeed, searching the online databases of local, national, and international newspapers yields helpful results, but as with every historical research, the documents have to be read critically and in context. The same applies to the accounts published by professional historians while the controversy was unfolding, such as the 1994 newsletter by the Organization of American Historians on the topic. It is thus left to the scholar to puzzle together events and timelines, but also the actual plans for this theme park as they changed over a span of several months in reaction to the backlash. Due to the sheer amount of material, Disney's America may thus be easier to research than most of the other never-built projects, but at the same time the analysis is complicated by the project's highly contested nature and the politically charged discussions surrounding it.

Other never-realized Disney projects, such as the addition of a land called "Discovery Bay" to Disneyland in Anaheim (see Mittermeier 2017), have become easier to research as the company itself has begun to reflect on its own past and has released some of its archival material—if only to a selected, paying audience via their fan club D23. Of course, the materials that have become available through, for instance, presentations at the D23 Expo or Destination D, only provide restricted access to the project, as the flow of information remains entirely governed by the company.

With respect to both Disney's America and "Discovery Bay," it is interesting to note that many designed concepts later found their way into existing theme parks in revised form—many of the ideas for Disney's America were eventually realized at Disney's California Adventure (Mittermeier 2016, 130), while "Discovery Bay" influenced the design of "Discoveryland" at Disneyland Paris (Mittermeier 2021, 122) as well as "Port Discovery" and "Mysterious Island" at Tokyo DisneySea (Mittermeier 2017, 183–84). Such palimpsests may provide an opportunity to trace never-realized projects through a quasi-archaeological process.

Yet such an abundance of material is not available for all projects. Nor have all failed projects reached the stage of planning and thus left traces in archives. In these cases, scholars have to rely on press releases, interviews, and newspaper articles. To generate publicity, park managements usually publish press releases about the construction of a new park, a new themed area, or a new attraction. The internet has made this kind of investigation much easier—not only because many articles, even from local newspapers, are searchable and available online, but also because of specialized sites that collect and publish news from theme and amusement parks all over the world. Interesting examples are themeparkinsider.com, with a news archive going back to 2004, or blooloop.com, which offers a specific section on theme park news.

Scholars working with articles and press releases about construction and expansion plans that ultimately faltered are obviously confronted with the question of why such projects did not come into being. Answering this crucial research question requires not only systematically and thoroughly consulting the available sources, but also paying attention to the political, economic, and cultural contexts in which the park projects were revealed and discussed. The best example is once again Disney's America: beyond the official explanations provided by Disney, the failure of this project must be located, as already noted above, in the cultural and political context of the Culture Wars during the early 1990s (see Mittermeier 2016 for on overview of the debate).

In other cases, researchers may wonder whether the announcement of a new project was really underpinned by concrete plans or, rather, whether it was simply published to attract attention and provoke discussion. Here, too, scholars must carefully consider the political, economic, and cultural contexts of the documents. As theme parks can often be direct vehicles of political statements and ideologies (see WORLDVIEWS), the announcement of a new theme park that bears a direct political significance can be particularly noteworthy. For example, in summer 2014, only a few months after the military annexation of Crimea by Russia, it was announced that Russian banker Konstantin V. Malofeev and French theme park owner Philippe de Villiers would realize two theme parks in Moscow and on the peninsula, both dedicated to Russian history—an announcement that caused substantial outrage (see WORLDVIEWS). In such a case, scholarly research needs to not only position the project within the general political situation and debates about the "Russian identity" of Crimea, but to also investigate the investors' involvement in politics and in other projects. In the case of de Villiers, for instance, this should include a look at his park Puy du Fou (Les Epesses, France) with its strongly nationalistic ideological underpinnings, and a thorough consideration of his activities as a writer and politician (he was president of the nationalist, Eurosceptic party Mouvement Pour la France). With the internet page dedicated to the proposed parks having since been deleted, it is only through such research that scholars can attempt to formulate theories about the possible intentions behind such announcements.

Another good example is the theme park dedicated to Alexander the Great that was to be built in Thessaloniki, Greece (see Carlà-Uhink 2020, 193–194). Here, too, the theme park scholar must reconstruct the origins of the idea and the source of funding (the Greek diaspora in the U.S.), as well as the political context, the so-called "Macedonia naming dispute" (1991–2019), which saw the Republic of Greece confronting the republic that, since 2019, has been called North Macedonia—a dispute in which the ancient heritage, and particularly the heritage of Alexander the Great, was a substantial issue.

Working on theme parks that never came into being, therefore, requires a substantial amount of work with archives, collections, newspapers, and press statements, as well as a thorough study of the cultural, political, economic, and social contexts. Of course, depending on the specific project and the scholar's approach, some aspects may be more salient than others—never built theme parks as political statements, economic reasons for not building a planned theme park, conflicts over a plan that lead to its abandonment, etc. Nonetheless, it is important that the theme park scholar—given the innate interdisciplinarity of the subject—remains open to the entire context in which an announced or planned theme park project was (supposed to be) embedded.

4 History and Archaeology

Another branch of theme park research that involves working with archives is historical theme park studies—research on parks, themed areas, and attractions that once existed but have since been removed. As with the theme park form as a whole (see ANTECEDENTS), each individual theme park has its own history. Technological innovation, marketing decisions, operational strategies, natural wear, changing legislation, or broader cultural changes constantly bring change to theme parks, sometimes in ways that are highly obvious, and sometimes barely noticeable (Bryman 1995, 83; Freitag 2021, 138–44). Theme parks, however, have different ways of dealing with their own history, ranging from complete silencing to self-celebration and functionalization for commercial purposes, especially in the case of parks with a longer history. As in every form of historical narrative, then, the theme park carefully selects which aspects of its history—which episodes, rides, characters, and moments—are memorialized or forgotten, whether in the park landscape itself or in theme park paratexts such as coffee table books (see PARATEXTS AND RECEPTION).

The representation and celebration of a park's history may take its cue from and/or revolve around three possible different topics: the park itself, or specific elements thereof; the pre-history of the park or of the company owning the park; and individual people connected to the park's or the company's development, most often their founders. Such a typology of memorialization constitutes, however, a heuristic idealization: more often than not, the three different types cannot be neatly divided. Disneyland in Anaheim, for instance, frequently makes references to past rides and attractions. Thus, the Many Adventures of Winnie the Pooh ride, opened in 2003, features somewhat hidden tributes to characters from the ride it replaced, Country Bear Jamboree, which had been located on that site since 1972. More prominently, the soundtrack of Light Magic, a short-lived nighttime parade, contained excerpts from the soundtrack of the Main Street Electrical Parade, which it was supposed to permanently replace in 1997. Other parks such as Luna Park on Coney Island, Brooklyn, or Kennywood in Pittsburgh dedicate themselves entirely (Luna Park) or partly (see Kennywood's "Lost Kennywood" section) to referencing their (pre-) history, namely, the turn-of-the-century amusement parks previously located on the site (Rabinowitz 2012, 173).

Rather than focus on a continuity of site, parks such as Mount Olympus (Wisconsin Dells, Wisconsin), or Europa-Park (Rust, Germany) focus on a continuity of theme park actors: in the former case, the object of celebration is Demetrios "Jim" Laskaris, a Greek immigrant who moved to Wisconsin Dells in 1970 to open a fast-food diner, to which go-kart tracks and roller coasters were later added. Following Jim's death in 2003, the family turned the property into a theme and water park, whose Greek theme celebrates the family's history. This is made explicit to park visitors: Jim's life and his heritage are not only memorialized on the park's website, but also through a commemorative monument in the park that tells visitors about Jim's life and achievements (Carlà-Uhink 2020, 117–121). At Europa-Park, in turn, park designers created an entire attraction around the history of the park's founders and of its parent company, the ride manufacturer Mack rides. Operating from 2010–2017 (and thus now itself a part of history), the Europa-Park Historama combined "talking busts" of the Mack patriarchs and old photographs with an animated timeline and other special effects into a multimedia celebration of the history of the Mack family and their business venture (see Freitag 2016, 143–44). As it also covered the history of Europa-Park itself, however—through models, old sketches, photographs, and footage, and once again through animated timelines—the Historama united all three different approaches to theme park memorialization.

Celebrating its own history can add to a park's emergent authenticity (see AUTHENTICITY), yet the celebration of the park's history is usually restricted to a limited set of memorial elements, lest its immersive potential should suffer because of an excess of musealization and self-referentiality (see IMMERSION). As with all aspects of theming, every form of memorialization is the product of a deliberate selection (see THEMING). This is also the reason why generally, replacements and new additions to the park are integrated as seamlessly as possible: to hide the changes the park underwent, to silence the park's historicity, and thus to avoid signs of the intrusion of external forces into the self-contained themed space.

Therefore, what is perhaps even more interesting are those cases in which silencing is chosen in order to hide events or aspects of the park's past that might tarnish its public image in the present. An extreme example is offered by Parc Astérix (Plailly, France): after a deadly accident on the river rapids ride La descente du Styx in the park's "Le Grand Lac" section in 2006, the ride was reopened two years later not only with new ride vehicles and better safety regulations, but also with a new location on the park's map (as part of the "Rome" section) as well as a new name (Romulus et Rapidus) that offered no reference to the Underworld and instead evoked the legend of the twins who, abandoned on the Tiber, were then deposited safely by the river where the she-wolf would find them (Carlà-Uhink 2020, 178–179). Such sudden adjustments in theming can also be triggered by events outside the park. A case in point is the GAZPROM Erlebniswelt, a multimedia exhibit on natural gas which from 2010 onwards constituted part of the waiting line of the Blue Fire coaster, sponsored by the Russian energy corporation. Renamed Nord Stream 2 Dome in 2020 after the proposed gas pipeline from Russia to Germany financed by Gazprom, the exhibit's name was quickly changed to Blue Fire Dome in 2022 after Europa-Park had discontinued its cooperation with the Gazprom daughter Nord Stream 2 in response to the Russian invasion of Ukraine.

Not only sudden events, but also slower-paced socio-cultural developments can induce re-themings of this kind. In 1970, Disneyland in Anaheim renamed the Quaker Oats-sponsored Aunt Jemima's Kitchen restaurant Magnolia Tree Terrace and removed all references to the notorious character of Aunt Jemima, who had welcomed visitors to her kitchen since the opening of the park in 1955. Following the Civil Rights movement, a plantation-themed restaurant hosted by a former slave who even after "the 'War Between the States'" could not resist "the opportunity to make so many families happy with the ease and satisfaction of serving her mouth-watering pancakes" would have created too much controversy. Hence, Aunt Jemima was erased from the park landscape as well as from park maps, histories, and promotional material (Freitag 2021, 150–151; 51 years later, Quaker Oats would themselves retire their mascot in the wake of the Black Lives Matter movement).

Yet not all changes can be silenced: major modifications that impact the park's layout, thematic coherence, or overall structure cannot be hidden and are immediately visible. The 2011–2013 expansion of Hong Kong Disneyland, for example, which added three "lands" to the park, seriously upset the original magic wand layout, as the new lands are not accessible from the main hub. The change is visible not only from the "irregularity" on the park map, but also through a comparison with the layout of all other Magic Kingdoms. In Terra Mítica (Benidorm, Spain), financial difficulties encountered by the park led to a series of dramatic changes which could not possibly be silenced, including the addition, on top of the original theme (the ancient Mediterranean world), of Vicky the Viking and Maya the Bee, the closure of a section of the park and thus the abandoning of the original loop pattern, and finally the division of the park into two different theme parks (Carlà-Uhink 2020, 80–82). Once specific changes have been identified, the task of the researcher is to motivate them. As the potential factors are varied and complex, ranging, as just stated, from economic to cultural and political shifts, such research questions necessarily involve different fields of study, and therefore require an interdisciplinary approach.

However, not all changes are so easily recognizable. Theme park scholars therefore need to pay close attention to minute details of the palimpsest of the park landscape in order to discover traces of previous phases and elements. The methods of landscape archaeology and archaeology of architecture, for instance, can help researchers to identify where elements have been added, removed, or modified. In Terra Mítica, for example, the ride La Cólera de Akiles was moved to what had previously been the sit-down restaurant Corfú. Two walls of the original building are still in place, and while the designers tried to insert them into the theming of the ride as ruins, they reveal to the observer that this was not the original purpose of this building (Carlà-Uhink 2020, 103–104). To confirm such hypotheses, researchers may interview regular visitors and staff, but also need access to old park maps, photographs, and promotional material, e.g. via fan sites dedicated to individual parks (see VISITORS). The next section therefore deals with the specific methods connected to archival work on theme parks.

5 Working with Archives

In order to provide an overview of the archival resources available to theme park scholars, but also to illustrate some of the difficulties one might encounter when working with theme park archives, we will discuss only one case study here: the material relating to Disney parks. In the case of Disney, the company houses its own archives on the Walt Disney Studios lot in Burbank, California, but access for scholars is notoriously hard to come by, as it is not openly advertised online and the lot is off-limits to visitors without a personal invitation. Meanwhile, scholars who have gained access report severe restrictions on the use of materials or the company's wish to have veto power over future publications, almost rendering access moot (see Budd 2005, 15–16). While by far not as extensive as the official company archives, there are three archives within the United States that have small holdings on Disney: the archives of the Anaheim Heritage Center (https://www.anaheim.net/2473/Anaheim-Heritage-Center), the Orlando Public Library (https://www.libguides.ocls.info/specialcollections/Disney), as well as the university archives of UC Irvine (as part of their special collections; https://www.special.lib.uci.edu/). These collections mostly consist of Cast Member publications or annual company reports, but also of documents from the time of the building of these parks (especially on Walt Disney World at the Orlando Public Library) or newspaper clippings across the decades.

Yet, the richer source for Disney theme parks history are fan-built online archives. One of the most extensive ones is "Daveland" (https://www.davelandweb.com/disneyland), an online photo archive that contains thousands of old photographs from both Disneyland and Walt Disney World Resort. The site is a great resource, especially on now-defunct attractions, as it also features extensive descriptions of the parks through the years. Many other bloggers also trace Disney history and often upload rare documents or photos, such as "Passports to Dreams and Magic" (http://www.passport2dreams.blogspot.com/2018/10/magic-kingdom-in-early-1972.html), which focuses on Walt Disney World, the "Disney History Institute" (http://www.disneyhistoryinstitute.com/), which also covers the company beyond the parks, or "Disney History" (http://www.disneybooks.blogspot.com), which also publishes books on the parks.

In addition to these older blogs and websites, social media have become increasingly important for Disney hobby historians and fans to bring together content that also holds potential scholarly value, in particular on Twitter, Instagram, and Tumblr. The bigger theme park news websites, whether they approach the topic more from an industry (see the section on databases below) or from a tourism perspective (https://www.insidethemagic.net), are also interesting when it comes to newer developments and occasionally past plans, especially as they also cover the Universal Theme Parks and many other smaller, non-franchised parks all over the world. Ultimately however, for such online archives, longstanding fan interest is vital to keeping them alive—which is also why content on Disney continues to dominate the landscape. Moreover, obtaining publishing rights for their content (e.g. for scholarly publications) may sometimes prove problematic (see the section on publishing below).

6 Working with Databases

Identifying, measuring, comparing, and analyzing theme parks at different geographical scales from global to local is not an easy task for researchers due to the lack of key and homogeneous information and the complexity of the industry. This has also been a problem for planners, designers, and consultants working in the industry. In fact, for many decades, the only publicly accessible source of benchmark data were the TEA/AECOM reports, which used, among others, statistics provided directly by operators, financial reports, investment banks, local tourist organizations and, when necessary, professional estimates to identify the attendance figures of the top parks worldwide and in different regions of the world.

More precisely, the annual *Theme Index and Museum Index: Global Attractions Attendance Report* (https://www.aecom.com/theme-index/) has been the result of a collaboration between the Themed Entertainment Association (TEA) and the consultancy firm AECOM (formerly Economics Research Associates or ERA) initiated in 2006 (see INDUSTRY). Prior to that year, the ERA estimates of theme park attendance had been annually published in the trade magazine *Amusement Business*, which folded, however, in 2006. Founded in 1991 and counting some 1,600 members worldwide in 2022, TEA (https://www.teaconnect.org/index.cfm), in turn, is an international non-profit association representing creators, developers, designers, and producers from all over the sector of themed places and immersive experiences. One of its goals is to become an industry leader in terms of information, education, and standards, which is why it has committed itself for many years to the compilation of attendance data for the industry.

Even though the TEA/AECOM information has been tremendously useful in terms of monitoring the annual performance of the main players, understanding the worldwide expansion of the industry, assessing the leadership of some companies, and measuring the role of specific theme parks in various regions of the world, the main limitation of the TEA/AECOM reports is precisely that they only inform about attendance and only report on the largest parks in an industry that is mainly formed by small- and medium-sized facilities. A second limitation is due to the diversity of types of facilities included in the reports and the lack of definitions and classifications of entertainment facilities, as TEA/AECOM for example does not distinguish between theme and amusement parks. Thirdly, and depending on the type of parks or the regions of the world, some gaps and inaccuracies can be observed. For example, the 2019 report offers no information on such retail theme parks as Cultural Village in Dubai, which has an estimated annual number of visits of more than 5 million; and several Chinese destinations—from the show-based Songcheng parks to the Chimelong Guangzhou wildlife park in Guangdong, which also have a large number of attendees—have simply been omitted. Therefore, when using the TEA/AECOM reports e.g. to create aggregate figures, it is important to remember that these figures may be partial, fragmented, or imprecise. Finally, regarding the attendance figures of the top ten theme park operating groups in the world, TEA/AECOM lists the total number of visitors of all the destinations and facilities owned by each corporation, including theme parks but also amusement parks, water parks, zoos, and other location-based entertainment facilities (see INDUSTRY).

Being aware of those limitations, authors such as Anton Clavé (2007) have used this data extensively and shown its usefulness and potential areas of application. Nevertheless, as Cornelis states,

> publicly accessible benchmark data that is available to our industry should increase in variety and reliability in the near future. In this way, an important step will be taken towards a greater degree of accountability in our industry; the small and medium sized parks, in particular, will benefit from this. These parks often lack the means to do extensive research, nor do

> they generally have the budgets with which to hire external consultants for the job. (Cornelis 2017, 38)

Since 2016, The Parks Database (https://www.theparkdb.com/) has sought to contribute to improving the range of available data by gathering the existing publicly available information about the global theme park industry, including benchmark data from any size and type of park. The website praises itself as "the market intelligence solution for the attractions industry" and is a valuable source for analysis and research of attractions and leisure real estate including, obviously, theme parks. Data is aggregated weekly from published public sources; uncredited data comes from the reported facilities themselves or has been estimated by the database's own analysts. In addition to attendance, information on each facility includes at least the opening year, the estimated value, size, and cost, as well as a Google Maps capture providing either a map or an aerial view of the venue. In some cases, more elaborate facts and figures supplement this basic information. For example, the file on Europa-Park (Rust, Germany) discusses the history, location, size, country, type of park, the opening year, capacity, URL, ticket prices for adults and children, and ratios of capacity/attendance, attendance/size, and size/capacity. Moreover, information about individual rides (name, type, and capacity) is offered and attendance data includes figures from 1975, 1976, 1993, and from 1998 to 2020 (taken from the *Amusement Business*-TEA/ERA-TEA/AECOM attendance reports).

As of February 2022, The Parks Database has gathered information on over 4,228 parks and 11,239 attractions from all over the world, which are classified into theme, indoor, water parks, and others. Although conceived as a tool for planners, designers, and consultants working in the attractions industry, it also constitutes a useful source for theme park researchers as it features rankings of the top 50 facilities according to attendance and estimated value, a blog with insightful informative articles about the business of theme parks, the different types of parks, and such issues as size, seasonality, and resorts, resources and case studies, as well as an online store offering reports that combine information with experience-based knowledge.

Founded in 1918, the International Association of Amusement Parks and Attractions (IAAPA), finally, is the largest international trade association for location-based amusement facilities worldwide and represents more than 6,000 facility, supplier, and individual members from more than 100 countries (https://www.iaapa.org/; see INDUSTRY). To support the industry, it has also developed a valuable series of resources that could be useful for researchers interested in the dynamics of theme park development in different regions of the world. Whereas the previously discussed sources are public and refer to each particular attraction, the resources provided by IAAPA are mainly addressed to their members and consist of reports with aggregated data on operational benchmarks for different types of attractions (themed attractions, museums and science centers, zoos and aquariums, amusement parks, family entertainment centers), economic impact studies by regions, and attendance projections. Among them, the annual global outlook for theme and amusement parks includes a five-year projection of spending and attendance by world region.

7 Quantitative Research

Together with the growing availability of industry data, research on theme parks shows an increasing interest in applying quantitative methods of analysis to theme park planning, development, and management issues. Two reasons may explain this evolution. First, the motivations for scholarly engagement with theme parks have diversified, moving from a more general critical understanding of the theme park as a cultural, societal, entrepreneurial, and economic hegemonic artifact to, e.g. the analysis of specific aspects of the industry as they are studied in other fields such as economics, marketing, or landscape studies. This has also been observed by Leask (2016) with respect to research on visitor attractions, of which theme park research forms a part. For the period from 2009 to 2014, she notes an increasing number of researchers active in the field, growing opportunities for inter- and multidisciplinary work, rising implications of research for managers and decision-makers, a growing range of methods applied, and, as a result, an increase in the quality of research published in journals. The growing role of Asian and particularly Chinese scholars (or scholars based at Chinese universities and research organizations) in theme park studies has also contributed to this development. Zhang and Shan (2016) report that among the 145 Chinese-language articles about theme parks published between 1990 and 2014 in 65 key journals "over eighty percent of journal articles analyse[…] the theme park from the economic and industrial perspective, such as operational costs, tourism income and land commercialization," whereas topics more favored by scholars publishing in English, such as cultural perspectives, are comparatively neglected.

Second, some industry operators have acknowledged that academic research on the industry generates new opportunities for knowledge (and value) creation. More specifically, as scholars and theme park companies continue to develop mutual trust, partnerships between the two have stimulated new areas of research and have eased the implementation of quantitative methods of research. In some cases such partnerships are based on making empirical data from particular parks available for academic analysis. This has been especially useful for medium and small parks, i.e. parks that—in contrast to the largest theme park and entertainment corporations—do not have in-house research facilities. This is also the reason why the number of non-Disney-related academic contributions has been growing, with the industry gaining insights into what can be known and learned that go beyond Disney. As previously stated, the availability of empirical data has stimulated transdisciplinarity and the use of quantitative approaches, and currently there are an increasing number of papers that discuss highly specialized topics ranging from sociology and mathematics to design and environmental sciences.

Traditionally, the lack of data made it difficult for scholars to get involved in theme park research and the most insightful studies based on empirical observations were those published by leading professional experts from the industry. A case in point is *Walt's Revolution! By the Numbers* (2003), written by one of the best-known industry consultants, Harrison Price. Nowadays, access to data from diverse sources, including theme park operators themselves, allows researchers to develop quantitative research related to the design, planning, development, operations, and general administration of theme parks. A few examples with different research characteristics and based on different data sets are discussed next.

Data on parks obtained through financial company reports is a useful source of information to develop quantitative research, especially from an economic perspective (see ECONOMIC STRATEGY and INDUSTRY). Even though they are only available for the largest operators and the information they provide depends on the location of the company, annual financial reports from, e.g. The Walt Disney Company, Cedar Fair, Six Flags Entertainment Corporation in the United States, Merlin Entertainment, Parques Reunidos, or Compagnie des Alpes in Europe, or Overseas Chinese Town Holding Limited and Haichang Ocean Park Holding Limited in the People's Republic of China, allow for research on the business models of different types of parks, facilities, and companies and the role of such factors as attractions mixes and markets. Basic financial indicators obtained from the reports are regularly compiled in specialized databases from where information can be extracted and processed. Based on data obtained from the FAME database, for example, Liu (2008) seeks to apply data envelopment analysis (DEA) to a sample that includes indicators related to 13 parks managed by the ten leading theme park operators in the UK in order to aggregate various financial ratios into a composite profitability index.

More interestingly, there is a growing number of cases in which parks provide data to researchers in order to facilitate studies that benefit both the scholar or academic research group as well as the operator. For example, for his doctoral thesis, Cornelis (2011) received financial data from many European theme parks on a daily and weekly basis. This rich source of information (in some cases reaching as far back as 25 years) allowed him to explore the impact of investments in new attractions on the parks' performance. In a similar study focused on Efteling (Kaatsheuvel, Netherlands), van Oest, van Heerde, and Dekimpe (2010) use a unique data set of weekly visitor figures over a period of 25 years to develop a response model of investment decisions for theme park attendance. Their return on investment calculations show that it is more profitable to invest in multiple smaller attractions than in one big one and, moreover, that even though thrill rides tend to be more effective than themed rides, there are conditions under which one should prefer investing in the latter (see ATTRACTIONS). Obviously, these research outputs have been of utmost interest for managerial decision-making. With an entirely different interest, Finnegan et al. (2018) monitored energy data provided by a UK safari park in order to apply the Greenhouse Gas (GHG) Protocol to understand the company's carbon footprint. In this case, park management sought to identify realistic carbon reduction solutions (see ECONOMIC STRATEGY). This is one case in which the name of the park was not disclosed, even though it is described in revealing detail in the study.

Further quantitative methods used in theme park research are based on information obtained through surveying visitors inside or outside of the parks (see VISITORS). In the latter case, research may be conducted offsite because there is no need to interview visitors during their visit or because, as often occurs, the park has not agreed to the survey (see also the section on private and public space below). In the former case, there is a growing interest in analyzing the spatial-temporal behavior of customers during their visit. This kind of research is particularly valued by park managers as it can provide useful information on the operation of attractions (see SPACE). Two exemplary studies in this vein are those by Birenboim et al. (2013) and Huang et al. (2020). Both integrate multiple data sources to analyze tourists' spatial-temporal behavior patterns on a micro scale at PortAventura World and Ocean Park Hong Kong (People's Republic of China), respectively. In both cases, information was gathered using GPS tracking devices delivered in the park. With a different aim, although more common, are studies that explore specific issues such as customer expectations, satisfaction, loyalty, or perceptions. A good example is the study by Ryan, Shih Shuo, and Huan (2010), who collected data from a sample of 402 respondents at Janfusan Fancyworld (Yongguang, Republic of China) when visitors were leaving the park during the main summer season in order analyze the satisfaction they derived from the visit. In this case, socio-demographic characteristics of the sample were checked with park management who confirmed they were consistent with their own records. Kao et al. (2008), in turn, surveyed 408 respondents at Farglory Ocean Park (Yanliao, Republic of China) over a period of two weekends in order to explore the effects of the theatrical elements of theme parks on consumers' experiential quality, experiential satisfaction, and loyalty intentions (see SPACE and THEMING). To study "Children's Experiences of Disney World," finally, Pettigrew (2011) measured the heart rates of children during their visit to the park.

Quantitative research based on surveying theme parks visitors but not related to any particular theme park venue has also been published in academic journals. For example, Yoonjoung Heo and Lee (2009) examine how customers perceive Revenue Management practices in the theme park industry compared to traditional Revenue Management in the hotel industry. In this case, to collect opinions about the perceived fairness in the theme park and hotel industry, 523 students at east coast universities in the United States were invited to participate in a survey by email (see ECONOMIC STRATEGY). Another example is the study by Torres et al. (2019), whose quantitative research approach sought to identify the various emotions customers experienced throughout the theme park visit by using the Positive and Negative Affect Schedule (PANAS). The survey they used as the primary means of data collection was distributed via Amazon Mechanical Turk (MTurk) and the authors set a selection criteria of valid surveys based on responses to two discriminant questions (see ATTRACTIONS, SPACE, and VISITORS).

Finally, quantitative research methods have also been implemented to assess the economic and environmental impact of theme parks. Anton Clavé (2010), for instance, examined the ability of a medium-sized theme park such as PortAventura to generate external economies, create a new image for a specific place, and catalyze the development of new economic activities. Research outcomes presented in this paper suggested that the park played a role in the restructuring of the destination, the attraction of new market segments, and the stimulus for private actors in the area to respond to the new demand for accommodation facilities. With an entirely different scope and objectives, Wang et al. (2017) investigated the greenhouse gas emissions of 26 amusement parks evenly distributed throughout the island of Taiwan.

Finding significant discrepancies in energy use among the parks, the study proposes two models for predicting annual carbon emissions and average carbon emissions per visit, along with several guidelines to assist park operators in reducing carbon emissions and improving energy efficiency (see ECONOMIC STRATEGY).

8 Private and Public Space

One of the reasons why onsite surveys in theme parks—and field research in theme parks in general—are difficult to conduct is that they require the approval of the park's management. Although theme park spaces frequently use theming (architecture, landscaping, signage, street furniture, soundtracks, performances) as well as paratextual discourse (see the names of such areas as "Main *Street*, U.S.A." or "New Orleans *Square*") to simulate public urbanity (see THEMING), the vast majority of theme parks and themed commercial spaces are nevertheless "*private* spaces. Privately owned, privately developed, privately operated" (Orvell 1993, 244)—simply put, "Main Street, U.S.A." is not a street, just as "Disneyland isn't a city" (Marling 1997, 176; see also Freitag 2021, 180–181). This has numerous consequences for researchers: unless specific arrangements have been made with the theme park prior to the trip, the field researcher is subjected to the same rules and regulations as any other visitor. Not surprisingly, these regulations will invariably prohibit visitors from entering the park's backstage areas and from conducting systematic surveys or interviews with employees and other visitors within the park. Consequently, several critics have resorted either to (surreptitious) interviews and surveys outside the park (see, e.g. Real 1977; Kuenz 1995; Bryman 1999; Raz 1999, 19; Wasko et al. 2001; Waysdorf 2021) or, preferring the situatedness and suggestive context of in situ conversations, to informal chats within the park (see, e.g. Fung and Lee 2009; Firat and Ulusoy 2011), also as part of collaborative field research (see the section on reflexivity below).

In addition, however, due to their often deliberately vague wording, park rules and regulations may even interfere with basic documentary work such as taking photographs or notes. Consider, for example, the "Disneyland Resort Park Rules," according to which the company reserves the right to

> deny admission, prevent entry or to require a person already admitted to leave the Disneyland Resort or any party [sic] thereof, without refund, liability or compensation, for failure to comply with any of these rules, for unsafe, illegal or offensive behavior, to ensure safety, security or order, or if we consider that the circumstances otherwise so require, in our sole and absolute discretion. (N.N. N.D.)

Indeed, as a private company owning and operating a private space, Disney has every right to decide what precisely the circumstances are under which a researcher—or any other person—may have to leave the park. And while the work of a researcher may not in itself compromise the "safety, security or order" of the park, its employees, and its visitors, it may occasionally be initially perceived as such: during a research trip to Disneyland in California, one of the co-authors of this book—a white male then in his mid-30s visiting the park by himself—had spent less than two hours systematically photographing all the buildings in "New Orleans Square" and taking notes before he was approached by two security guards who asked him what he was doing there. He never learned whether it was one of the employees or visitors who thought his behavior suspicious and decided to call security or whether it was park security itself that spotted him through CCTV; while he perfectly understood how his behavior could be misconstrued, he was also worried he would be asked to stop his work and perhaps even leave the park. Yet the encounter turned out to be amicable, and after the researcher had explained his project and provided the security guards with his business card and a link to his professional website, they let him go on with his work and he was not approached again throughout the rest of his stay (perhaps because, as he found out later, he had made it onto the briefing sheet that is distributed to the park's individual operational managers every day).

This anecdote may not be representative of Disneyland or of theme parks in general—during other trips to Disneyland and to other parks, including other Disney parks, the same researcher had never been approached by security—but it illustrates that due to their work, scholars stand out from the crowd and may attract attention that, ultimately, could harm their work (in fact, Lukas tells anecdotes of encounters with security guards that were apparently much less amicable and after which he had to stop filming or taking pictures; see Lukas 2016a, 159–160). Students of theme parks on a research trip may thus be well advised to remember that theme parks are private spaces, to familiarize themselves with the park regulations, and to bring appropriate documents and material in order to explain their work.

9 Reflexivity

Apart from park regulations, another factor that may impede field research in theme parks is the parks' very nature as immersive leisure spaces. In fact, from the point of view of rhetorics, the term "theme park studies" constitutes an oxymoron: whereas "theme park" connotes immersion and imaginative or emotional involvement, "studies" rather evokes reflection and critical distance. To be sure, "immersion" and "reflection" have to be considered extreme points on a spectrum rather than mutually exclusive categories, with some scholars arguing that some residual awareness of the outside world, some "latent rational distance" (Wolf 2013, 51) or a "split loyalty" (Ryan 2015, 68), is even constitutive of the experience (see IMMERSION). Nevertheless, and whether research on theme parks involves short-term field research or long-term participant observation (see the section on participant observation below), the nature of theme parks as commercial immersive spaces—as restricted and physically engulfing environments that actively seek to diminish their visitors' critical distance—poses certain challenges for the scholar. On the one hand, and in contrast to other immersive media, theme parks simultaneously constitute a mediated space and a space of medial reception—there is literally no "outside" space for reflection. To use Janet H. Murray's famous comparison of a "psychologically immersive experience" to "a plunge in the ocean or swimming pool" (1997, 98), doing field research in theme park studies requires the scholar to breathe under water. On the other hand, time and other constraints may seriously diminish critics' willingness to "let go" or "take the plunge," with the result that they may miss out on a crucial, if not constitutive aspect of the theme park experience.

To answer both of these challenges, theme park scholars have suggested collaborative field research. For her research on Moomin World (Naantali, Finland), for instance, Kirsti Hjemdahl solicited the help of "Aksel, my research assistant who is six years old" (2003, 129). While during a previous visit on her own, Hjemdahl found herself "on the periphery of the big events," experiencing the park with a child made her come "to a totally different understanding about how theme parks happen, how they speak and move, and how you get in touch with their senses" (2003, 135, 136). Rather than remaining critically distanced, as during her first visit, the visit to Moomin World with Aksel allowed, even forced, Hjemdahl to become immersed in the experience: "Perhaps the theme park contributes to creative imagination and the ability to dream without scrutinising everything too closely" (2003, 145). In "Research Dialogue: The Ways of Design, Architecture, Technology, and Material Form" (2016), in turn, designer Gordon Grice has defined "Creative Research" as "the examination of a subject, approached simultaneously from a variety of directions, with the intent of stimulating new insights—in a manner of speaking, serious research conducted as a creative act" (Carlà et al. 2016a, 109). Conceived during a visit to a theme park with three theme park scholars from different disciplines, "Creative Research" invites transdisciplinary teams to leisurely stroll through the theme park and spontaneously comment on individual phenomena until the "collective mind" fixes itself on one particular item, which is then selected for closer examination through a free and open-minded exchange of ideas. "Creative Research" thus combines

> two disparate elements into a single exercise that is spontaneous and deliberate, carefree and rigorous, cerebral and visceral, mindful and mindless—all at the same time. The tension that exists between the two parts—the disorder of creativity and the methodical orderliness of research—is what provides the motive force, as well as the unpredictable results. (Carlà et al. 2016a, 109)

Note that in both cases, the diversity of the research team seems to play a central role, as this allows for the inclusion of new, different perspectives on the theme park (see Fig. 1).

Even during the precious time of field research, theme park scholars should thus strive to keep an open mind and body to both the theme park's assault on their senses as well as the various theoretical reflections these sensations may engender. Of course, engaging in field research together with other people may, for various reasons, not always be an option. Yet reading a few pages of Jean Baudrillard's semiotic study of Disneyland (Baudrillard 1994) immediately after riding Space Mountain as part of a phenomenological approach to theme parks may have a similar effect.

Fig. 1 A geographer and an American Studies scholar from the *Key Concepts in Theme Park Studies* team at work. *Photograph* Salvador Anton Clavé

10 Phenomenological Approaches

Indeed, yet another way to productively use the tension between immersion and reflection during field research in theme parks is to draw on phenomenological approaches, as is practiced in theater and performance studies. In fact, theme park visits may be considered events that are characterized by a form of co-presence (Fischer-Lichte 2008) and may thus be thought of as theater or performances. Kokai and Robson (2019, 13–14) even go so far as to define theme parks and so-called immersive theater as one and the same thing, and to see Disneyland as a precursor of today's immersive theater. In their introduction to performance analysis, Christel Weiler and Jens Roselt note that experiences made during a visit to a performance are above all subjective (2017, 12–14) and that the first step for researchers should be to determine what significance the event had "for me" (13)—indeed, even though theme parks are designed to provide standardized experiences, different visitors react differently to individual attractions and situations (see Schwarz 2017), and this applies to theme park researchers as well. While the researcher's individual reaction, Weiler and Roselt continue, should not necessarily be discussed in the analysis of the event, it should be kept in mind that other people may have responded differently (16).

Noting that a semiotic approach cannot do justice to the large variety of responses or meanings that an event may elicit within different people (2017, 82), Weiler and

Roselt suggest a phenomenological approach. On the one hand, such an approach allows for a theoretical understanding of experience that makes spectators' or visitors' perceptions and reactions describable and analyzable; on the other hand, it allows spectators, visitors, and researchers to distance themselves from the event and thus makes new, different insights possible (85). In a phenomenological approach, the experience of the analyzing recipient becomes the starting point of a process of reflection and cultivates a practice of noticing, marveling, and wondering (100). Analyzing a performance phenomenologically thus implies making experiences while watching and afterwards describing and reflecting on them.

This can also be applied to field research at theme parks. As has been noted above, theme parks simultaneously constitute a mediated space and a space of medial reception: in the "performance" of the theme park, stage and auditorium are one and the same, and due to the sometimes highly intense ways in which the body is affected on this stage—e.g. on a roller coaster—it may be difficult to develop a distanced or reflective attitude. While some researchers may therefore decide to postpone the reflective phase until after the field trip, one may nevertheless try to incorporate it into the theme park visit—either as part of collaborative or creative field research (see above) or individually, in a more remote corner of the park. For the authors of this article, smaller restaurants and snack bars have often provided a suitable atmosphere for reflection, especially before or after the lunch and dinner rushes: the seating area across from Boiler Room Bites in Tokyo Disneyland's "Adventureland," Rutmor's Taverne in the "Klugheim" area of Phantasialand, Pica Picae in the Roman section of Terra Mítica, and the Café de la Brousse in Parc Disneyland's "Adventureland" are our top recommendations. In any case, it is important for researchers to take their feelings of fun or fear seriously and to ask themselves how these feelings have been produced. As Weiler and Roselt point out, after a performance it is possible to think that you have not understood it, but it is impossible not to remember anything (2017, 103). It is these individual memories, feelings, and physical impressions that can serve as the starting point for an analysis and that can help us better understand how theme parks work.

11 Landscape Analysis

One's reactions to the park—as well as the reactions of other people—may be at least partly explained through landscape analysis. When applied by social scientists to theme parks, landscape analysis mainly helps researchers understand how space and place make sense—or not—to visitors. According to landscape semiology, the landscape constitutes a text whose meaning is encapsulated in signs—mostly visuals, although smells, sounds, and tastes may be involved in the process as well (see SPACE and THEMING). In the case of a landscape designed as such, these signs have been purposely created so as to make sense for visitors according to their culture and the latter's references and narratives. Hence, landscape analysis is twofold: first of all, researchers need to identify the signs inscribed in the landscape. Listing these signs requires paying close attention to details, because everything—from the texture of the ground and the color of the electric light to the font used on the restaurant's menus and the species of plants chosen for landscaping—may be used to give sense to the place (see MEDIA and THEMING). Particular attention should be paid to central and iconic landmarks that give meaning to the entire themed area, or even an identity to the park itself (such as the castle at Disney parks), but also to what is hidden and kept out of view in order to avoid cognitive dissonance (such as the urban landscape surrounding the park). Listing all the signs is useful but difficult. Taking pictures certainly helps—and looking at them after the field trip is often more effective than doing observations directly on site—but non-visual signs need to be listed as well (e.g. through video recordings or notes; see Fig. 2). However, signs rarely make sense individually, but rather collaborate to convey a specific style (e.g. Orientalism, steampunk, etc.) or to tell a story (e.g. Asterix fighting the Romans, the conquest of the Wild West, etc.), which calls for a lexical and syntactical analysis.

Secondly, the landscape is an open text: its meanings are plural and created individually by each visitor. Some patrons may ignore some of the signs inscribed in the landscape, have trouble understanding their meaning, or read them in radically different ways. This particularly applies to culturally specific signs and landscapes such as Disney's "Main Street, U.S.A.": while most American visitors will understand the nostalgic evocation of a Midwestern turn-of-the-twentieth-century small

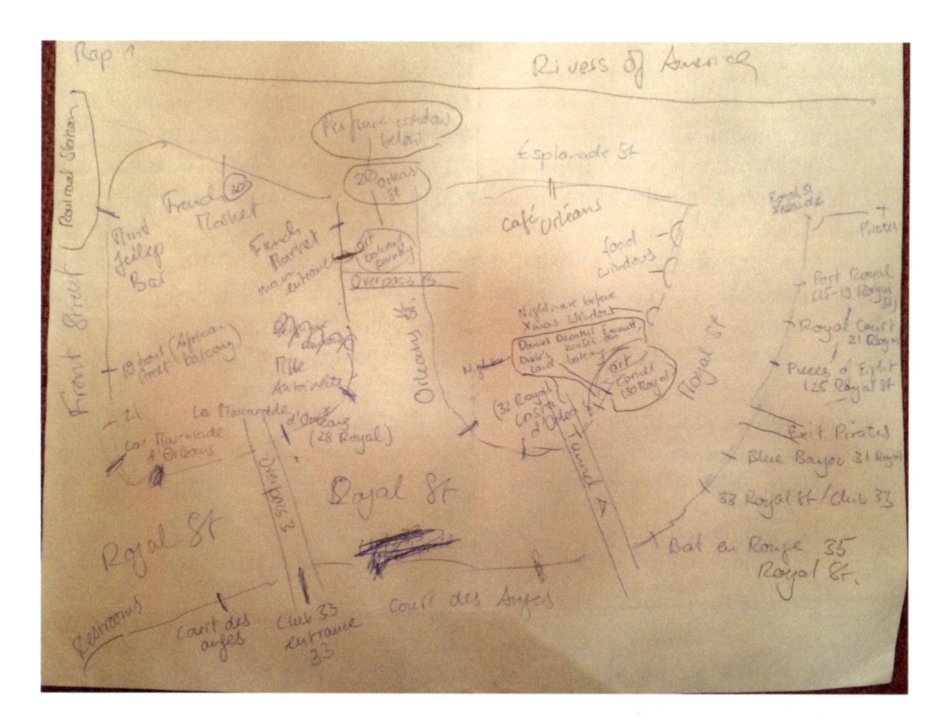

Fig. 2 From the field notes of a theme park researcher: hand-drawn map of "New Orleans Square" at Disneyland (Anaheim) with notes. *Drawing and photograph* Florian Freitag

town in this area, overseas visitors (or young children) who are less familiar with American culture and Disney's narratives and ideology may read this landscape in a different way (see THEMING). Therefore, landscape analysis should not be limited to the intended or assumed meaning of the landscape text. Instead, visitors need to be observed and questioned about what they actually see, hear, smell, and taste, and about whether or how this makes sense to them. For visitors, reading the landscape is usually an unconscious process, and they may frequently have difficulty explaining how they do it and struggle to put into words what they get from this reading. For example, an informal interview with a family of German tourists to Terra Mítica conducted by two of the co-authors of this book within the area dedicated to ancient Egypt (with obelisks and pyramids) revealed not only that these visitors had not grasped that the entire park was themed to ancient Mediterranean civilizations, but also that they were not in the least troubled by that—even if the children did recognize, after further questions and hints, the connection between the themed area they were in and what they learned in history class at school. There is probably always some discrepancy or transmission loss between the message as written by theme park designers and as read by visitors. But of course, the point is not to identify the "mistakes" made by visitors or blame their ignorance, but to understand how, on the basis of their perception, culture, and (often limited) attention, they meaningfully connect to the surrounding landscape—or not.

At other times, the process is very self-conscious and even laborious. For instance, when people get lost in the park, they stop and look around, trying to figure out where they are, searching for signs to find out whether this is (fake) France or (fake) Italy, and trying to make relevant connections between the map they have in their hands and the surrounding landscape. Visitors frequently also pay close attention to the themed area they are in when they sit down to eat, wait in a line, or simply when they enjoy the place and the moment, and perhaps take a picture of some element of the landscape they have noticed and appreciated. In addition to interviews on site, analyzing photographs taken and shared by visitors may thus also give researchers a good idea of what other visitors actually "see," and photo-elicitation (i.e., asking visitors to comment on their own photographs) constitutes an entryway into other people's readings of the landscape. Hence, landscape analysis involves cataloguing not only the signs inscribed in the landscape, but also the various ways in which different people have read and interpreted this landscape, whether these interpretations come in verbal (interviews), textual (reviews, trip reports, scholarly analyses), or visual and multimedia forms (photographs, videos, or even recreations).

12 Participant Observation

Participant observation has not only been conducted during the brief period of the field trip (see, e.g. Waysdorf 2021), but also over longer stretches of time, from travel journals and comparative descriptions of multiple parks (see, e.g. Hendry 2000; Paine 2019) to retrospective analyses by researchers who used to work as employees in the sector (see, e.g. Lukas 2013; Ron and Timothy 2019). This method allows researchers to gain much deeper insights into the design processes or into the ideological framing of human interactions in the theme park. As Bender and Zillinger have pointed out, it is "through the observation of routines that elude predication, through the situated embodiment of the researcher himself, and last but not least through the conflicts and coincidences of stationary field research, that new and unexpected knowledge emerges" (2015, xxxi; our translation).

Participant observation is especially interesting for those who seek to move away from "etic" approaches to theme parks—i.e. approaches that focus more on the researcher's analytical and critical views and sensibilities (see Fjellman 1992; Lukas 2018)—and instead are interested in interpretations of these spaces that emerge from the "emic" (or insider's) understandings of those individuals who work in theme parks, design them, or have a professional or corporate affiliation with theme parks. As has been noted,

> in the literature on themed, immersive, and consumer spaces, there is a notable contrast (and distance) between the interpretations, experiences, and attitudes of the person who visits these spaces (as well as the designers and operators connected to them) and the academic researcher or cultural critic who encounters these same environments. (Lukas 2016b, 261)

This divide is increased by perceptions of some professionals within the theme park industry that academic researchers are condescending in their critiques of theme parks, their designs, and those who work in them; similarly, academics may share the opinion that those who frequent theme parks or who design and operate them may do so without challenging the problematic consumerist, hegemonic, and political issues common to these spaces (Lukas 2018, 172). This notable distance between academic and professional interpretations of theme parks will hopefully be lessened in future research. Collaborations between members of these groups—for example, through "Creative Research" (see above)—could result in powerful impacts on both the academic study and the professional design and operation of theme parks (see Carlà et al. 2016a, 2016b; Lukas 2016a, 168).

However, if long-term participation or participant observation is to be applied as a research method, the biggest hurdle is to get the formal permission to do so. Usually, theme park owners have little interest in letting researchers stick their noses into their affairs and risk negative press, except if there is a benefit for them—which in turn contradicts the rationale of the objectivity of research. And while employees usually have to sign non-disclosure agreements, being a long-term visitor to a park is quite expensive and offers little room for investigation. Therefore, studies using participant observation in the classical anthropological sense are still rare.

Religious or ideologically loaded theme parks were the first to attract the interest of anthropological long-term approaches. Brosius' case study of the Akshardham Cultural Complex in Delhi (India) is one example of in-depth anthropological research in a themed environment. Brosius focuses on the opening ceremony of the structure and a series of rituals that took place there, thus taking into account the complexity of the social interactions and the "ambivalent modes of switching codes (e.g. between religion and secularism)" (2010, 149) in visitor and employee behavior. Among other methods, Callahan (2010) also uses participant observation as a frequent visitor to the Holy Land Experience (Orlando, Florida) for her in-depth study of this particular park. Bielo (2016; 2018) has found a way to conduct classical participant observation in the form of a workplace anthropology by accompanying a team of designers as they produced content for Ark Encounter (Grant County, Kentucky) when the park was still under construction. Although his access was sometimes restricted, he could gain important insights into the life world of these cultural producers and their challenges in bringing the biblical world to life—as artists, inspired by Disney, Hollywood and the like, but also as believing creationists. Ultimately, participant observation challenges researchers to reflect on their own theme park experience and the multiple contrasts that working in hybrid leisure spaces brings with it.

13 Publishing

The challenges connected to theme park studies continue even when the research is done, the data have been gathered, and the conclusions have been written down. The ultimate goal of theme park studies—as with every form of scholarship—is, of course, to share one's work with fellow scholars, professionals, students, and a broader audience through presentations at workshops and conferences, outreach activities, and, perhaps most importantly, publications. Yet, publishing confronts the scholar with new difficulties and challenges. This begins with the question of which journal or publisher the work should be submitted to. The inter- and transdisciplinary nature of theme park studies can make this decision highly difficult: is a publication organ specializing in cultural studies "better" than one focusing on media studies? Or should the authors go for an established journal in the field of tourism studies or architecture? This depends of course very much on the specific research questions asked, but as we have already noted, many research questions connected to theme parks transcend the borders of established disciplines, fields, or research areas—at least until theme park studies establishes itself as a discipline of its own (the launch of the *Journal of Themed Experience and Attractions Studies* in 2018 may be a sign pointing in this direction; see https://www.stars.library.ucf.edu/jteas/). Answering the question of where to publish is even more challenging in the case of works with multiple authors from different disciplines and/or if the authors are emerging scholars: in this case, the choice of publication outlet becomes a highly strategic decision that may have an impact on the researcher's career—a geographer, for example, might not benefit from a publication in an internationally renowned journal for public history, while the visibility of the co-author from history would immensely profit from it.

Once a publication venue has been chosen, the challenges continue. In "See You in Disneyland," the final chapter of his 1992 collection *Variations on a Theme Park: The New American City and the End of Public Space*, Michael Sorkin includes a picture of "the sky above Disney World" (207). As the caption notes, the picture

> here substitutes for an image of the place itself. Disney World is the first copyrighted urban environment in history, a Forbidden City for postmodernity. Renowned for its litigiousness, the Walt Disney Company will permit no photograph of its property without prior approval of its use. (1992, 207)

This illustrates—quite literally so—some of the further challenges that confront theme park scholars. Indeed, depending on the particular sources the research draws upon, rights and releases for interviews and image permissions, for example, may result in some theme park research not being published at all, or lead to severe and context-changing alterations of the work before it goes to print (Lukas 2016a). This applies to material gathered from the parks as well as from private (fan-run) websites: as some publishers today also require publishing permissions for any kind of material (including text) originally published on websites, some manuscripts may need substantial editing before they see the light of day.

Theme park researchers who also work as consultants in the theme park industry further face the requirements and limitations posed by NDAs (Non-Disclosure Agreements). In the case of some NDAs required by some theme park corporations, all work conducted as a consultant for the corporation may be deemed as non-publishable in any form for the life of the consultant-researcher (Lukas 2016a). While contacts with the industry may generally seem desirable, productive, and fruitful (see above), some forms of collaboration also impose limits on the work of the scholar.

These challenges are by no means merely a matter of the private nature of theme park spaces and archives: even if the theme park management agrees, e.g. to the use of photographs for publication, it is still almost impossible to get pictures of a theme park without people—nor would that be particularly useful, as such a picture would give readers a false impression of what theme parks look like and how they work. But if the picture shows visitors, these people have to be asked whether they agree to the publication of the picture—making it virtually impossible to overcome the bureaucratic difficulties. And yet, discussing a ride, a waiting area, the structure of a park, signs, etc. is hard without providing an illustration. The best solution—or rather the only viable one—in these cases is to recur to stock photo collections, although this implies that researchers only have a limited choice and are unable to use the photo they would ideally want in their publication.

And it is not only a matter of illustrations: the analysis of an immersive environment refers to a multisensorial experience that is extremely hard to reproduce via text (and illustrations)—how can you address in a publication the role played by the soundtrack of a ride without adding a sound sample? Beyond copyright problems, it is to be hoped that the new digital technologies and the ongoing change in scholarship and its dissemination will allow researchers to find better and more effective ways of commenting upon and analyzing the theme park than the written word of the printed article or book (or their pdf equivalent, for that matter).

References

Anton Clavé, Salvador. 2007. *The Global Theme Park Industry*. Wallingford: CABI.
Anton Clavé, Salvador. 2010. Leisure Parks and Destination Redevelopment: The Role of PortAventura, Catalonia. *Journal of Policy Research in Tourism, Leisure and Events* 2 (1): 67–79.
Banet-Weiser, Sarah. 1999. *The Most Beautiful Girl in the World: Beauty Pageants and National Identity*. Berkeley: University of California Press.
Baudrillard, Jean. 1994 [1978]. The Precession of Simulacra. In *Simulacra and Simulation*, 1–42. Trans. Sheila Faria Glaser. Ann Arbor: The University of Michigan Press.
Bender, Cora, and Martin Zillinger, eds. 2015. *Handbuch der Medienethnographie*. Berlin: Reimer.
Bielo, James S. 2016. Materializing the Bible: Ethnographic Methods for the Consumption Process. *Practical Matters Journal* 9. http://practicalmattersjournal.org/2016/05/04/materializing-the-bible/. Accessed 28 Aug 2022.
Bielo, James S. 2018. *Ark Encounter: The Making of a Creationist Theme Park*. New York: New York University Press.
Birenboim, Amit, Salvador Anton Clavé, Antonio Paolo Russo, and Noam Shoval. 2013. Temporal Activity Patterns of Theme Park Visitors. *Tourism Geographies* 15 (4): 601–619.
Brosius, Christiane. 2010. *India's Middle Class: New Forms of Urban Leisure, Consumption and Prosperity*. London: Routledge.
Bryman, Alan. 1995. *Disney and His Worlds*. London: Routledge.
Bryman, Alan. 1999. Global Disney. In *The American Century: Consensus and Coercion in the Projection of American Power*, ed. David Slater and Peter J. Taylor, 261–272. Oxford: Blackwell.
Budd, Mike. 2005. Introduction: Private Disney, Public Disney. In *Rethinking Disney: Private Control, Public Dimension*, ed. Mike Budd and Max H. Kirsch, 1–33. Middletown: Wesleyan University Press.
Callahan, Sara B. Dykins. 2010. *Where Christ Dies Daily: Performances of Faith at Orlando's Holy Land Experience*. Diss., University of South Florida.
Carlà-Uhink, Filippo. 2020. *Representations of Classical Greece in Theme Parks*. London: Bloomsbury.
Carlà, Filippo, Florian Freitag, Gordon Grice, and Scott A. Lukas. 2016a. Research Dialogue: The Ways of Design, Architecture, Technology, and Material Form. In *A Reader in Themed and Immersive Spaces*, ed. Scott A. Lukas, 107–112. Pittsburgh: ETC.
Carlà, Filippo, Florian Freitag, Gordon Grice, and Scott A. Lukas. 2016b. Research Dialogue: The Place of the Future. In *A Reader in Themed and Immersive Spaces*, ed. Scott A. Lukas, 301–304. Pittsburgh: ETC.
Cornelis, Pieter C.M. 2011. *Attraction Accountability: Predicting the (Un)Predictable Effects of Theme Park Investments?!* Breda: NRIT Media.
Cornelis, Pieter C.M. 2017. *Investment Thrills: Managing Risk and Return for the Amusement Parks and Attractions Industry*. Nieuwegein: NRIT Media.
Finnegan, Stephen, Steve Sharples, Tom Johnston, and Matthew Fulton. 2018. The Carbon Impact of a UK Safari Park: Application of the GHG Protocol Using Measured Energy Data. *Energy* 153. https://doi.org/10.1016/j.energy.2018.04.033. Accessed 28 Aug 2022.
Firat, A. Fuat, and Ebru Ulusoy. 2011. Living a Theme. *Consumption Markets & Culture* 14 (2): 193–202.
Fischer-Lichte, Erika. 2008. *The Transformative Power of Performance: A New Aesthetics*. Trans. Saskya Iris Jain. London: Routledge.
Fjellman, Stephen M. 1992. *Vinyl Leaves: Walt Disney World and America*. Boulder: Westview.
Freitag, Florian. 2016. Autotheming: Themed and Immersive Spaces in Self-Dialogue. In *A Reader in Themed and Immersive Spaces*, ed. Scott A. Lukas, 141–149. Pittsburgh: ETC.
Freitag, Florian. 2021. *Popular New Orleans: The Crescent City in Periodicals, Theme Parks, and Opera, 1875–2015*. New York: Routledge.
Fung, Anthony, and Micky Lee. 2009. Localizing a Global Amusement Park: Hong Kong Disneyland. *Continuum* 23 (2): 197–208.

Hendry, Joy. 2000. *The Orient Strikes Back: A Global View of Cultural Display*. Oxford: Berg.

Hjemdahl, Kirsti Mathiesen. 2003. When Theme Parks Happen. In *Being There: New Perspectives on Phenomenology and the Analysis of Culture*, ed. Jonas Frykman and Nils Gilje, 129–148. Lund: Nordic Academic Press.

Huang, Xiaoting, Minxuan Li, Jingru Zhang, Linlin Zhang, Haiping Zhang, and Shen Yan. 2020. Tourists' Spatial-Temporal Behavior Patterns in Theme Parks: A Case Study of Ocean Park Hong Kong. *Journal of Destination Marketing & Management* 15. https://doi.org/10.1016/j.jdmm.2020.100411. Accessed 28 Aug 2022.

Kao, Yie-Fang, Li-Shia Huang, and Cheng-Hsien Wu. 2008. Effects of Theatrical Elements on Experiential Quality and Loyalty Intentions for Theme Parks. *Asia Pacific Journal of Tourism Research* 13 (2): 163–174.

King, Margaret J. 1981. Disneyland and Walt Disney World: Traditional Values in Futuristic Form. *The Journal of Popular Culture* 15 (1): 116–140.

Kokai, Jennifer A., and Tom Robson. 2019. Introduction: You're in the Parade! Disney as Immersive Theatre and the Tourist as Actor. In *Performance and the Disney Theme Park Experience: The Tourist as Actor*, ed. Jennifer Kokai and Tom Robson, 3–20. Cham: Palgrave Macmillan.

Kuenz, Jane. 1995. Working at the Rat. In *Inside the Mouse: Work and Play at Disney World*, ed. Jane Kuenz, Susan Willis, Shelton Waldrep, and Stanley Fish, 111–162. Durham: Duke University Press.

Leask, Anna. 2016. Visitor Attraction Management: A Critical Review of Research 2009–2014. *Tourism Management* 57: 334–361.

Liu, Yi-De. 2008. Profitability Measurement of UK Theme Parks: An Aggregate Approach. *International Journal of Tourism Research* 10: 283–288.

Lukas, Scott A. 2013. *The Immersive Worlds Handbook: Designing Theme Parks and Consumer Spaces*. New York: Focal.

Lukas, Scott A. 2016a. Research in Themed and Immersive Spaces: At the Threshold of Identity. In *A Reader in Themed and Immersive Spaces*, ed. Scott A. Lukas, 159–169. Pittsburgh: ETC.

Lukas, Scott A. 2016b. Judgments Passed: The Place of the Themed Space in the Contemporary World of Remaking. In *A Reader in Themed and Immersive Spaces*, ed. Scott A. Lukas, 257–268. Pittsburgh: ETC.

Lukas, Scott A. 2018. The Value of Research. In *CLADbook*, ed. Liz Terry, 172–176. Herts: Leisure Media.

Marling, Karal Ann. 1997. Imagineering the Disney Theme Parks. In *Designing Disney's Theme Parks: The Architecture of Reassurance*, ed. Karal Ann Marling, 29–177. Paris: Flammarion.

Mittermeier, Sabrina. 2016. Disney's America, the Culture Wars, and the Question of Edutainment. *Polish Journal for American Studies* 10: 127–146.

Mittermeier, Sabrina. 2017. Utopia, Nostalgia, and Our Struggle with the Present: Time Travelling through Discovery Bay. In *Time and Temporality in Theme Parks*, ed. Filippo Carlà-Uhink, Florian Freitag, Sabrina Mittermeier, and Ariane Schwarz, 171–187. Hanover: Wehrhahn.

Mittermeier, Sabrina. 2021. *A Cultural History of the Disneyland Theme Parks: Middle Class Kingdoms*. Chicago: Intellect.

Murray, Janet H. 1997. *Hamlet on the Holodeck: The Future of Narrative in Cyberspace*. New York: Free Press.

N.N. N.D. Disneyland Resort Park Rules. *Disneyland*. https://disneyland.disney.go.com/park-rules/. Accessed 26 Dec 2020.

Orellana, Alicia, Joan Borràs, and Salvador Anton Clavé. 2016. Understanding Emotional Behaviour in a Theme Park: A Methodological Approach. In *Consumer Behavior Tourism Symposium Book of Abstracts: Experiences, Emotions and Memories. New Directions in Tourism Research*, ed. Serena Volo and Oswin Maurer, n.p. Bruneck: Competence Centre in Tourism Management and Tourism Economics (TOMTE).

Orvell, Miles. 1993. Understanding Disneyland: American Mass Culture and the European Gaze. In *Cultural Transmissions and Receptions: American Mass Culture in Europe*, ed. Rob Kroes, Robert W. Rydell, and D.F.J. Bosscher, 240–253. Amsterdam: VU University Press.

Paine, Crispin. 2019. *Gods and Rollercoasters: Religion in Theme Parks Worldwide*. London: Bloomsbury.
Pettigrew, Simone. 2011. Hearts and Minds: Children's Experiences of Disney World. *Consumption Markets & Culture* 14 (2): 145–161.
Price, Harrison. 2003. *Walt's Revolution! By the Numbers*. Orlando: Ripley.
Rabinovitz, Lauren. 2012. *Electric Dreamland: Amusement Parks, Movies, and American Modernity*. New York: Columbia University Press.
Raz, Aviad E. 1999. *Riding the Black Ship: Japan and Tokyo Disneyland*. Cambridge: Harvard University Press.
Real, Michael R. 1977. *Mass-Mediated Culture*. Englewood Cliffs: Prentice-Hall.
Ron, Amos S., and Dallen J. Timothy. 2019. *Contemporary Christian Travel: Pilgrimage, Practice and Place*. Bristol: Channel View.
Ryan, Chris, Yeh (Sam) Shih Shuo, and Tzung-Cheng Huan. 2010. Theme Parks and a Structural Equation Model of Determinants of Visitor Satisfaction: Janfusan Fancyworld. *Taiwan Journal of Vacation Marketing* 16 (3): 185–199.
Ryan, Marie-Laure. 2015. *Narrative as Virtual Reality 2: Revisiting Immersion and Interactivity in Literature and Electronic Media*. Baltimore: Johns Hopkins University Press.
Schwarz, Ariane. 2017. Staging the Gaze: The Water Coaster Poseidon as an Example of Staging Strategies in Theme Parks. In *Time and Temporality in Theme Parks*, ed. Filippo Carlà-Uhink, Florian Freitag, Sabrina Mittermeier, and Ariane Schwarz, 97–112. Hanover: Werhahn.
Sorkin, Michael. 1992. See You in Disneyland. In *Variations on a Theme Park: The New American City and the End of Public Space*, ed. Michael Sorkin, 205–232. New York: Hill and Wang.
Steinkrüger, Jan-Erik. 2013. *Thematisierte Welten: Über Darstellungspraxen in Zoologischen Gärten und Vergnügungsparks*. Bielefeld: Transcript.
Torres, Edwin N., Wei Wei, Nan Hua, and Po-Ju Chen. 2019. Customer Emotions Minute by Minute: How Guests Experience Different Emotions within the Same Service Environment. *International Journal of Hospitality Management* 77: 128–138.
Van Oest, Rutger, Harald van Heerde, and Marnik Dekimpe. 2010. Return on Roller Coasters: A Model to Guide Investments in Theme Park Attractions. *Marketing Science* 29: 767–780.
Wang, Jen Chun, Yi-Chieh Wang, Li Ko, and Jen Hsing Wang. 2017. Greenhouse Gas Emissions of Amusement Parks in Taiwan. *Renewable and Sustainable Energy Reviews* 74: 581–589.
Wasko, Janet. 2020. *Understanding Disney: The Manufacture of Fantasy*, 2nd ed. Cambridge: Polity.
Wasko, Janet, Mark Phillips, and Eileen Meehan, eds. 2001. *Dazzled by Disney? The Global Disney Audience Project*. London: University of Leicester Press.
Waysdorf, Abby S. 2021. *Fan Sites: Film Tourism and Contemporary Fandom*. Iowa City: University of Iowa Press.
Weiler, Christel, and Jens Roselt. 2017. *Aufführungsanalyse: Eine Einführung*. Tübingen: UTB.
Williams, Rebecca. 2020. *Theme Park Fandom: Spatial Transmedia, Materiality and Participatory Cultures*. Amsterdam: Amsterdam University Press.
Willis, Susan. 1995. The Problem with Pleasure. In *Inside the Mouse: Work and Play at Disney World*, ed. Jane Kuenz, Susan Willis, Shelton Waldrep, and Stanley Fish, 1–11. Durham: Duke University Press.
Wolf, Werner. 2013. Aesthetic Illusion. In *Immersion and Distance: Aesthetic Illusion in Literature and Other Media*, ed. Werner Wolf, Walter Bernhart, and Andreas Mahler, 1–63. New York: Rodopi.
Yoonjoung Heo, Cindy, and Seoki Lee. 2009. Application of Revenue Management Practices to the Theme Park Industry. *International Journal of Hospitality Management* 28 (3): 446–453.
Zhang, Wen, and Shilian Shan. 2016. The Theme Park Industry in China: A Research Review. *Cogent Social Sciences* 2 (1). https://doi.org/10.1080/23311886.2016.1210718. Accessed 28 Aug 2022.

Paratexts and Reception: Images of Theme Parks in Art, Popular Culture, and Discourse

Abstract This chapter reflects on the depiction of theme parks in various media by the theme parks themselves ("paratexts") and by others ("reception")—two topics that have so far been utterly neglected by scholars. Consisting of advertisements, visitor guides, souvenir maps, guidebooks, and videos, self-reflexively or autothemed rides and attractions, as well as (more recently) websites and apps, paratexts have had a major impact on the theme park experience, shaping it before, during, and after the visit itself. Mainly intended to encourage repeat visits, paratexts have taken theme parks' representational politics of inclusion and exclusion to a new level, highlighting specific aspects of the parks (e.g. their photogenic vistas) while deemphasizing others (e.g. costs and waiting times). This is in contrast to independently created novels, documentary and fictional movies, and TV series about theme parks, where the theme park often serves as a setting for stories on disorder, terrorism, crime, deviance, dystopia, social satire, and existentialism. Here, the theme park becomes a "text" and serves to both express the realities (and contradictions) of the world while also providing an evocative setting for fictional media worlds and their stories. The final section of this chapter reflects on the metaphorical use of the term "theme park" to describe a large variety of places.

1 Introduction

This chapter surveys how theme parks are depicted in mediated versions of themselves, in fictional and documentary media artifacts, as well as the broader use of the theme park as a "metaphor" for various places. Applying literary critic Gérard Genette's term "paratexts" (from Greek *pará-*, "beside, adjacent to," and "beyond, or distinct from, but analogous to") to such media as film and television, Jonathan Gray has argued that media paratexts like ads, previews, trailers, posters, billboards, bonus materials, and spinoffs "establish frames and filters through which we look

This work is contributed by Florian Freitag, Scott A. Lukas, Céline Molter, Salvador Anton Clavé, Astrid Böger, Filippo Carlà-Uhink, Thibaut Clément, Sabrina Mittermeier, Crispin Paine, Ariane Schwarz, Jean-François Staszak, Jan-Erik Steinkrüger, Torsten Widmann. The corresponding author is Florian Freitag, Institut für Anglophone Studien, Universität Duisburg-Essen, Essen, Germany.

F. Freitag et al., *Key Concepts in Theme Park Studies*,
https://doi.org/10.1007/978-3-031-11132-7_12

at, listen to, and interpret the texts that they hype" (Gray 2010, 3). In Gray's study, theme parks merely function as paratexts for other media texts (namely, in the shape of "a new ride at an amusement park" that promotes e.g. a movie; see Gray 2010, 29), but of course theme parks also have their own paratexts: ads, signs, announcements, maps, leaflets, videos, websites, apps, and other representations of the theme park or its parts that are produced by the park itself and that serve as a medial interface between the park landscape and its visitors, providing the latter with "frames and filters" or "scripts" for how to experience the park. Theme park paratexts sometimes become souvenirs and collectors' items (see Fig. 1) and have frequently been used as sources in theme park research (see METHODS; see also Williams 2021, 142) but have rarely featured as objects of research themselves.

However, as this chapter will show, as well as offering practical information they also support the theme park's underlying economic and aesthetic strategies and thus shape the theme park experience in important ways. By contrast, fictional and documentary representations of theme parks that are produced by others generally take a much more critical look at both individual theme parks and the form in general. As in the case of theme park paratexts, however, scholars have so far shown surprisingly little interest in these artistic critiques of and responses to theme parks. Finally, this chapter also offers a first look at the reception of theme parks in critical discourse,

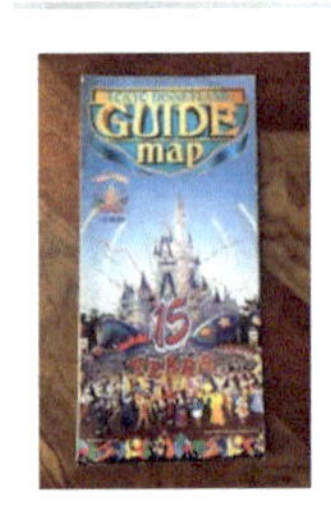

Fig. 1 Once handed out for free, some old theme park guide maps have become collectors' items and are traded on the internet. *Screenshot* Florian Freitag

where the term and the concept have also been used in a decidedly critical way to call attention to specific developments in the tourism industry.

2 Theme Park Paratexts: Mediated Landscapes

Theme park paratexts are distributed in a variety of ways to visitors, are consumed by them at various points in time, and fulfill several functions. With respect to distribution, they may be sold in the park's shops (e.g. postcards and other souvenirs, including souvenir maps, videos/DVDs, and coffee table books) or displayed and distributed for free either on site (e.g. physical and virtual guide maps, promotional posters and videos, and augmented reality features), or elsewhere (e.g. print and TV ads, brochures and leaflets, and websites). Theme park visitors may thus consume them before, during, and/or after their visit. Finally, some theme park paratexts primarily seek to advertise the park and familiarize visitors with its various offerings, whereas others mainly intend to provide orientation to visitors, add to their immersion, or simply entertain them (see Younger 2016, 271). Yet regardless of how they are distributed, when they are consumed, and what their primary function is, theme park paratexts always serve as a medial interface between the visitor and the park landscape. As such, they support the theme park's general "politics of inclusion/exclusion" (Lukas 2007, 277; see also INCLUSION AND EXCLUSION), stressing certain aspects of the park while deemphasizing others, and thus have a major impact on how visitors anticipate, experience, and remember the park. Theme park maps, which scholars have discussed in detail, are a case in point (see also SPACE), but this observation also applies to other paratexts, and notably those which take a diachronic approach—as coffee table books, websites, and self-reflexive events, exhibitions, and rides often do—and tell the history of the theme park.

From a historical perspective, it is also significant that the theme park was promoted as a mediated landscape long before it became a real, physical landscape. In September 1953, almost two years before the opening of Disneyland, Walt Disney had studio artist Herbert Ryman draw an aerial-view map of the park in order to attract investors to the project. A year later—while Disneyland was still under construction—another studio artist, Peter Ellenshaw, created a four-foot-by-eight-foot overhead view of the park that was, thanks to the use of fluorescent paint, even more dramatic than Ryman's rendering. Ellenshaw's painting was prominently featured on yet another pre-opening paratext of Disneyland—namely Disney's weekly show on ABC, which not only took its name from the park but also used an anthology format to familiarize viewers with Disneyland's themes and layout, and provided Disney with the necessary funding to realize the project (see Telotte 2004, 8–12; see also MEDIA). Even more importantly, during its first season the show also regularly discussed the park and offered "progress reports" from the construction site. The premiere episode of *Disneyland*, for instance, broadcast on 2 October 1954 and entitled "The Disneyland Story," showed host Walt Disney standing in front of the Ellenshaw painting and proclaiming: "'That's it, right here, Disneyland, seen from about 2,000 ft in the air, and ten months away'" (Florey and Jackson 1954, 03:53–04:00).

After it was opened ten months later, in July 1955, however, Disneyland would never quite look as depicted by Ellenshaw (or Ryman, for that matter)—that is, with the corners of the park filled with rides and neatly landscaped areas rather than backstage facilities and maintenance roads, with the backsides of "Main Street" buildings featuring detailed facades rather than looking like warehouses, with light crowds evenly dispersed throughout the site rather than gathering at bottlenecks and in waiting lines, and with not a single employee in sight. Of course, Ellenshaw was not a naïve artist who had failed to anticipate the realities and necessities of designing and operating a theme park; with its deliberate additions, omissions, and edits, Ellenshaw's painting showcased the park exactly as Disney wanted visitors to perceive and experience it. Yet while the real park designers and operators could merely do their best to deemphasize certain aspects of Disneyland—architectonically hiding backstage areas, rhetorically disguising employees as hosts and work as performance (see LABOR), always with the risk that visitors might accidentally glimpse the inner workings of the machine—in his map, Ellenshaw was free to entirely erase these aspects. As such he could, as Stephen Yandell has argued, support Disneyland's "simulation more effectively than the park itself" (2012, 25).

In fact, according to Yandell, all of Disneyland's maps—including those by Ryman and Ellenshaw, but also (and especially) the regularly updated, intricately labeled "fun maps" first drawn by Sam McKim and sold as souvenirs at the park from 1958 onwards—rely on what he terms "carto-reality." While their use value for the purposes of navigation and orientation might be limited, with elements omitted and added, and distances and scales adjusted at liberty, they depict the park "with a level of control and perfection not achievable by moving through a hands-on world, even through a simulation like Disneyland" (2012, 25). Yet this "agenda of perfection," as Yandell also calls his concept (2012, 22), is not only supported by the Disneyland maps, it informs virtually every other sort of theme park paratext, too: tellingly, Ellenshaw's painting was not only featured on the *Disneyland* TV show but also served as the motif of Disneyland's first ever postcard and was reprinted on the cover of the 1955 Disneyland souvenir book (see Strodder 2012, 154). Thematically odd sightlines and other visual intrusions, construction sites, overcrowded areas, long waiting lines, unhappy visitors, and any reminder of the theme park as a workplace and a commercial enterprise are either entirely omitted from or at least seriously downplayed in all theme park paratexts. The Disneyland preview and souvenir maps may have set the "agenda of perfection," but other, later paratexts have adapted this agenda to their individual formats and respective functions.

This is well illustrated by one of the closest "relatives" of the theme park souvenir map and, simultaneously, one of the most common forms of theme park paratexts: namely the guide map (i.e. the paper leaflet or brochure usually given out for free to visitors at park entrances and elsewhere). Throughout its history, this paratextual genre has undergone several dramatic shifts in both form and function. Early

Disneyland guide maps from the 1950s, for instance, were portrait-oriented, four-page paper leaflets in black, white, and green that folded out to landscape orientation and appeared to be primarily concerned with communicating the basic park concept and layout to visitors. While the front dedicated two entire pages to written descriptions of Disneyland's five individual "lands," each illustrated with a small line drawing, the back featured Walt Disney's dedication speech, a brief explanation of the park's general layout, as well as a large, illustrated park map, also line drawn and labeled with letters that allowed visitors to locate the park's various subsections (though not its individual attractions, restaurants, or shops). Later, logos of Disneyland's "lands" were added to the front page of the guide map, presumably in an attempt to even more firmly "locate the park's [...] overall layout in the minds of the customers" (Lukas 2008, 104).

By the mid-1960s, Disneyland had switched from folded leaflets to stapled, 30-page brochures printed in multiple colors and illustrated with black-and-white or tinted photographs. The guide maps now offered detailed descriptions of general park services, the Disneyland A-B-C-D-E ticket system, individual rides, exhibits, restaurants, and shops, as well as an entertainment schedule and schematic land maps that were decorated with simple line drawings and used numbers and shapes to allow visitors to precisely locate each ride, restaurant, shop, etc. within a given "land." They also featured a highly abstract park map, however, that employed color-coding and labels in thematically appropriate fonts to identify the location of each of Disneyland's "lands" within the park, thus testifying to Disney's continuing concern with familiarizing visitors with the general layout and concept of the park.

The general structure of the Disneyland guide maps would be left more or less untouched until around the mid-2000s; the late 1970s, however, saw an interesting innovation in terms of theme park paratexts: sponsored by GAF (a manufacturer of photographic equipment), the guide maps now not only featured full color photographs and three pages of "GAF Photo Tips," the general park map also indicated the various stops of the "GAF Photo Trail," a series of spots that had been "carefully researched and selected as the best locations for taking well-composed photographs" (N.N. 1977, 14–15) and which were marked in the actual park landscape by "Photo Trail" signposts. Hence, Disneyland sought to extend its "agenda of perfection" even to such visitor-produced paratexts as souvenir photographs and movies, prescribing exactly what should be photographed in the park and how it should be done (with the introduction of "roving" photographers and the "Photopass" system in the 2000s, Disney later attempted to gain even more control over visitors' pictorial souvenirs).

The new additions were kept even when sponsorship of the guide maps switched from GAF to Polaroid and, later, to Kodak; yet in the mid-2000s, Disneyland completely redesigned its guide maps, reverting to the old "folded leaflet" format. As in the 1950s, the front part once again offers general information about the park in textual form, while the back features a line-drawn, illustrated park map. But instead of explaining Disneyland's individual themed "lands," the front part now introduces Disney's various "pass" systems (Photopass, Fastpass), while the map on the back no longer just allows visitors to locate the individual themed areas (via color-coding and labels) but also each of their attractions, services, and restaurants (via numbers, letters, and icons that correspond to a detailed legend). Clearly, Disney assumes that by now the park's general concept and overall layout have been firmly etched into visitors' minds and instead uses the guide map to further drive its "agenda of perfection," hiding backstage facilities and construction sites behind "innocuous clumps of trees" (Yandell 2012, 34) on the map, deemphasizing the park's essentially commercial agenda by not even listing the large number of shops and merchandise carts, and advertising "solutions" to such potential imperfections or problems as long waiting lines (Fastpass).

Most theme parks today have adopted Disney's "folded leaflet" format, with several interesting variations: some guide maps include entertainment schedules, dispensing with the need for an additional show times guide (e.g. Europa-Park or Hansa-Park, both located in Germany); some include road maps as well as opening hours for the park, indicating their use as not just guide maps during the park visit, but also as advertising brochures (e.g. Belantis, Germany; or E-Da, Taiwan); some list attractions and restaurants not according to themed areas, but according to target groups, like families, thrill-seekers, small children, etc. and price ranges, respectively, thus emphasizing practical visitor concerns over the thematic organization of the park (e.g. Parc Astérix, France); others have replaced the intricately detailed, painted reality style of Disney's park maps with more thematically appropriate art styles, thus seeking to carry over the park's distinct thematic identity into its paratextual representation (see Younger 2016, 272). Perhaps even more importantly, many guide maps now include links or QR codes to direct visitors to the park's website or digital guide maps, which will be discussed in more detail below.

It should be noted that there are more and more guide maps and guidebooks for popular theme parks such as Disney and Universal that are not produced by the respective parent companies. Indeed, numerous free fan sites and online fora are dedicated to planning strategies for visitors (see also VISITORS). Increasingly, however, these planning guides and tools have become their own industry of paratexts—the *Unofficial Guide* book series, for instance, is updated every year, includes titles on Walt Disney World Resort, Disneyland Resort, as well as Universal Orlando Resort and Universal Studios Hollywood, and is available as a print publication in bookstores and online. The same group behind the books also runs an online service called Touring Plans that predicts crowds and makes full crowd calendars available for paying customers months ahead of time. The DisneyFoodBlog, meanwhile, in addition to regular blog posts, self-publishes paid e-books dealing with all aspects of dining in Walt Disney World, ranging from snack and restaurant recommendations and reviews to how-tos for scoring popular dining reservations, as well as coverage of Epcot's popular seasonal festivals.

A much less common, though with respect to the "agenda of perfection" no less interesting and relevant paratextual genre is the self-reflexive or "autothemed" (see Freitag 2016) theme park ride or show. Just like the guide map, the autothemed space or show serves as a medial interface between the physical park landscape and the visitor during the park visit, but does so using the medium of the theme park itself. Autothemed elements are often introduced on the occasion of park anniversaries:

thus Disneyland's Parade of Dreams, which featured references to several of the park's iconic buildings and attractions, premiered in 2005 as part of Disneyland's 50th anniversary celebrations, and for Europa-Park's 35th anniversary in 2010, the park opened the Historama, a self-reflexive revolving theater show. The latter is notable for being probably the first ever autothemed ride at a theme park: in its four main acts, the "Historama" used animatronics, projections, models, sounds, lights, and other special effects to not only tell the history of Europa-Park and its parent company (Acts 1 and 2) but also to present the park's main rides (Act 3) and shows (Act 4)—which are located right outside and, in some cases, just a three-minute walk away from the Historama itself. Yet similar to the Historama's portrayal of Europa-Park's history, its depictions of the park's attractions plainly demonstrated the effects of theme park paratexts' "agenda of perfection": the collage of point-of-view shots of Europa-Park's main coasters used in Act 3, for instance, only offered visitors the highlights of the various rides (e.g. the "big splash" in the case of water coasters) and cut from one coaster to the next without any segue. The collage showed neither visitors waiting in line nor people getting on and off the vehicles, nor people walking from one coaster to the next, let alone breakdowns or accidents. Until its closure in 2017, the "Historama" thus gave visitors a pure yet simplified coaster experience, without any of the negative "side effects" that actually riding the coasters may entail.

Incidentally, the point-of-view shots used for the Historama have also been featured on the souvenir DVDs sold in the park's shops as well as on the park's website, YouTube channel, TV ads, and mobile app. Electronic media have become central to communicating and promoting the park landscape to visitors before, during, and after the visit. Perhaps most importantly, digital theme park paratexts frequently include filter options that, in addition to the selections already made by the creators, allow customers to further tailor the paratextual representation of the park landscape to their personal needs and preferences. Hence, the "agenda of perfection" is partly handed over to customers, as they turn the theme park into their own, personalized theme park. Once again, this is perhaps best illustrated by (virtual) guide maps, which are now available for most of the larger theme parks in the form of mobile apps for visitors' electronic devices. Park apps generally offer a large number and variety of features, including targeted advertising, opening hours, weather forecasts, waiting times for individual attractions, online ticket and souvenir shops, entertainment schedules (and changes), reservation services for hotels and restaurants, links to services such as Fastpass or Photopass, social media features, and even thematically appropriate games.

Most importantly, however, mobile apps employ GPS technology to position visitors directly on the park map (generally the illustrated park map also used for the print version of the guide map, not the high-resolution satellite imagery available on Google maps, which would work directly against the "agenda of perfection"; see Yandell 2012, 35 and Younger 2016, 271). Via multiple filter options, visitors can look for the nearest restrooms, interactive rides suitable for small children, restaurants offering gluten-free products, or shops selling specifically themed merchandise—and can even get directions to arrive as quickly as possible. In addition, search results usually offer links to detailed descriptions of the respective space, including pictures and videos, technical data and safety restrictions (for attractions), as well as review options, and can be put on a personal "favorite" list. Hence, the already highly selective paratextual representation of the park landscape on the printed guide map becomes even more selective in the electronic version, as elements that do not match the visitor's filters are not even indicated—indeed, on the electronic map, elements that have been filtered out by customers become as invisible to them as closed attractions and backstage areas. By putting customers on the map, both literally and metaphorically, mobile apps thus not only address the problems of visitors being distracted and the park's immersive potential being reduced by personal electronic devices (see Younger 2016, 228), they also directly involve customers in theme park paratexts' "agenda of perfection."

Since theme park experiences are ephemeral, souvenirs serve as a means to store the memories and take them home, and thereby form a paratextual genre of their own. Souvenirs "carry the magic of place" (Löfgren 1999, 87) and while one reason for their purchase might be the being-there-proof (especially for Disney pilgrims), they are also symbolically charged public relations tools that store core messages or values of a park. The signature souvenirs from Disney parks are costumes, the most popular of which are the mouse ears: wearing them visibly turns the visitor into a member of the Disney community (see VISITORS). The souvenirs sold at Ark Encounter, a creationist Noah's Ark theme park in Grant County, Kentucky, focus on truth and authenticity: the shops sell artisanal products, fudge, and books, but a signature souvenir is "Noah's cubit," a wooden stick of the assumed length of a cubit, the Biblical measurement unit for the Ark. Accordingly, "Noah's cubit" symbolically condenses Ark Encounter's message and mission: to prove the Bible true.

Theme park souvenirs also frequently use mascots that embody or personify the parks' messages. For example, Phantasialand's mascots are six different dragons, one for each of the themed areas (the symbolism is telling though—"Berlin" has a mighty golden dragon, whereas "Deep in Africa" has a lazy but cheerful crocodile). Apart from being used as souvenir branding in the central House of the Six Dragons shop, however, the dragons do not play much of a role in Phantasialand. Efteling (Kaatsheuvel, the Netherlands), meanwhile, has numerous mascots, but introduced the wizard Pardoes as the "charming host" for the park in 1989. Ten years later, his girlfriend Pardijn was added, and in 2017 the dark ride *Symbolica* was opened, which narrates Pardoes' background story. Since then, his LED-lighted magic wand has been a popular souvenir, allowing visitors to literally carry the magic of Efteling home with them.

Theme park paratexts not only show the parks themselves, however. Print, TV, and online ads, as well as souvenir postcards, videos, and books also frequently depict theme park actors, both visitors and employees (including costumed characters; see also LABOR and VISITORS). A quantitative study of the still images used on the official websites of the five Disney "castle parks" in Anaheim, Orlando, Tokyo, Paris, and Hong Kong, found that with respect to gender, depictions of male and female visitors were more or less balanced across the parks, but that Tokyo Disneyland showed almost nine times as many female as male employees (excluding characters; see Auster and Michaud 2013, 8). With respect to race, whereas images of the parks in

Anaheim and Paris as well as Tokyo and Hong Kong showed predominantly white and exclusively Asian visitors, respectively, pictures of Orlando's Magic Kingdom showed significantly more Black and Asian visitors (12). Additional research on other forms of paratexts, as well as involving other variables (age, disabilities, etc.), may determine whether theme park paratexts are also driven by the "agenda of perfection" in this context, or whether they reflect actual instances of inclusion and exclusion (see INCLUSION AND EXCLUSION).

3 Theme Park Texts: Theme Parks in Popular Culture and the Arts

The theme park has a rich and interesting history in terms of its portrayal in popular culture and the arts—films, fiction, music, video games, paintings, photography, performances, and art installations have frequently depicted or referenced theme parks. Such representations of theme parks within media worlds draw as much on existing theme parks and their paratexts (the 1977 movie *Rollercoaster*, for instance, was filmed at several "real" American parks and visual artist Blain Hefner's "Arctic World" references the style and layout of Disneyland's attraction posters) as they do on theme park criticism (with, for example, a "simulation" of Jean Baudrillard appearing as a character in Julian Barnes's 1996 novel *England, England*). In fact, theme park criticism itself (see INTRODUCTION) arguably constitutes a "theme park text": a media portrayal in which the theme park "moves from being a themed landscape full of thrill rides to a place of the mind—a mental image" (Lukas 2008, 216). In its state as a text, the theme park not only provides an evocative setting for fictional media worlds and their stories, but also functions to express realities (and contradictions) of the world.

There are numerous documentary films that focus on the myriad issues connected to theme parks. The most notable film is *Blackfish* (2013), which focuses on the issues connected to the captivity of orca whales at theme parks like SeaWorld. The popularity of the film led to a public relations scandal and, in 2016, the ultimate end to the orca breeding program and live orca performances at SeaWorld theme parks. Other documentary films about theme parks, including those that have aired on travel- and science-related television channels, have focused attention on the dynamics of theme park development, including new and innovative rides. The spread of videos on social media fora has resulted in additional attention being given to theme parks and their attractions. A number of popular YouTube channels are composed of content created entirely on park rides, often coordinated with their grand openings. Additional YouTube and social media theme park content often relates to notable ride accidents, fires, or other such spectacular situations. One site, Theme Park Review, had 1.76 million subscribers as of 2022.

Another topic popular on social media fora and websites are abandoned theme parks. Numerous so-called urban explorers have sought to capture, in photographs and videos, the atmosphere of defunct parks from Japan's Nara Dreamland (see ANTECEDENTS) to the amusement park in Pripyat (Ukraine), the latter abandoned, along with the rest of the city, after the 1986 Chernobyl disaster. The material published by Florian Seidel on his "Abandoned Kansai" website, for instance, suggests a dystopian, post-apocalyptic (theme park) world and thus contrasts sharply with the imagery of theme park paratexts and their "agenda of perfection" (see above). At the same time, urban theme park exploring also points to some of the practical problems of conducting research in theme parks, as well as in commercial spaces in general (see METHODS), as most (abandoned) theme parks are private spaces and therefore, technically, urban exploring constitutes trespassing.

Theme park fans and aficionados have also imagined their own theme parks, however. On fan websites, message boards, and other social media, they engage in "online memorialization" (among other things; see VISITORS) by archiving material about theme park elements (rides, shows, restaurants, shops) that have been closed or replaced (see Williams 2020, 211–242). They also employ media such as drawings and models to visualize what the park(s) would look like if they were in charge. Using existing parks as starting points and combining, for example, currently operating and long-gone elements, as well as elements from various parks of the same chain, fans can thus create semi-fictional depictions of parks that occasionally reflect selected principles of theme park planning and management (see ECONOMIC STRATEGY), but which mainly respond to their own, personal vision of their favorite park.

Entirely fictional films and novels have also been a popular foundation for narratives related to theme parks (see Lukas 2008, 212–245). Some of the themes considered in such stories include: the contrast between order and disorder; terrorism, crime, and deviance; the contrast between dystopia and utopia; social satire and existentialism; and futuristic visions. In terms of order and disorder, novels like *Utopia* (Child 2003) and films and television series like *Westworld* (2016–) use the theme park as a foundation for considering the human condition among situations of entropy. Theme parks are often modeled on order, and considerations like those in these films and novels suggest an existential theme when humans are unable to control their own theme park creations. The film *Escape from Tomorrow* (2013), which is famous for being filmed surreptitiously at Walt Disney World and Disneyland, takes the traditional trope of the orderly, family-friendly theme park and turns it on its head: the protagonist's trip to a Disney theme park with his family is interrupted by his bizarre hallucinations of theme park rides, lusting after young French women, drunkenness and spats of vomiting, and encounters with sinister corporate robots. Pat Rocco's *Ron and Chuck in Disneyland Discovery* (1969), covertly filmed at Disneyland, by contrast, upsets the park's heteronormative order (see INCLUSION AND EXCLUSION) by telling the story of two young men falling in love at Disneyland.

Crime, deviance, and terrorism are often the subject of fictional theme park representations. Numerous young adult novels, such as *Disneyland Hostage* (Wilson 1995), imagine young people battling criminals and terrorists in Disney theme parks. One of the earliest examples of this genre is *Rollercoaster* (1977). In the film, a terrorist who seeks ransom money from theme park owners places bombs on roller coasters, instilling fear in the theme park-going public. A similar theme is explored in Lincoln Child's novel *Utopia* (2003). Numerous other novels and films, including *Slayground* (1983), *Final Destination* 3 (2006), and *House of 1000 Corpses* (2003), use theme park rides as situational contexts for violence, murder, and crime.

Because theme parks have been considered as utopian visions of society, a number of popular novels and films have reimagined the theme park as a dystopian entity. *The Florida Project* (2017) is a film set in the outskirts of Kissimmee, Florida, near Walt Disney World. The film imagines the harsh and chaotic family dynamics of working-class people who struggle with the day-to-day realities that contrast with the fantasy worlds of Walt Disney World visitors. Social satire has been a popular theme in many theme park novels. British writer Julian Barnes uses the fictional theme park of *England, England* (1996) to interrogate themes of authenticity, nationalism, nostalgia, and tradition in British society. A few novels and films have also used the setting of the theme park to tell existential tales. Stanley Elkin's *The Magic Kingdom* (1985), in which terminally ill English children make a pilgrimage to Walt Disney World, and the Canadian film *Rollercoaster* (1985), in which themes of teen angst, suicide, and depression are realized in an abandoned amusement park, are reminders that theme parks may be the subject of more serious social and philosophical issues.

Another variety of theme park novels and films extends theme parks into the realm of fantasy. The popular *Jurassic Park* and *Jurassic World* films (and their related media) experiment with the idea of a theme park based on genetically reincarnated dinosaurs. The HBO series *Westworld*—and its two previous film versions, *Westworld* (1973) and *Futureworld* (1976)—similarly imagines a future of theme parks in which it is impossible to tell the robot workers from the human visitors—a reversal of the familiar front- and back-stage trope of theme parks. Very few fictional films have focused on life inside a theme park as their direct subject; one exception is *The World* (2004), which follows the lives of two theme park workers at Beijing World Park.

Like their filmic counterparts, video games with theme park content often rely on familiar social, cultural, and existential tropes. Two general types of theme park video games include those focused on ride simulations (the *RollerCoaster Tycoon* series) and those emphasizing the organization and running of a virtual theme park (Theme Park World, Sim Theme Park). A new genre of such videos games involves the branding of the familiar rides and park operations into more specific re-presentations of real theme parks, as is the case with *Efteling Tycoon* (2008), which took the familiar simulations of the *RollerCoaster Tycoon* series and branded them to the landscapes and attractions of Efteling. While they are simulations of real theme parks, such games are notable for their highlighting of many of the actual day-to-day contexts of theme parks. Thru-put and efficiency of rides and attractions, the dynamics of guest queuing, guest satisfaction, and ride and attraction breakdowns are a few of the issues that such games profile (Lukas 2008, 227–229). Notably, they introduce the contrast between order and disorder as a primary concern. In such games, the player's success is determined by the extent to which he or she can balance the variety of park rides, attractions, decorations, and services with guest interests and desires, staffing limitations, and other issues. Like their real-world equivalents, theme park video games emphasize that park disorder or entropy is just around the corner.

Apart from theme park management games, others use the theme park as a narrative trope or to transmediate theme park landscapes into the gaming world. In the Monkey Island adventure game series, the hero Guybrush Threepwood must riddle his way through a seventeenth century Caribbean pirate world full of clichés. At the end of part 2, after hours of pirate adventures, the puzzled player is confronted with a plot twist: Guybrush suddenly exits the pirate world, which in fact is only part of an amusement park. Looking back, there had been some hints, like strangely modern vending machines, a rubber chicken, and other tools which were clearly out of place in the seventeenth century setting. However, to the player who has spent so much time on Monkey Island, this twist is somewhat appalling—the strenuous journey was all fake. Monkey Island transmediates "*the cultural logic of immersion* in theme parks" into the video game genre (Makai 2019, 165), simultaneously mocking and praising the theme park medium: the pirate world was make-believe, a rip-off of popular tropes, stories, and even Disney rides (for more on this, see Makai 2019), just like a theme park world.

Theme parks have also been represented in popular music, notably in *Tragic Kingdom* (1995), the breakthrough album of Anaheim, California-based rock band No Doubt. Written by the band's founding member Eric Stefani and performed by lead singer Gwen Stefani, both frequent visitors to Disneyland in their youth in the 1970s and 1980s, the album's title track is a play on "Magic Kingdom," the park's unofficial nickname. Opening with audio recorded at Disneyland, including announcements heard on Disneyland's Matterhorn ride, the song takes a rather cynical view of what Stefani believes to be the increasing commercialization of the park and denounces Disneyland's new leadership after the death of Walt Disney in 1966. The song ends with the refrain "Welcome to the Tragic Kingdom" repeated with increasing speed and overlapping of instruments, evoking an image of a ride spinning out of control. Despite the album's cynical take on Disneyland, the band has performed at the park many times throughout its career, including Grad Night in 1995, just months before the release of *Tragic Kingdom*, and again in November 2002, where the band was awarded the Key to Anaheim City. In addition to the band's music, No Doubt's marketing imagery in the 1990s was also influenced by Disneyland. This was especially apparent on the band's official website, which was designed to look like a theme park guide map, with the image of a gondola providing a link to the visitor chatroom and the image of a train circling the park linking to the band's fan club membership.

The design of No Doubt's website provides us with a segue to the depiction of theme parks in visual art—drawing, painting, photography—as well as forms of "high art" such as installation and performance. Like theme parks themselves (see Marling 1991; Jodidio 1992; Neuman 1999, 254; and Sotto 2017), concept art produced in the context of theme park design and marketing has sometimes been received as art in its own right. In 1964, for instance, Herbert Ryman, who had already drawn the first map of Disneyland (see above), created a large opaque watercolor and pen & ink painting entitled "The Square," which depicts Disneyland's then-soon-to-be-opened "New Orleans Square" section. "The Square" has been regularly printed in Disneyland ads, souvenir books, and other paratexts; reproductions of the painting can also be found in several Disney resorts (e.g. the lobbies of Disney's Port Orleans—French Quarter at Walt Disney World and the Disneyland Hotel at Disneyland Paris), but it has also achieved distinction as the "first piece of Disney artwork to inaugurate the State Department's Art in Embassies Program" and was displayed in the U.S. Embassy in Paris in the late 1980s (Gordon and Mumford 2000, 180).

Whereas "The Square" and many other artworks by Ryman have a sketchy quality and show the parks thickly peopled with visitors and diegetic inhabitants (in the case of "The Square," for example, two nuns and a painter), the primitivist paintings of Orlando, Florida-based artist Larry Dotson usually depict the Disneyland and Walt Disney World landscapes as entirely devoid of visitors, or only populated with Disney characters. Though marketed as souvenirs via the artist's own website and via the shops at Walt Disney World, where Dotson also holds regular signing events, Dotson's works thus assume, despite their bright colors and cheerful titles (e.g. "Springtime Waltz"; "Figment Celebrates Epcot"), a somewhat eerie, even post-apocalyptic quality.

Much more openly associating Disney's theme parks with catastrophic events and dystopian scenarios are the works of Southern California-based artist Jeff Gillette, whose "slumscapes" combine depictions of urban wastelands and residential slums with iconic symbols of Disney parks, including the Disneyland marquee, Sleeping Beauty Castle, and the Matterhorn Mountain roller coaster. While the latter are sometimes portrayed in pristine condition and hover as a hazy background over the slums in the foreground to mark an ironic contrast between "First" and "Third World," they also at times resemble decrepit leftovers from a different time. In 2010, Gillette's paintings were exhibited at the CoproGallery in Santa Monica, California, under the title "Dismayland" and caught the interest of British street artist Banksy, who eventually invited Gillette to feature his work in "The Galleries" section of Banksy's (2015) avant-garde "bemusement park" Dismaland.

Dismaland is perhaps the best known of the many art installations that have drawn on the iconography and the somatic and immersive strategies of (Disney) theme parks. Some of these, such as the various haunted house-inspired art installations examined by Tina Klopp (2014) or Julijonas Urbonas's "Euthanasia Coaster" (2010), have been inspired by individual theme park elements and primarily focus on offering sensorial and bodily experiences—in the case of Urbonas's work, for instance, the euphoric experience that would precede death by prolonged cerebral hypoxia caused by the massive G-forces that the riders of Urbonas's coaster would be exposed to, if it were ever built. Other installations draw on entire parks and use them as conceptual devices as a way of dealing with social, political, and aesthetic issues. In the 1990s, for instance, the artists Laurie Anderson, Peter Gabriel, Brian Eno, and filmmaker John Waters had planned an avant-garde theme park known as The Real World (Lukas 2007, 285; Rawlins 1994). This would have featured rides and attractions—sixty-foot-high working tornadoes and attractions designed by controversial filmmaker John Waters, among others—that were deliberately envisioned as contrasting with traditional theme park design. Building on the artist's previous engagement with theme parks (see his documentary *Exit through the Gift Shop*, 2010) and using the familiar iconography of theme parks such as Disney's

Cinderella Castle and SeaWorld's orcas, Banksy's Dismaland confronted visitors with surreal, critical, and explicitly political narratives. As the artist himself stated in an interview, Dismaland proclaims that "theme parks should have bigger themes" (Banksy 2015) and thus "engages with the medium to creatively extend its heretofore rather limited range of expression and to give it a social and cultural relevance that matches its enormous worldwide popularity" (Freitag 2017, 2). In 2014, a third project was announced. This theme park, if built, will be located in Ohio and is the idea of filmmaker Wes Anderson and musician Mark Mothersbaugh (Berman 2014). One overarching foundation of The Real World, Dismaland, and similar projects is their discontent with the traditional orderly and conservative foundations of theme parks and a shared interest in what has been classified as "dark theming" (the representation of controversial, surreal, and dark topics and contexts in immersive spaces; see Lukas 2007, Lukas 2016; see also THEMING).

Dismaland featured not only two-dimensional artworks, sculptures and dioramas, and even working rides, but also performances in the shape of the performative labor (see LABOR) of its employees (who were asked to do anything *but* smile) and live concerts. Other performance art inspired by theme parks has included Ruby Lerner and George King's play "Bananaland: A Central American Theme Park" (1988) and Lea Moro's choreography "FUN!" (2017). Involving more than 20 visual, media, and performing artists engaged in cabaret, video, installation pieces, and puppetry, "Bananaland" premiered at the Seven Stages Theatre in Atlanta in July 1988, just as the National Convention of the U.S. Democratic party also took place in the city (see Evans 1989, 95). Visitors could enjoy the various tours, shows, restaurants, and shops of the eponymous fictional park, which used fake palm trees, rhumba music, a Caribbean yellow, red, and green color scheme, and numerous references to bananas and former Nicaraguan dictator Anastasio Somoza ("Somoza e Hijos Bar y Grill") to evoke a Latin American theme. While also offering a critique of the imperialism of U.S. foreign policy, Paul Evans has argued that "Bananaland" is primarily about "confusion: the confusion of good intentions and perverse payoffs, progress and domination, truth and PR, art and propaganda, seriousness and humor, information and entertainment—even, theatre and theme park" (1989, 7), but also the creative potential of confusion. Thus, the play harkens back to early twentieth century artistic renderings of amusement parks such as Joseph Stella's painting "Battle of Lights, Coney Island, Mardi Gras" (1913–1914).

With its reduced stage and costumes, Swiss-born Lea Moro's one-hour choreography "FUN!," which premiered at the Tanzhaus Zürich in 2017 and has since toured throughout Switzerland and Germany, may at first appear to be in sharp contrast with the "assaultive circus" (Evans 1989, 7) of "Bananaland." Yet "FUN!" also creates a fictional theme park, relying almost exclusively on the bodies of its five performers to alternately depict visitors (cautiously sliding down a slide), employees (chanting "happiness is our business"), and the mechanics of the park's rides (spinning like cogwheels and, in one stunning scene, joining hands to depict a moving Ferris wheel). But the choreography also transcends the idea of "fun" as an "imaginary, reproducible, and marketable commodity" or an "informed consent constantly prescribed by the entertainment industry" (as defined in the accompanying "FUN! Glossary"), for instance when the individual performers soliloquize about what brings them fun (e.g. the feeling of the wind on their legs during a bike ride). These "confessions"—"narratives of pleasure [...], which can be acquired for (nearly) nothing," as the "Glossary" defines them—may very well spring to mind at the theme park, but they may be rooted somewhere entirely different.

In addition to referring to real theme parks and imagining their own fictional parks, some artists have also drawn on the "fake" theme parks of popular culture: curated by illustrator Chogrin (Joseph Game) in 2018 for Gallery 1988 in Los Angeles, the "Fake Theme Parks" art show displayed the works of over 50 artists who were inspired by "theme parks that only exist in movies, television, & video games" (see Gallery 1988). In fact, the show's poster, itself designed to look like a theme park map, shows several of the "fake" theme parks featured in the exhibit (see Fig. 2); others included Monkey Island (from the eponymous video game) and Brisbyland (from the animated TV series *Venture Bros.*).

Many of the featured artworks—whether realized as prints, oil and acrylic paintings, drawings, dioramas, sculpture wall art, or enamel pins—also took up and reworked well-established genres of theme park paratexts, particularly the theme park map, the attraction poster, and the souvenir pin. Scott Balmer's "Big Whoop—Monkey Island" depicts a labelled treasure/theme park map of Monkey Island and includes numerous references to the video game (e.g. the Big Whoop, the Carnival of the Damned, the Voodoo Swamp, and the Scumm Bar). Digital artist Nemons's "Pleasure Island" is in turn reminiscent of the Victorian-style attraction posters designed for the Toy Story Midway Mania! ride at Disney California Adventure (Anaheim, California), although it actually refers to Pleasure Island from the animated movie *Pinocchio* (1940). Miranda Dessler's "Jurassic Fun Park" enamel pin, showing dinosaurs on a carousel, is almost indistinguishable from the myriad of pins sold for as much or even more at (Disney) theme parks (the gallery website lists the price at $12).

"Isla Nublar," finally, which depicts a detailed and intricately labelled map of Jurassic Park, brings us back to Herbert Ryman and "The Square," as the artwork was originally created by *Jurassic Park*'s art director John Bell as concept art for the 1993 movie. Hanging "Isla Nublar" next to art works inspired by the movie in the 2018 "Fake Theme Parks" show offers an interesting comment on Baudrillard's idea of theme parks as simulacra, as copies without originals.

4 The Theme Park as Metaphor

"VENICE is the first urban theme park. Like any other theme park, it is full of attractions, but impractical for everyday living," author John Kay states in a 2008 *Times Online* article.

> Aesthetes might be appalled by the comparison between Venice and Disneyland, but Venice is as artificial as Disneyland. The city ceased to be a significant commercial and political centre more than 200 years ago. (Kay 2008)

Fig. 2 Jurassic Park and Jurassic World (from the *Jurassic Park* novels and movie adaptations), Itchy & Scratchy Land, Krustyland, and Duff Gardens (from the animated TV series *The Simpsons*), and Walley World (from National Lampoon's "Vacation" film series) are among the parks featured in Luke Flowers's poster for Gallery 1988's "Fake Theme Parks" show. *Poster design* Luke Flowers

To preserve Venice, Kay suggests turning it into a "proper" tourist venue with a corporate management (like Disney, or the U.S. national parks) and to charge visitors an entrance fee, which would finance urgently needed conservation work—and make no difference to the tourists, who he notes are nowadays "constantly ripped off" anyway. In 2019, the Venice city council actually voted to impose such an entrance fee (Bastianello 2019).

Since Bryman's concept of the "Disneyization" of society (see Bryman 1999, 2004), the discursive practice of naming places a "theme park" has become quite commonplace. And at least in Western reception, it often bears the more or less silent reference to Disney, which is also partly indebted to Bryman's long-time dominance of the discourse. Bryman sought to describe a development in global society, which he considered to become more and more shaped by the Disney theme park ideology and business model. Whereas in Bryman's theory, the theme park has become a cultural form that can be applied to any other social space, Lukas refers to the cultural application of theme parks in a broader sense as a cultural text, which exists not only in material forms, but also as an immaterial "life form, […] a text, that can be recalled in nearly anyone's mind" (2008, 216). By understanding the theme park as a text, Lukas elaborates on the possible narratives that are produced by it (the emotional nuances, from prescribed happiness to the uncanniness of potential disorders) and on the associated moral questions, thus focusing on themed spaces and remediations of the theme park concept. This section will stick to a more linguistic perspective, looking at the theme park as a metaphor for cultural and social spatial practice: who calls something a theme park, in which context, and what does this imply? If a metaphor is understood as a container that is filled with meaning and can be deciphered by a group of people who share the same cultural knowledge, this section tries to dismantle the contents of the container, citing different examples of places being described as theme parks and analyzing the contextual frame.

The Venice example stands for one meaning of the theme park metaphor: to allege that the place thus described was staging itself—or rather, an image of itself. Often, this image stems from an episode of the past, which has become remembered as the "glorious days" or "good old times" of this place. In *Retroland* (2018), historian Valentin Groebner, as many others before (e.g. in Schlehe et al. 2010), vividly describes how staging the past is a common feature in urban and rural architecture around the globe, for example through the nineteenth century (re-)construction of pseudo-medieval "Old Towns" in Luzern, Basel, or Danzig (Groebner 2018, 82). Hence, naming these places a "theme park" carries a second meaning: the extent to which the place is aesthetically, economically, and socially shaped by the tourist industry.

In consequence, many other municipalities share Venice's fate of being called a "theme park": the Bavarian village of Oberammergau, for example, which is famous for staging a passion play every ten years that attracts up to 500,000 visitors per season and many more all year round. In the early days of nineteenth century mass tourism, Thomas Cook chose the village as one of his first destinations for holiday package tours, thereby opening the path for extensive infrastructural measures in guest accommodation and the service sector. In his volume on the 2010 Oberammergau Passion Play, theatre scholar Kevin Wetmore describes how nowadays the whole village is shaped by Passion Play-tourism:

> Also of significance in 2010 are the epiphenomena of the production: a light show in the village church, a lecture series [...], artistic displays, a museum, the environment of the theatre itself, with a series of displays around the building, not to mention dozens if not a hundred shops filled with wood carvings, Passionsspiel memorabilia [...], and other souvenirs. [...] It has become like attending a festival or, to the cynic, a theme park. (2017, 8)

"Theme park" is in this context a clearly pejorative term, attributing to Oberammergau a loss of authenticity and an increase in commercialization. If the goal of theming in theme parks is to create a coherent image by masking elements that are potentially disruptive to the overall picture (see THEMING), then by contrast it seems unwanted, or perceived as masking the authentic backstage, in places like Oberammergau that live on selling authenticity. Other places which center their tourism around a singularity as a unique selling point face the same controversy that theming for tourists produces.

Nicola J. Watson (2007) describes how Stratford-upon-Avon developed into "Shakespeare-town," a process whose beginning she dates to the early eighteenth century with the publication of Nicholas Rowe's Shakespeare biography and play edition in 1709. According to Watson, the text collection transferred Shakespeare's work from the collective stage experience to the individual reading experience which, together with the first dissemination of detailed biographical information, sparked interest in the places of his life, where a personal relationship with the playwright could be enhanced through shared experience. Although Stratford had not been of great interest to Shakespeare and lacked any particular remains of his life, the sudden tourist interest attracted investors, who transformed the place into what it was expected to be: "Stratford represented, and still largely represents, a Victorian dream of Englishness, an energetic dreaming that turned Stratford into the world's first theme-park" (2007, 215). What does Watson mean by using this comparison? In both cases, Stratford and Venice, "theme park" is the etic frame from the visitor's perspective. Attracted by travel accounts, paintings, pictures, movies, or celebrities, tourists make their pilgrimage (see VISITORS) to pre-imagined destinations, which in turn become modified to match their expectations. Seeing that, critical tourists (and academics) start to sense a lack of authenticity and call the place a theme park. Although this is a simplified summary, it describes the popular reception of places as diverse as Venice, Stratford, the Bavarian village of Oberammergau, or even the Holy Land. The pejorative implication of the term theme park is in these cases linked to the notion of authenticity (see AUTHENTICITY).

In travel literature, there is a classical reaction to artificial theme park places: individuals who feel deterred by what they regard as staged representations tend to turn to nature to experience "true" authenticity. Watson cites Christian Tearle's American tourist Mr. Fairchild, who, when visiting Shakespeare's birthplace, flees from the unloved representation to the countryside to take a bath in "Shakespeare's river." He throws his souvenir plaque into the Avon and plucks some leaves of ivy from the birthplace instead (see Watson 2007, 215). Based on his experience as a tourist guide in the Holy Land, Amos Ron (2010) observes similar tourist behavior: Protestant visitors tend to disfavor the reconstructed, "themed" sacred sites and instead wish to be taken to the countryside, preferably on high vantage points, to experience the landscape, which they regard as the most authentic remnant from biblical times. In this contextual setting, the term theme park is employed for (re-)constructed places, meaning the opposite of untouched nature.

Yet the idea of untouched nature, especially if enjoyable for tourists, is based on the same wishful fantasy as the constructed tourist destinations are, and is consequently not immune to theme park framing. "The Grand Canyon offered early visitors nothing less than a theme-park landscape and a promise of redeeming spiritual and traditional values through contact with nature and people who lived in a 'natural' world," writes Neumann (1999, 12). A theme park landscape, for Neumann, is a landscape that triggers in its visitors a spectacular sensual experience. This idea is best exemplified by his account of an evening lecture, held by a national park ranger who pointed out similarities between the Grand Canyon and Disneyland in Anaheim before coming to the conclusion that, as the title of his presentation announced:

> "Gee … This Place Is Just Like Disneyland." [...] But then he came to his senses, he told us, when he returned to the canyon after visiting the theme park and found himself watching the sunrise over the chasm. "I found myself screaming to no one in particular that this place is not like Disneyland!" he screamed at us, and then he related his epiphany to all of us seated in neat rows on benches. "To compare Grand Canyon to anything man-made—Disneyland or otherwise—would be to trivialize what this canyon is all about. It's so special, that there's a special name for it. It's called Grand Canyon National Park. It's so special, we protect it." (Neumann 1999, 303)

In Neumann's story, the ranger's emotional outburst illustrates the painful paradox that there is no absolute opposite to "theme park," no pure nature for man to consume. Experience is always framed. Löfgren (1999, 16) illustrates how travelling and "seeing" landscapes became linked to aesthetic values in eighteenth century travel literature, and how the "picturesque" turned into a mirror that constantly reflected the perceived landscape and memories of its reproductions. For the twenty-first century (in discussing Disney's "Animal Kingdom"), Stephanie Rutherford even argues that

> Disney commodifies our vision and understanding, and through that commodification, governs what we come to know as nature. [...] this works in large part because nature has already become a theme park, or perhaps a mausoleum. Nature is where people are not. It is the journey to find nature at a zoo, a theme park, an ecotour, or a museum that has allowed for it to be consumed in normalized ways by certain people. (2011, 88)

In the Grand Canyon park ranger's understanding, a theme park is man-made. For Rutherford, nature turns into a theme park through the presence of biased humans. In the ranger's world, a "true" experience means to free oneself from the theme park frame of vision; following Rutherford, this is impossible. In both cases, "theme park" describes a layer of artificiality that is virtually placed onto an imagined "real" thing beyond. It is a political term, used in an arena of competing interpretational frames for certain places.

Judging by these findings, the struggle over "theme park" framing seems to occur especially in places that have been referenced or duplicated elsewhere, notably in other theme parks. The Grand Canyon inspired Disney's "Frontierland," Stratford inspired the urban architectural fashion of "mock-Tudor pastoral" (Watson 2007, 215) along with a series of "birthplace copies" all over the world (222), the Venetian casino in Las Vegas is almost as famous as the city of Venice. The Oberammergau Passion Play, meanwhile, has been duplicated many times, especially in the U.S., while itself referencing the Holy Land, which is probably the most referenced and copied place in the world, even though it is more of a virtual concept than a physical place. Even "nature" is a blueprint for zoos, parks, and ecotours. In all these cases, "theme park" is not only a lens, but also a mirror, reflecting back onto the source of reference.

When theme parks entered public discourse as a concept beyond their spatial boundaries, their potential as objects for intellectual examination was soon discovered. The academic debate on theme parks began in the 1970s, when Eco and Baudrillard (see ANTECEDENTS) famously created the still predominant public image of theme parks as a Western capitalist phenomenon, initiated by Disney, that transformed into the metaphor discussed above. This opened the topic for lower-level education on arts, culture, society, and other issues:

> art educators and students could develop understandings of the value and complexity of the theme park as a cultural vortex whose swirling forces collapse boundaries between Western dualities of place and space, myth and reality, and work and play. (Jeffers 2004, 222)

Some theme parks have reacted to this new form of interest by issuing their own learning material, with background information on ride physics, history, and culture (referring to the "cultural" themes employed in the park). These materials comprise another type of theme park paratext, employing the excitement of the theme park experience as a tool to inspire students. Davies and Knivett (2008) suggest the use of "exciting" theme park maps for geography classes. Bouzarth, Harris, and Hutson (2014) describe their teaching of the mathematical "traveling salesman problem" by challenging their students to complete the 18 rides of Walt Disney World in the shortest possible time. Tho, Chan, and Yeung (2015) have developed an educational program for Hong Kong's Ocean Park, using the newly built thrill rides for "experience-based learning" in physics. Since these materials rely on hands-on experiences in the park, they are also a means to attract new visitors (see ECONOMIC STRATEGY). On the other hand, teaching cultural and historical issues based on theme park theming is a rather sensitive issue, especially if the themed worlds are used as models for real places and epochs (see WORLDVIEWS).

5 Conclusion

From its very beginnings, the theme park has been an extraordinary success—not only in terms of the spread of theme parks themselves throughout the Western world and beyond (see ANTECEDENTS), but also in terms of the theme park's global impact on architecture and urban planning. Through processes variously referred to as "Disneyfication" (Francaviglia 1981, 146) or "theme-park-ification" (Dear and Flusty 2002, 417), theme parks have contributed to such "variations on a theme park" (Sorkin 1992), "urbanoid environments" (Goldberger 1996), "non-place urban realms" (Carosso 2000), "public-esque" (Wells 2004), or "Disneyfied places" (Kolb 2008) as themed commercial environments, shopping malls, festival marketplaces, revitalized downtowns, and New Urbanist communities (see THEMING). Within these, the theme park not only provided ideas and acted as a catalyst, but also often functioned as an active agent (see, for instance, the Disney Development Company, founded in 1984, and its involvement in various projects outside of the Disney resorts; see Lukas 2019; Warren 1994, 2005; see also SPACE).

It is not only in the realm of built environments that theme parks have left a trace, however. As this chapter has shown, cultural artifacts realized in various media—from maps and apps to movies, television, video games, and fictional and critical discourse—have depicted and engaged with the theme park, sometimes reproducing and even intensifying its representational strategies, sometimes offering critical perspectives on the phenomenon. The reception of theme parks in para- and other texts constitutes a rich and relevant field for future research, not least because representations of theme parks—even highly critical ones—have occasionally found their way back into the parks. For instance, The Simpsons Ride, opened in 2008 at Universal Studios Hollywood (California) is not just—like so many other rides and attractions at this park—a ride adaptation of a movie or a TV series. Taking visitors to Krustyland, the fictional theme park from the TV series, the ride reproduces the series' satirical references to Disneyland (e.g. "Sleeping Itchy's Castle"), but also incorporates *The Simpsons*' critique of issues of labor, safety, and commercialization at theme parks. Visitors to Universal Studios are thus confronted with a theme park adaptation of a fictional—and not at all flattering—portrayal of theme parks, which will in turn be represented in Universal Studios paratexts, as well as in visitor pictures and YouTube videos of the ride. In this and many other cases, the study of theme park paratexts and theme park reception forms an integral part of theme park studies in general.

References

Auster, Carol J., and Margaret A. Michaud. 2013. The Internet Marketing of Disney Theme Parks: An Analysis of Gender and Race. *SAGE Open* 3.1: 1–16. https://doi.org/10.1177/2158244013476052. Accessed 28 Aug 2022.

Banksy. 2015. I Think a Museum Is a Bad Place to Look at Art. *The Guardian*, August 21. http://www.theguardian.com/artanddesign/2015/aug/21/banksy-dismaland-art-amusements-and-anarchism?CMP=twt_gu. Accessed 28 Aug 2022.

Bastianello, Riccardo. 2019. Italy's Venice to Charge Admission Fees for Tourists. *Reuters UK Online*, February 27. https://uk.reuters.com/article/uk-italy-venice/italys-venice-to-charge-admission-fees-for-tourists-idUKKCN1QG1F4. Accessed 27 July 2020.

Berman, Eliza. 2014. Wes Anderson Might Create a Theme Park with Devo's Mark Mothersbaugh. *Time*, November 4. https://time.com/3556593/wes-anderson-mark-mothersbaugh-theme-park/. Accessed 28 Aug 2022.

Bouzarth, Liz, John Harris, and Kevin Hutson. 2014. Math and the Mouse. *Math Horizons* 22 (2): 12–14.

Bryman, Alan. 1999. The Disneyization of Society. *Sociological Review* 47: 25–47.

Bryman, Alan. 2004. *The Disneyization of Society*. London: Sage.

Carosso, Andrea. 2000. America's Disneyland and the End-of-the-Century Cityscape. *Revue française d'études américaines* 83: 64–75.

Child, Lincoln. 2003. *Utopia*. New York: Doubleday.

Davies, Richard, and Louise Knivett. 2008. Using Theme Parks in Geography. *Teaching Geography* 33 (1): 15–18.

Dear, Michael J., and Steven Flusty. 2002. The Spaces of Representation. In *The Spaces of Postmodernity: Readings in Human Geography*, ed. Michael J. Dear and Steven Flusty, 415–418. Oxford: Blackwell.

Elkin, Stanley. 1985. *The Magic Kingdom*. New York: Dutton.

Evans, Paul. 1989. Bananaland: PR, Propaganda, and Infotainment. *The Drama Review* 33 (3): 95–102.

Florey, Robert, and Wilfred Jackson, dir. 1954. The Disneyland Story. *Disneyland* 1 (1), October 27. Burbank: Walt Disney Productions.

Francaviglia, Richard V. 1981. Main Street U.S.A.: A Comparison/Contrast of Streetscapes in Disneyland and Walt Disney World. *The Journal of Popular Culture* 15.1: 141–156.

Freitag, Florian. 2016. Autotheming: Themed and Immersive Spaces in Self-Dialogue. In *A Reader in Themed and Immersive Spaces*, ed. Scott A. Lukas, 141–149. Pittsburgh: ETC.

Freitag, Florian. 2017. Critical Theme Parks: Dismaland, Disney and the Politics of Theming. *Continuum* 31 (6): 923–932.

Gallery 1988. 2018. Fake Theme Parks. https://nineteeneightyeight.com/collections/fake-theme-parks. Accessed 28 Aug 2022.

Goldberger, Paul. 1996. The Rise of the Private City. In *Breaking Away: The Future of Cities. Essays in Memory of Robert F. Wagner, Jr.*, ed. Julia Vitullo-Martin, 135–147. New York: Twentieth Century Fund.

Gordon, Bruce, and David Mumford. 2000. *A Brush with Disney: An Artist's Journey, Told through the Words and Works of Herbert Dickens Ryman*. Santa Clarita: Camphor Tree.

Gray, Jonathan. 2010. *Show Sold Separately: Promos, Spoilers, and Other Media Paratexts*. New York: New York University Press.

Groebner, Valentin. 2018. *Retroland: Geschichtstourismus und die Sehnsucht nach dem Authentischen*. Frankfurt: S. Fischer.

Jeffers, Carol S. 2004. In a Cultural Vortex: Theme Parks, Experience, and Opportunities for Art Education. *Studies in Art Education* 45 (3): 221–233.

Jodidio, Philip E., ed. 1992. *Euro Disney*. Special Issue of *Connaissance des Arts*.

Kay, John. 2008. Welcome to Venice, the Theme Park. *The Times Online*, March 1, https://www.thetimes.co.uk/article/welcome-to-venice-the-theme-park-nj8hzntbdpf. Accessed 28 Aug 2022.

Klopp, Tina. 2014. *Die Geisterbahn als Modell und Mode in der zeitgenössischen Kunst*. Diss., Hochschule für Bildende Künste Hamburg.

Kolb, David. 2008. *Sprawling Places*. Athens: University of Georgia Press.

Löfgren, Orvar. 1999. *On Holiday: A History of Vacationing*. Berkeley: University of California Press.

Lukas, Scott A. 2007. A Politics of Reverence and Irreverence: Social Discourse on Theming Controversies. In *The Themed Space: Locating Culture, Nation, and Self*, ed. Scott A. Lukas, 271–293. Lanham: Lexington.

Lukas, Scott A. 2008. *Theme Park*. London: Reaktion.

Lukas, Scott A. 2016. Dark Theming Reconsidered. In *A Reader in Themed and Immersive Spaces*, ed. Scott A. Lukas, 225–235. Pittsburgh: ETC.

Lukas, Scott A. 2019. Between Simulation and Authenticity: The Question of Urban Remaking. In *The New Companion to Urban Design*, ed. Tridib Banerjee and Anastasia Loukaitou-Sideris, 327–336. London: Routledge.

Makai, Péter Kristóf. 2019. Three Ways of Transmediating a Theme Park: Spatializing Storyworlds in Epic Mickey, the Monkey Island Series and Theme Park Management Simulators. In *Transmediations: Communication Across Media Borders*, ed. Niklas Salmose and Lars Elleström, 164–185. London: Routledge.

Marling, Karal Ann. 1991. Disneyland, 1995: Just Take the Santa Ana Freeway to the American Dream. *American Art* 5 (1–2): 168–207.

N.N. 1977. *Your Guide to Disneyland: Summer 1977*. Burbank: Walt Disney Productions.

Neumann, Mark. 1999. *On the Rim: Looking for the Grand Canyon*. Minneapolis: University of Minnesota Press.

Neuman, Robert. 1999. Now Mickey Mouse Enters Art's Temple: Walt Disney at the Intersection of Art and Entertainment. *Visual Resources* 14 (3): 249–261.

Rawlins, Melissa W. 1994. Peter Gabriel's Real World Experience Park. *Entertainment Weekly*, May 20. https://www.ew.com/article/1994/05/20/peter-gabriels-real-world-experience-park/. Accessed 28 Aug 2022.

Ron, Amos S. 2010. Holy Land Protestant Themed Environments and the Spiritual Experience. In *Staging the Past: Themed Environments in Transcultural Perspectives*, ed. Judith Schlehe, Michiko Uike-Bormann, Carolyn Oesterle, and Wolfgang Hochbruck, 111–134. Bielefeld: Transcript.

Rutherford, Stephanie. 2011. *Governing the Wild: Ecotours of Power*. Minneapolis: University of Minnesota Press.

Schlehe, Judith, et al., eds. 2010. *Staging the Past: Themed Environments in Transcultural Perspectives*. Bielefeld: Transcript.

Sorkin, Michael. 1992. Introduction: Variations on a Theme Park. In *Variations on a Theme Park: The New American City and the End of Public Space*, ed. Michael Sorkin, xi–xv. New York: Hill and Wang.

Sotto, Eddie. 2017. Thoughts on the (Accidental?) Art of Theme Parks. *Blooloop*, November 30. https://www.blooloop.com/art-theme-parks/. Accessed 28 Aug 2022.

Strodder, Chris. 2012. *The Disneyland Encyclopedia*, 2nd ed. Salona Beach: Santa Monica.

Telotte, J.P. 2004. *Disney TV*. Detroit: Wayne State University Press.

Tho, Siew Wei, Ka Wing Chan, and Yau Yuen Yeung. 2015. Technology-Enhanced Physics Programme for Community-Based Science Learning: Innovative Design and Programme Evaluation in a Theme Park. *Journal of Science Education and Technology* 24 (5): 580–594.

Warren, Stacy. 1994. Disneyfication of the Metropolis: Popular Resistance in Seattle. *Journal of Urban Affairs* 16 (2): 89–107.

Warren, Stacy. 2005. Saying No to Disney: Disney's Demise in Four American Cities. In *Rethinking Disney: Private Control, Public Dimensions*, ed. Mike Budd and Max H. Kirsch, 231–260. Middletown: Wesleyan University Press.

Watson, Nicola J. 2007. Shakespeare on the Tourist Trail. In *The Cambridge Companion to Shakespeare and Popular Culture*, ed. Robert Shaughnessy, 199–226. Cambridge: Cambridge University Press.

Wetmore, Kevin J., ed. 2017. *The Oberammergau Passion Play: Essays on the 2010 Performance and the Centuries-Long Tradition*. Jefferson: McFarland.

Wells, Kathryn J. 2004. *Easton Town Center: Creation of the Retail Public-esque*. Diss., Ohio State University.

Williams, Rebecca. 2020. *Theme Park Fandom: Spatial Transmedia, Materiality and Participatory Cultures*. Amsterdam: Amsterdam University Press.

Williams, Rebecca. 2021. Theme Parks in the Time of the Covid-19 Pandemic. In *Pandemic Media: Preliminary Notes toward an Inventory*, ed. Philipp Dominik Keidl, Laliv Melamed, Vinzenz Hediger, and Antonio Somaini, 137–142. Lüneburg: Meson.

Wilson, Eric. 1995. *Disneyland Hostage*. New York: HarperCollins.

Yandell, Stephen. 2012. Mapping the Happiest Place on Earth: Disney's Medieval Cartography. In *The Disney Middle Ages: A Fairy-Tale and Fantasy Past*, ed. Tison Pugh and Susan Aronstein, 21–38. New York: Palgrave Macmillan.

Younger, David. 2016. *Theme Park Design & the Art of Themed Entertainment*. N.P.: Inklingwood.

Space: Representing, Producing, and Experiencing Space in Theme Parks

Abstract This chapter is dedicated to the spatial dimension of theme parks and investigates their being within and structuring space. It first conceptualizes the theme park as a relation between two spaces—one representing a theme and the other one represented by a theme—and discusses how this "doubled" space provides visitors with an immersive experience of virtual tourism. The chapter then goes on to examine how the park's space is both materially produced by the planners and engineers involved in its design and performatively constructed by visitors according to how and where they act in the park. The material production of space includes such issues as spatial organization and transportation, landmarks and landscaping, control of and the dichotomy between front stage and backstage, as well as mapping and materiality. With respect to visitors' affective and subjective spatial experiences, the chapter discusses their perception and imagination, their performance and agency, and the ways in which they connect and contribute to the park's atmosphere.

1 Introduction

A theme park is a place that visitors and employees go to and a space in which they move. As a place, a theme park can be identified by a dot on a map, within a city, a region, a country, a continent, and on the planet. While such a map may, for example, tell visitors how to get there by car, train, or plane, it usually informs them about neither the park's size nor what they will find there. As a space, a theme park is mapped to help visitors find their way and locate the main gates, subsections, pathways, attractions, etc. This map may also include the parking lot, the subway station, or hotels outside of the park, but its main purpose is to represent the space within the park's limits, to give patrons an idea of the distance between the different elements, and to help them plan their visit. Such a map is usually handed to visitors as they enter

This work is contributed by Jean-François Staszak, Salvador Anton Clavé, Jan-Erik Steinkrüger, Ariane Schwarz, Astrid Böger, Filippo Carlà-Uhink, Thibaut Clément, Florian Freitag, Scott A. Lukas, Sabrina Mittermeier, Céline Molter, Crispin Paine, Torsten Widmann. The corresponding author is Jean-François Staszak, Département de Géographie et Environnement, Université de Genève, Genève, Switzerland.

F. Freitag et al., *Key Concepts in Theme Park Studies*,
https://doi.org/10.1007/978-3-031-11132-7_13

the park or can be accessed virtually via an app. Large mural maps can also be found at strategic locations, for example at so-called distribution points where visitors are expected to stop and plan the next steps of their visit. For employees, similar (albeit usually more functionally designed) maps are available, which additionally indicate the locations of employee services such as transportation systems, breakrooms, or cafeterias. Maps, as well as signs and signals materially placed inside the park, are part of and simultaneously document the strategies developed by park planners and designers for handling large amounts of people as efficiently as possible (see also PARATEXTS AND RECEPTION).

At the same time, theme parks are also maps that visitors can walk in. This is perhaps most obvious in the case of miniature parks such as France Miniature (Elancourt, France), whose outlines represent—at a reduced scale but as accurately as possible—the spaces they refer to, although this also applies to other parks that are themed to geographical areas. Generally, the theme park constitutes a microcosm, divided into several "lands," and the map given to visitors is a map of this world. A case in point is Belantis (Leipzig, Germany), which reproduces almost the entirety of the Western world, including Europe, North Africa, the Americas, the Atlantic Ocean, and the Mediterranean Sea, both in the park and on the maps of the latter (see Carlà-Uhink 2020, 62). The theme park picks up on the idea of enclosing a meaningful universe in an ordered space, capitalizes on the narrative symbolism of the place, and instrumentalizes it to generate business opportunities (see also WORLDVIEWS).

From a geographical point of view, theme parks are not so different from such other *places*—or non-places (see Augé 1995)—as shopping malls or touristic resorts. The location of these venues is a matter of town planning and profitability: in this regard, production costs, taxes, accessibility, local market, and competition are key issues. Their internal layout is also a matter of urban planning in the sense discussed by Eyssartel and Rochette, who define theme parks as urban spaces where architecture, art, and technology are at the service of a coherent "projet culturel" ("cultural project"; 1992, 37). A theme park is a place in which every architectural shape, volume, sign, or any other form might have an evocative (suggesting a certain image, story, or setting) and invocative (suggesting a certain action or response) power, but it is also a place where everything is rationalized in order to guarantee the fulfilment of visitor needs and expectations, from the location of the toilets to the photographic souvenir system (either by facilitating the purchase of photographs at the park's rides or via the classic "picture spot" signposts; see PARATEXTS AND RECEPTION).

However, theme parks are very special *spaces* because what matters is the story that is conveyed through the design elements that shape the landscape recreated—in the case of Disney parks, for example, the Victorian storefronts of "Main Street, U.S.A.," the spires of Sleeping Beauty Castle in "Fantasyland," the cartoonish angles of "ToonTown," or the rough surfaces of "Frontierland" (see MEDIA and THEMING). Indeed, the clash of various scales and the mix of fiction and reality result from and illustrate the complexity and ambivalence of space in theme parks, as is illustrated by a directional sign at Europa-Park (Rust, Germany) informing visitors that the "Wikingerland," the Netherlands, Austria, Spain, the United Kingdom, and Portugal are to the left and the exit to the right (see Fig. 1).

Fig. 1 The clash of various scales and the mix of fiction and reality in theme park spaces crystallize in this directional sign at Europa-Park in Rust (Germany). *Photograph* Peter Bischoff/Getty Images

Seeking to explain how theme parks make sense spatially, this chapter explores where we are when we are in a theme park. The chapter first analyzes the so-called doubling of space and the articulation between the themed space and the theming space in the park, then goes on to examine how the park's space is in turn produced by planners and experienced by visitors.

2 The Doubling of Space: Themed and Theming Space

Visiting a theme park without accepting the rules of make-believe would be an instructive but also very boring experience. Indeed, a visitor who experiences the park for what it is rather than what it pretends to be would probably notice things that more naïve or enthusiastic visitors usually miss—and yet this visitor would not understand what is going on. What theme parks do is use shapes, volumes, performances, etc. (see MEDIA) to materialize and thus bring to life narratives that seek to create a complete world subordinated to a script. Visitors must accept this script for the mechanisms of immersion to take place (see IMMERSION). The world created in the theme park is thus organized as a sequence of scenes and represents a harmonious spectacle that can be seen, heard, and lived according to its own rules, but always in a different way to how life is lived conventionally.

Similar to going to the movies, then, visiting theme parks is all about pretending. While nobody believes that there are cats that can talk or that Mickey Mouse actually exists, there is no point watching Disney's *The Aristocats* (1970) and doubting cats could talk or meeting Mickey Mouse at Disneyland with the suspicion that it might not be the actual Mickey Mouse. What distinguishes these two experiences, however, is the distance between the audience and the fiction: spectators watch a movie from the outside, as there is an irreducible distance and (virtually) no interaction between the screen and the audience. Likewise, botanical and particularly zoological gardens keep visitors at a distance from their exhibits. In a theme park, by contrast, spectators willingly and necessarily go on stage. Theme parks thus provide an immersive and interactive experience because they are neither a screen nor an exhibit created for spectators, but a space of representation created for visitors. The latter thus enter inside the fiction and become part of the show (see VISITORS). In this sense, the "invented" landscape of parks becomes a product that is "experienced" and produces "effects of recognition" among visitors.

But fiction is not the only issue. Theme parks bring visitors to such imaginary places as Neverland (at Disney parks) or Hogwarts (at Universal parks) but also to places that existed in the past (e.g. Jerusalem at the time of Jesus at Holy Land Experience in Orlando, Florida, as well as at Tierra Santa, Buenos Aires, Argentina) or that exist today (Paris, e.g. at Walt Disney Studios Paris). The difference between tourists enjoying the "real" Paris and visitors enjoying Ratatouille's Paris at Walt Disney Studios in Paris is less about fakery than about distance and location: tourists are in the Paris they visit, whereas visitors in Disneyland are not—they are in a theme park. Visitors to Walt Disney Studios' Ratatouille ride know that their Paris is not Paris (even if they are within the space of Greater Paris), whereas tourists in central Paris might believe that they are experiencing the "real" Paris without realizing that the authenticity of the place is staged (MacCannell 2013; see AUTHENTICITY). Moreover, the distance between themed space and theming space can also be constructed around time. At Disneyland, for example, visitors are in Anaheim, California, but simultaneously they are in the future (e.g. in "Tomorrowland") or in the past (e.g. in "Frontierland"). Indeed, most themed spaces have a specific temporal setting that creates a distance between themed space and theming space, even when the physical geography of places overlaps (see TIME).

Hence, the distance between the theming and the themed space is both material and symbolic. The themed space is very often far away, belonging to a different continent or an imaginary world, but the symbolic distance also applies when the geographical, material one is narrow: even if Disney's Ratatouille is in the suburbs of Paris, and Old Hong Kong (in Ocean Park, Hong Kong) is just a few miles from the actual city, there is still a recognizable gap between the themed space and its local environment (and between the themed and the theming space). The themed space is located on a cultural map where imagination and well-known fictions are the main resources, whereas the theming space is located on an economic map concerned with costs, benefits, and efficiency.

Therefore, the themed space "is not only a landscape but the *relation between* two landscapes" (Steinkrüger 2017, 90; emphasis original). To a large extent, theming has to do with establishing this relation: the doubling of space ("the doubling of landscape," according to Steinkrüger) is operated by superimposing the theming onto the themed space, through simulacrum, replica, pastiche, recreation, reconstruction, etc. (see THEMING). Hence, analyzing the geography of theme parks is less about mapping the theming and/or the themed space than about mapping how one space is articulated through the other, or how visitors travel back and forth between them. Mapping theme park space is both a geographical and a semiotic operation. In this sense, everything starts with a story. The story inspires an architectural program in which the solid and empty spaces are ordered with the idea of shaping the place as a whole and creating a sense of place. This sense of place allows visitors to establish a relationship of empathy and identification with the planners' and designers' spatial purposes. Travelling through the material landscape serves to give coherence to the variety of attractions within the park and to highlight the universe created rather than each of the emotions.

Theme parks could thus be described as metatouristic places: their visitors (physically) travel to the park in order to (imaginatively) travel somewhere else (Freitag 2018, 166–167). They enjoy the journey because it takes them from their everyday local world to a distant and exciting place, satisfying a taste, or longing for escapism (Tuan 1998). This escapism is based on the visitors' desire to leave—even for the short time of a visit to a theme park—their daily lives and go to a different world. At Disney's Animal Kingdom (Orlando, Florida), Pandora—based on James Cameron's movie *Avatar* (2009; see MEDIA)—is an area that allows visitors to not only queue and walk in the exotic and pristine landscapes of the Valley of Mo'ara, to meet a Pandora Ranger, to eat and drink like a Na'vi in the Satu'li canteen, but also to have their face painted in Na'vi designs and ultimately to get their own avatar, become a Na'vi, and thus connect to a dragon on the back of which they enjoy a 3D simulated flight over Pandora's spectacular landscapes. In Pandora, visitors not only escape Earth, gravity, the present, and the real world; thanks to their avatar, they also escape their own bodies.

In the sense that they are in a different time and a different space, theme parks are efficiently designed utopias, providing visitors with the happy experience of an imaginary world within a carefully designed space that has been created exclusively to fulfill their dreams and desires. Yet, etymologically, utopias are places that do not exist, except in the imagination. Contrary to theme parks, they are not located anywhere on earth. Theme parks might rely on imagination, but they nevertheless exist for real; therefore, one could apply to them the concept of heterotopia. Foucault, who first developed this concept, defines heterotopias as

> real places—places that do exist and that are formed in the very founding of society—which are something like counter-sites, a kind of effectively enacted utopia in which the real sites, all the other real sites that can be found within the culture, are simultaneously represented, contested, and inverted. (1986, 24)

Examples are prisons, brothels, and colonies, but also theaters and gardens, as they are "capable of juxtaposing in a single real place several spaces, several sites that are in themselves incompatible" (25). Theme parks clearly also fall within this category.

To be sure, one could argue that Foucault's definition of a heterotopia is so broad and his examples so diverse that more or less any place could be considered a heterotopia. It could be argued, however, that this is the case precisely because the doubling of space is no longer confined to theme parks; indeed, it would be naïve to assume that there could be a space not referring to another space: this is very clear in the case of postmodern architecture (which is all about pastiche), but is also true for neoclassical architecture (inspired by Roman and Greek monuments) and even negatively for modern architecture (in its opposition to the ornamental tradition, for instance). No garden or landscape is designed without referring to other gardens and landscapes. Hence, the relevant question is not to identify or locate heterotopias but rather to characterize the heterotopian dimension of a given place. In this sense, theme parks should not be considered as unusual places or exceptions to the rule. Instead, as spaces that are not to be taken seriously and that are conspicuously separated from the space (and time) of "regular" life (see Lukas 2007), theme parks are ideal-types, models for architecture and (new) urbanism that perfectly exemplify and demonstrate the double nature of the space we live in (cf. the theming or Disneyfication of urban space and touristic places; see Gottdiener 1997; Lukas 2007).

3 Designing and Mapping Theme Park Space

From the perspective of designers and engineers, theme parks are "total design spaces" that trans- or disfigure the existing geography of places. Their design is based on formulas of planning that seek to evoke an emotional response among their visitors in order to provide pleasant experiences and that seek to ensure efficiency regarding operational needs. The spatial design first focuses on the location of each attraction, show, restaurant, shop, animation, etc. according to the available space and its topography (which is sometimes also altered from its natural state in order to produce the desired experience). The spatial design is also instrumental for the management of flows and must consider attendance levels, visitors' preferences, their perceptions (for example, concerning whether to queue up), and finally, their decision-making process as consumers.

The discontinuity between the themed space and its local environment (see above) is marked by the park's gate, usually a monumental and spectacular border post which signals to visitors that they are about to enter a different world (namely, the themed space). Engraved brass plaques over the entry tunnels to the Disneyland parks clearly announce: "Here You Leave Today and Enter the World of Yesterday, Tomorrow and Fantasy." The berm around the park is a physical and visual border that isolates the themed space and protects it against any form of intrusion (for instance, modern buildings in the background of the medieval castle). "I don't want the public to see the world they live in while they're in the park. I want them to feel they're in another world," Walt Disney said (qtd. in Younger 2016, 187). Spatial representations, aerial photographs, or satellite images clearly show this inside-outside dichotomy, the parks' discontinuity with their surrounding areas, and the relevance of their internal spatial organization (Anton Clavé 2007). A case in point is the collection of Disney parks maps (Neary et al. 2016), which charts 60 years of the inside of the parks as "snapshots of a place and time"—according to the endorsement of the book—without any connection with their surroundings.

According to Rebori (1993), three design dimensions ensure that the spatial organization of parks is intelligible, comfortable, and satisfactory: order, enclosure, and context. Limits such as berms or fences around the parks, organizational axes, directional alternatives, and landmarks organize the space and direct movement. The interrelation between solid (buildings) and empty spaces (streets and squares) in a continuous distribution of sizes, scales, forms, heights, shapes, subdivisions, and changes of ground level create feelings and emotions (see also below). Finally, although many parks use copy-paste solutions (see Carlà-Uhink and Freitag 2022), the unique grouping of common elements providing distinctive meaning to every place ensures that two theme park places never really feel the same. Walls, gates, viewpoints, paths, landmarks, and gardens define the specific dimensions of order, enclosure, and context that guide visitors toward magnet landmarks, must-see attractions, and business opportunities for the operators. Additionally, these are also elements that help to organize the transitions between themed areas, giving specific sensorial signals to visitors through e.g. colors, trees, edges, views, paths, water, sounds, vegetation, furnishings, fittings, and hedges.

Landmark attractions are of particular importance: they magnetize and provoke journeys to each themed area because, thanks to word-of-mouth and paratexts, visitors already know about the symbolic elements of reference that "must not be missed" even prior to coming to the park (see PARATEXTS AND RECEPTION). In Disney's terminology, a Weenie is an eye catcher, "something that would beckon the customer to go inside a space" (Lukas 2007, 199) and thus has the power to influence visitor behavior and lead them in a certain direction—for instance, a building at the end of a corridor that appears so enticing it makes you want to reach the end. Landmarks have effects that are the equivalent of unification, central focalization, landscape image creation, and entertainment values communication. What is important is that via its landmark attractions, each park should be perceived as unique, entertaining, and worthy of being remembered. This is also why many parks tend to identify themselves with an iconic landmark that they have but their competitors do not. Examples include the Dragon Khan roller coaster in the "China" area of the PortAventura World theme park near Barcelona (Spain), the iconic sphere located at the entrance to Epcot (Orlando, Florida), or the huge and dynamic Whale shark icon in Chimelong Ocean Kingdom in Zhuhai (China).

The spatial organization of the park may be upset by intrusions from the outside world or from within the park itself, for instance at Disneyland when a "Frontierland" employee in a cowboy costume walks through "Tomorrowland." This may cause cognitive dissonances and challenge the make-believe atmosphere. The backstage thus needs to be kept strictly separated from the stage; at Magic Kingdom (Orlando, Florida), the underground utilidor system allows employees to go from one place to the other without being noticed by visitors, while also hiding storage, sewer lines, garbage disposals, etc. The themed space must be preserved to protect the illusion, not because the illusion could be ruined (after all, who really believes that a theme park castle is medieval?) but because it has to be as perfect as possible; the attraction is less in the illusion itself than in the technological challenge and attention to detail (to the risk of overdetailing, when too much unnecessary information is provided to overwhelmed visitors). Good theming might play on a "revered gaze": "a response marked as much by recognition of the labor and effort involved in creating the spectacle as in the spectacle itself" (Griffiths 2008, 286; see THEMING).

The doubling of space occurs in theme parks when visitors suspend their disbelief and accept the themed space. For a themed space to be identified as such, it must be recognizable and located on the map of popular culture. Theming employs iconic landmarks, architecture, landscaping and gardening, props and costumes, shows and performance, food, sounds, smells, etc. as signs that visitors are able to read and understand according to their knowledge and imagination (about history, geography, or fictions they have learned in books, comics, or movies). This understanding allows them to know and enjoy the place they virtually travel to. Hence the importance of identifying a target audience: a theme park dedicated to a national hero might puzzle foreign visitors, whereas Disney's characters are more or less universal. However, as the themed space is usually much larger than the theming space, designing theme parks is not only about theming space, but also about "creating space." Numerous techniques, many of them used in theater and playing with perspective, are used: disorientation, forced perspective, negative space, haze, false portals, infinity or mirror walls, open-plan layouts, peaking, artificial skylights, stratification, trompe l'oeil, windowing, video screens, etc. (see Younger 2016, 168–173; 510–512).

The four main ingredients of what Ritzer (1996) has called McDonaldization—namely efficiency, quantification, predictability, and control—are also four of the basic ingredients of theme parks' strategic spatial operation processes (see Anton Clavé 2007, 170–177). There are two reasons for their importance: the need to control the internal processes around the park, and the need to give visitors—who are permanently moving inside the park—guarantees. Theme parks thus implement strict systems of control over visitors (by directing flows, surveying behaviors via

CCTV, designing creative queue management, etc.), workers (via codes of appearance, the obligation to "appear to be" happy, differentiation and stratification of the work posts, etc.; see LABOR), and the physical space (through recreated, themed, sanitized, and ordered spaces; see Fjellman 1992; Archer 1997). Theme parks can be characterized as engines of control, or even as panoptic systems in the sense of Bentham (2017), as total institutions in the sense of Goffman (1961), or as closed spaces in the sense of Foucault (1977). However, within the parks control is exercised subtly and tenuously, and visitors must remain unaware of the fact that they are being controlled. In this sense, the systems for handling visitor flows, the key issue of theme park management, are commonly designed following the McDonaldization principles. They relate to mechanical transportation, pedestrian flows, and queuing management.

Mechanical transportation systems are conceived both as attractions that capture visitors' attention and improve their experience of the park and as means to move visitors (as well as employees) from one themed area to another, usually through land-based transport systems (either on wheels or on rails) or on water. A designed, customer-oriented layout directs pedestrian flows, breaking them into increasingly smaller groups as they enter each themed area and through the use of distribution points or loops within the area (Anton Clavé 2007, 372–373). This layout generally starts in the parking area, moves to the admissions zone (ticket sales and validation), customer services, the common services areas (cloakroom, the hire of wheelchairs and a shopping pickup point, among other services), to the entrance of the park (with shops, restaurants, and access to internal transport systems) and then to the inner parts of the park and its themed areas, where visitor flows are managed through visual magnets, colors, maps, and signs (see above). The smallest scale at which theme park flow management takes place is the queue management of individual attractions. This is dealt with in the layout design of the attraction itself, which includes the design of the queue, the loading structure, and the loading system. A good management of these flows is one of the key elements for visitor satisfaction.

More recently, many parks have installed systems that allow visitors to reserve a time slot for an attraction at a precise moment of the day (see TIME). Efficiency, quantification, predictability, and control are further improved over the course of a day by setting timetables for entertainment opportunities in accordance with the behavior observed among visitors according to age, seasonality, and other sociocultural characteristics. In order to do so, theme parks can use central control and real-time control systems (RTCS) management software to interconnect operations processes and to work smoothly and efficiently. These systems track how guests explore the theme park and how visitor flow is moving, thus allowing the park to schedule the timing and position of park workers depending on the real time needs of the facility. As is well-known, this has repercussions on both customer satisfaction and the park's income and profit (Thompson 1999).

The "doubling of landscape" does not apply equally to every place in the theme park, however. The backstage, accessible to employees only, is not themed: it is not supposed to "mean" anything and is basically governed by efficiency. The backstage includes such fundamental facilities as the human resource buildings (recruitment, candidate selection, contracting, and specific training); maintenance buildings (which contribute not just to visitor satisfaction but also to safety); gardening services areas and warehouses; electrical and electronic installations; the distribution of gas and water; installations for the production of shows with systems, logistics, and stage elements; waste and environmental management installations; common facilities and specific installations for food and beverage; facilities for administration, management, and new projects; roadways and technical facilities; general storage facilities; warehouses for goods to be sent out to the shops and play areas; food stores; wardrobe stores; maintenance workshops; greenhouses; and fuel pumps (Anton Clavé 2007, 354). Invisible to the customer, all these services and installations are incorporated in the park design and, therefore, in the park's layout, but there is no need for doubling there. Even some spaces that are accessible or even dedicated to visitors, such as toilets, paths, etc. are sometimes less themed, providing visitors with a kind of pause (a space and time-out zone) that allows them to fully appreciate the theming elsewhere.

4 The Experience of Space in Theme Parks

While the preceding section of this chapter has focused on the production of space by the makers, this section concentrates on the role of visitors in producing the theme park space. In theme parks, the visitor's experience is highly affective—not only because of rides and rollercoasters, but also because of the experience of space in general. This section will use a phenomenological approach to visitor experiences, starting from the assumption that visitors create space on different levels, or, conversely, that space influences visitor behavior and the visit to a theme park in different ways. Conceptually and methodologically, this section thus draws on the paradigm shift that took place in Social Sciences and in the Humanities in the 1980s and that has been conceptualized as the "spatial turn." Most importantly, the spatial turn has shown that space should not be understood as a "container," a natural or objective element within which people move and act, but that it is socially and culturally constructed.

When looking at the theme park, it is primarily the perception and imagination of the visitor that creates space. According to de Certeau (1984, 91–93), space can be perceived from two different positions: that of the "voyeur" and that of the "flaneur" (walker). The former views space in a "theoretical" fashion, "from above," either physically (in a theme park, most notably, thanks to attractions such as rotating towers that allow a broader, "external" view of the park and its spatial structures) or mentally (by looking, for example, at the park's map). The "flaneur," on the contrary, perceives and constructs space through practice, in an "operational" fashion by moving through it—and thus for instance by turning towards the Weenie. While both the "voyeur" and the "flaneur" must be understood as ideal types and the theme park visitor perceives the theme park space from both positions, it must be stressed that the visitors' role as walkers makes their perception radically dynamic.

Indeed, most of the time visitors are in motion, whether on rides and rollercoasters or in the rest of the park. The different three-dimensional images that visitors perceive are connected through their movements and thus relate to an overall experience which creates the theme park space. The flow of visitors is thus not only central to the production of theme park space (see above), but also to its perception. In contrast to television or the movies, visitors to a theme park can bodily enter the imaginary world, which becomes a physical experience in what is actually a fictional space. The theme park experience is thus "like walking into a movie" (Freitag 2017), with the different parts of the park or the themed areas connected not via the movement of a camera or via filmic cuts but through the movements of the visitors through the imaginary (but nevertheless physical) space. Thus, space in theme parks has a special connection to authenticity and to the authenticity of feelings in space, as it often consists of rebuilt sceneries or refers to existing cultures (see AUTHENTICITY).

At the same time, the theme park's design and layout to a certain extent also manage and control visitors' movements (see above). Hence, space and visitors' movements influence each other: on the one hand, the visitors' experience in and with space is highly individual as they can choose their own routes and itineraries, with the result that the connections between and perceptions of the different parts of the park are unique for each visitor. On the other hand, as the experience is to a certain extent already prescribed by the park's layout and (spatial) conditions, visitors can only react to them and thus space determines what is perceived. The water coaster Poseidon at Europa-Park (Rust, Germany) provides a good example: at a certain point on the way to the loading station of the coaster, a few steps lead visitors to a point where they decide whether to move on to the ride's entrance or turn left to a viewing platform where they have a perfect view of the final splash of the coaster. Whether they go to the platform or directly to the coaster, visitors automatically slow down due to the steps and have a longer view of the climax of the attraction, which increases their anticipation. Through the design of the path, the park thus seeks to intensify the experience. For their part, visitors unconsciously respond to the design of the paths, which guide both their movements and their emotions (see Schwarz 2017).

The viewing platform is not only a space where visitors connect with other spaces, but also a place where visitors visually connect with each other. This kind of interaction is highly relevant to visitors' experiences in theme parks. The platform shows how encounters between visitors are staged, and once again how the visitor's feelings and positions in space are prescribed by the park. As a spatial arrangement, the platform allows for some actions while prohibiting others—which is a characteristic of space in general, and of the theme park space in particular. When visitors enter this spatial arrangement, they are confronted with a limited selection of roles from which they may choose. For example, visitors can be mere onlookers who stand outside the actual action, simply watching the scenery and the boats dropping down, or—and especially when there are riders who they know—they can take a more active part in the game.

Yet visitors become part of the attraction whether they want to or not, as watching is a reciprocal process and the visitors on the platform are not only taking in the scenery, but are also being watched by the riders in the boats: as Poseidon is located in the "Griechenland" ("Greece") section of Europa-Park, which resembles a Greek tourist village (Carlà-Uhink 2020, 47–50), the visitors on the platform fit perfectly into the scenery as they play what they actually are: tourists. Hence, by going on the platform, exchanging gazes with the riders, and interacting with the space, visitors play a role that has been created for them through the space of the park. Of course, there are always ways to avoid these roles, for example by subversive behavior, but often the creators of the theme park take this into account as well (see ATTRACTIONS and VISITORS).

Paratextual representations of space in the shape of maps or apps (see PARATEXTS AND RECEPTION) assist visitors in finding their roles and ways, as do the various sorts of waiting lines for the various attractions ("single rider" lines, for example). Hence, theme parks permit various types of visits, as some patrons may prefer to plan ahead with the help of fast passes, whereas others may prefer to explore the park at their whim. The experience of space in theme parks as a whole fluctuates between these two poles of freedom and control (see ATTRACTIONS). This means that visitors must always orientate themselves and choose how to experience a theme park. And there is always the possibility of getting lost: for example, due to lack of space, Phantasialand (Brühl, Germany) is organized extremely densely, with the result that some visitors may find it difficult to orientate themselves. In addition, unlike e.g. Disneyland with its "Magic Wand pattern" that allows visitors to access various sections via a central platform (Mitrasinovic 2006, 128), Phantasialand forces visitors to choose between one of two central paths that both lead to dead ends: to the left are the "Wuze" and "Rookburgh" sections, to the right are "Mexiko" and, somewhat hidden behind it, "Deep in Africa," "Chinatown," "Klugheim," and "Mystery." Generally speaking, disoriented visitors may become frustrated, though it depends on their attitudes and expectations towards the visit and how they prefer to organize it. Nevertheless, theme parks generally try to avoid such reactions and organize space in ways that allow for both orientation and control even without maps, for instance through landmarks (see above).

This influence works to a large extent unconsciously and is thus another example of how space influences visitors' movement through space. However, not all theme parks have large corridors that lead to landmarks. In Phantasialand, for example, the organization of space differs from one area to the next. Whereas "Berlin" consists of one straight street and a large square evoking a city at the turn of the century, the "Deep in Africa" section represents a rural village and is structured into narrow, winding paths and small huts. Visitors move differently and thus also perceive these areas differently, unconsciously noting the atmosphere created by the spatial arrangement and the differences in interaction with the space. Moreover, according to Gernot Böhme (2013), atmosphere is not only created by the particular spatial arrangement but is also connected to the subjects being in this space, their feelings, their sensations, and their perceptions. Hence, it is not only the layout of the theme park space which engenders a specific atmosphere, but the interplay between all the elements of the built space, be it material, sound, light, smell, weather, etc. In the "Berlin" area in Phantasialand, for instance, materials like bricks and mortar are used for the facades, whereas the roads are asphalted or paved, and the overall color scheme is dominated by pastels. "Deep in Africa" differs from "Berlin" not only in its spatial structure, but also in the materials and their character. Here, the dominant colors are brown, red, and beige and the materials are wood, plants, or loam. Together with the spatial structure (as well as the different sounds and music), a completely different atmosphere is created.

Finally, as Böhme (2013) notes, not only must the individual subject want to be involved in and have to commit themselves to a certain atmosphere, but the presence of other people also has an impact on the prevailing atmosphere of a given space. Imagine the difference between a very crowded day during the peak summer season, a rainy weekday in fall, or a visit during the extra hours allocated for hotel guests (see TIME). A space consists not only of the built environment, but also includes the dimension of experience and of encountering people in a space. Therefore, the issue of crowds also needs to be considered. A crowded park requires visitors to slow down, perhaps allowing them to focus more on the theming, whereas an empty park will allow them to move faster at the risk of missing some details. Park designers must think about both possibilities, making sure the space engenders the desired responses under all sorts of different circumstances.

5 Conclusion

As a designed space, a theme park can be mapped as an artifact with gates, subsections, pathways, attractions, restaurants, shops, and other components relevant to the visit. As an experienced space, however, it can also be understood through the resulting emotional cartographies enacted by visitors. As is well known, individual emotional responses correlate with the current bodily state, and the bodily state directly influences the perceptions, interactions, movements, feelings, and sensations of visitors. This has been illustrated in a project by Orellana et al. (2016) that correlates changes in electrodermal activity (i.e. changes in sweat gland activity as a physical manifestation of emotional responses) with visitors' spatial behavior in PortAventura World. According to this analysis, the emotional cartography of a theme park may include a variety of maps that chart the ability of park areas and elements (e.g. rides, shows, games, or shops) to evoke specific emotional responses within visitors. Therefore, it is not only that the theme park space engenders a specific atmosphere, but also that the interaction between all the elements of the built space and the bodily (physical and mental) states of visitors results in individual and, by aggregation, collective emotional theme park maps.

References

Anton Clavé, Salvador. 2007. *The Global Theme Park Industry*. Wallingford: CABI.
Archer, Kevin. 1997. The Limits to the Imagineered City: Sociospatial Polarization in Orlando. *Economic Geography* 73 (3): 322–336.
Augé, Marc. 1995. *Non-Places: Introduction to an Anthropology of Supermodernity*. London: Verso.
Bentham, Jeremy. 2017 [1791]. *Panopticon*. Whithorn: Anodos.
Böhme, Gernot. 2013. *Atmosphäre: Essays zur neuen Ästhetik*. Berlin: Suhrkamp.
Carlà-Uhink, Filippo. 2020. *Representations of Classical Greece in Theme Parks*. London: Bloomsbury.
Carlà-Uhink, Filippo, and Florian Freitag. 2022. Theme Park Imitations: The Case of Happy World (Happy Valley Beijing). *Cultural History* 11 (2): 181–198.
De Certeau, Michel. 1984 [1980]. *The Practice of Everyday Life*. Trans. Steven Randell. Berkeley: University of California Press.
Eyssartel, Anne-Marie, and Bernard Rochette. 1992. *Des mondes inventés: Les parcs à thème*. Paris: Editions de la villette.
Fjellman, Stephen M. 1992. *Vinyl Leaves: Walt Disney World and America*. Boulder: Westview.
Foucault, Michel. 1977. L'œil du pouvoir: Entretien avec Michel Foucault par Jean-Pierre Barou et Michelle Perrot. In *Le panoptique*, Jeremy Bentham, 9–31. Paris: Belfond.
Foucault, Michel. 1986. Of Other Spaces. *Diacritics* 16 (1): 22–27.
Freitag, Florian. 2017. "Like Walking into a Movie": Intermedial Relations between Disney Theme Parks and Movies. *The Journal of Popular Culture* 50 (4): 704–722.
Freitag, Florian. 2018. "Who Really Lives There?": (Meta-)Tourism and the Canada Pavilion at Epcot. In *Gained Ground: Perspectives on Canadian and Comparative North American Studies*, ed. Eva Gruber and Caroline Rosenthal, 161–178. Rochester: Camden House.
Goffman, Erving. 1961. *Asylums: Essays on the Social Situation of Mental Patients and Other Inmates*. New York: Anchor.
Gottdiener, Mark. 1997. *The Theming of America: Dreams, Visions and Commercial Spaces*. Boulder: Westview.
Griffiths, Alison. 2008. *Shivers Down Your Spine: Cinema, Museums, and the Immersive View*. New York: Columbia University Press.
Lukas, Scott A., ed. 2007. *The Themed Space: Locating Culture, Nation, and Self*. Lanham: Lexington.
MacCannell, Dean. 2013 [1976]. *The Tourist: A New Theory of the Leisure Class*. Berkeley: University of California Press.
Mitrasinovic, Miodrag. 2006. *Total Landscape, Theme Parks, Public Space*. Burlington: Ashgate.
Neary, Kevin, Susan Neary, and Vanessa Hunt. 2016. *Maps of the Disney Parks: Charting 60 Years from California to Shanghai*. New York: Disney Editions Deluxe.
Orellana, Alicia, Joan Borràs, and Salvador Anton Clavé. 2016. Understanding Emotional Behaviour in a Theme Park: A Methodological Approach. In *Consumer Behavior Tourism Symposium Book of Abstracts: Experiences, Emotions and Memories: New Directions in Tourism Research*, ed. Serena Volo and Oswin Maurer, n.p. Bruneck: Competence Centre in Tourism Management and Tourism Economics (TOMTE).
Rebori, Stephen J. 1993. *Theme Parks: An Analysis of Disney's Planning, Design and Management Philosophies in Entertainment Development*. Master Thesis, University of Tennessee.
Ritzer, George. 1996. *The McDonaldization of Society*. Thousand Oaks: Pine Forge.

Schwarz, Ariane. 2017. Staging the Gaze: The Water Coaster Poseidon as an Example of Staging Strategies in Theme Parks. In *Time and Temporality in Theme Parks*, ed. Filippo Carlà-Uhink, Florian Freitag, Sabrina Mittermeier, and Ariane Schwarz, 97–112. Hanover: Wehrhahn.

Steinkrüger, Jan-Erik. 2017. Other Times and Other Spaces: Themed Places and the Doubling of Landscapes. In *Time and Temporality in Theme Parks*, ed. Filippo Carlà-Uhink, Florian Freitag, Sabrina Mittermeier, and Ariane Schwarz, 83–95. Hanover: Wehrhahn.

Thompson, Gary M. 1999. Labor Scheduling: Part 4. Controlling Workforce Schedules in Real Time. *Cornell Hotel and Administration Quarterly* 40: 85–96.

Tuan, Yi-Fu. 1998. *Escapism.* Baltimore: Johns Hopkins University Press.

Younger, David. 2016. *Theme Park Design & the Art of Themed Entertainment*. N.P.: Inklingwood.

Theming: Modes of Representation in Theme Parks and Themed Environments

Abstract Without theming, there is no theme park. Therefore, most of the chapters in a book on theme park studies address theming, or at least examples of theming, in one way or another. This chapter is, however, a central intersection within the book, explaining what theming is, as well as why and how it works. Following Gottdiener, theming can be seen as a symbolic bonus that gives additional meaning to objects or functions and unites them under one common roof. With theming, a theme park becomes more than a conglomeration of rides, shops, and restaurants. To be accessible to an audience and understood as a coherent unity, themes often rely on already established motives, narratives from history, geography, mythology, or media. In this context, the chapter also shows examples of failed theming and addresses the practical realization of themes by showing what elements—rides, shows, food, merchandise, furniture, landscaping, etc.—are themed, and how themes appeal to several senses, nowadays even to olfaction and thermoception.

1 Introduction

Theming—"the use of an overarching theme or key concept (like Western) to organize a space" (Lukas 2013, 68)—is central to the theme park industry and theme park research. Not only does it help us distinguish theme parks from closely related location-based entertainment spaces (as well as from other theme parks), but theming also provides a link between theme parks and other (commercial, public non-commercial, and even private) themed and immersive spaces. What differentiates the theme park from the amusement park or the fun fair is precisely the fact that it employs theming to connect its individual offerings. As the share prospectus for Disneyland Paris emphasizes,

This work is contributed by Scott A. Lukas, Filippo Carlà-Uhink, Florian Freitag, Salvador Anton Clavé, Astrid Böger, Thibaut Clément, Sabrina Mittermeier, Céline Molter, Crispin Paine, Ariane Schwarz, Jean-François Staszak, Jan-Erik Steinkrüger, Torsten Widmann. The corresponding author is Scott A. Lukas, Department of Sociology and Anthropology, Lake Tahoe Community College, South Lake Tahoe, USA.

F. Freitag et al., *Key Concepts in Theme Park Studies*,
https://doi.org/10.1007/978-3-031-11132-7_14

> Rather than presenting a random collection of roller coasters, merry-go-rounds and Ferris wheels in a carnival atmosphere, [theme parks] are divided into distinct areas called 'lands' in which a selected theme (such as exotic adventures, childhood fairy tales or the frontier life of the nineteenth century American West) is presented through architecture, landscaping, costuming, music, live entertainment, attractions, merchandise and food and beverage. (qtd. in Bryman 2004, 20)

Likewise, it is the specific theme(s) used in a given themed "land" or park that differentiates it from other lands or parks with similar attraction mixes (see ATTRACTIONS). For example, one of the design principles of Tokyo DisneySea, the second park added to Tokyo Disneyland Resort in 2001, was to avoid redundancy or duplicating the theming of the neighboring Tokyo Disneyland park: "Do not poach, or do not cannibalize on, any of the experiences in Tokyo Disneyland," was the direction given to designer Steve Kirk (qtd. in Younger 2016, 198). Therefore, Younger notes, "elements such as a pirate themed land were considered, but dropped due to the redundancy that would have existed with Pirates of the Caribbean" (198).

Since the 1990s, scholars have discussed the use of theming in historical and contemporary commercial and public (Sorkin 1992; Gottdiener 1997), as well as private spaces (Lukas 2007a, b, c), thus establishing theming as a more general cultural practice that is by no means restricted to theme parks. As Lukas has noted,

> Outside the corporate and consumer world, everyday people have reacted to the widespread nature of theming by using it as a means of personalizing their own home worlds. […] The theming of the home and of intimate spaces that are not used by the public or determined by a corporation suggests a movement of theming into a more personalized realm. (2007a, 4)

"Perhaps the most significant aspect of the themed space," Lukas concludes, "is its growing ubiquity" (15; see also below). The concepts of the "experience society" (Schulze 2005) and the "experience economy" (Pine and Gilmore 2011) provide the broader background for this development, as both have been recognized as characteristic of the Global North since the 1970s and have subsequently expanded to other regions of the world.

In addition to offering a general definition of theming as a design technique, the following section of this chapter will introduce the notion of the "duplicity" of theming, according to which visitors should perceive themed spaces as simultaneously strange and familiar. The next two sections will then take a closer look at the concepts of "externality"—i.e. the ways in which designers produce "strangeness" by establishing a notable contrast between the themed space and its surroundings—and "recognizability"—i.e. the strategies designers employ to produce "familiarity" by allowing visitors to connect to the theme cognitively and emotionally and thus enjoy its strangeness as a form of escapism. Next, the chapter will provide an overview of the most popular themes used in theme parks, as well as address such special cases and trends as autotheming and IP-based theming. The final three sections will trace the phenomenon of theming beyond theme parks, explore the limits of theming by analyzing examples where, for one reason or another, theming failed to work, and venture an overview on the future of theming.

2 Theming: A Definition

In very general terms, theming may be described as a design technique that takes all aspects of a delimited space—from the "material attributes of the environment (scale, color, layout, costumes)" and the "sensory environmental stimuli (visual, aural, tactile, olfactory)" to the "practices of all constituents (both on frontstage and backstage)" as well as the "commodities sold (arts and crafts, food, souvenirs)" (Mitrasinovic 2006, 36)—and consciously organizes them around a pre-defined topic (the "theme") in order to produce specific cognitive, emotional, and behavioral effects within its visitors. In their various definitions, scholars have highlighted different aspects of theming, including its spatiality, its mediality, its commercial function, as well as its ideological implications and behavioral effects. According to geographer Jan-Erik Steinkrüger, for instance, the act of theming constitutes a "doubling of landscape," in which a landscape represented by a theme (i.e. the theme) is superimposed onto or articulated through a landscape representing a theme (i.e. the themed space; see Steinkrüger 2017; see also SPACE). For media scholars Jay Bolter and Richard Grusin, in turn, theming joins the "hypermediacy" of "sights and sounds from various media [that] recall and refashion the experience of vaudeville, live theatre, film, television, and recorded music" with the "immediacy of presence," the "exhilaration of being physically surrounded by media" (2000, 169, 172; see MEDIA).

Other researchers have been more concerned with the functions and effects of theming. For Mark Gottdiener, for instance, theming primarily constitutes a marketing appeal of "symbolic differentiation" that "reduce[s] the product to its image and the consumer experience to its symbolic content" (Gottdiener 1997, 2, 74; see ATTRACTIONS). This image or symbolic content inevitably comes, as Scott A. Lukas has pointed out, loaded with certain values and worldviews: "Theming, as much as it is an entertainment practice, is an ideological undertaking that highlights certain values and underplays others" (2007b, 198; see WORLDVIEWS). Whereas Lukas thus emphasizes the efficacy of theming in promoting and perpetuating specific values and ideologies, Miodrag Mitrasinovic and Brian Lonsway are more interested in the immediate in situ effects of theming—namely, how the theming of a space impacts the behavior of its visitors. Theming, Mitrasinovic notes, "is de facto a deliberate attempt at designing an environment configured to produce specific behavioral outcomes in visitors" (2006, 36). Similarly, Lonsway argues that themed spaces "script a set of constraints on our spatial behavior" (2009, 175–176; see VISITORS).

Yet whatever aspect of theming individual researchers may be particularly interested in, most scholars agree that it is characterized by (1) its "multisensoriality" as well as (2) by an inherent "duplicity." With respect to the former point, theming is by no means only limited to visual (architectural and decorative) elements, even if they constitute a major component of it; independent of its function in a particular setting, an architectural form recognizable as a "Greek temple" constitutes a central element of a "Greek theme" (on the "Greek theme" and for many of the following examples, see Carlà-Uhink 2020). Next to such architectural forms, interior design, and more generally décor, also play a crucial role: to continue with the example of a

Greek theme, geometric decorative forms such as meanders, figures recalling ancient Greek vase painting, or the colors blue and white, are evocative of such a theme.

Material culture as a crucial ingredient to theming is by no means limited just to visual elements, however: typical food, for instance, produces smells that can be immediately recognized as related to the chosen theme (gyros, for example, or fish, in the case of Greece); touching the cool stone and marble of "Greek" columns or statues also heightens the believability of what can be seen. To all this, add performance (forms of acting and aesthetics that enhance the design, architecture, and material elements of a space), including the clothes worn by the staff—togas in an ancient setting, for example. The sounds of pre-recorded soundtracks, usually consisting of instrumental music and sound effects, further enhance the theme—in the case of a Greek theme, for example, the first two notes of Michail Theodorakis's "Zorba's Dance" (from the soundtrack of the 1964 movie *Zorba il Greco*) may evoke an association with Greece in many visitors.

Spoken or sung language may also be part of the soundtrack. In fact, whether spoken or written (e.g. on signs and inscriptions), language constitutes a very important and yet frequently neglected aspect of theming: in the Greek section of Europa-Park in Rust, Germany, for example, the soundtrack of the madhouse Cassandra contains a curse in modern Greek; in the same area, an imitation of an ancient Greek inscription decorates the waiting area of the water coaster Poseidon. More generally, the Greek alphabet is often evoked, although not used, through so-called simulation typefaces, or by replacing Latin letters with similar-looking Greek ones (e.g. sigma, Σ, for E; see Carlà-Uhink 2020, 49). Occasionally, themed spaces also work with thermoception—as, for instance, at Happy Valley Beijing (China), where a subsection of the Greek-themed area called "FantaSea" is located indoors to provide shelter from rain and low temperatures, neither of which visitors are likely to associate with Greece (see Carlà-Uhink 2020, 136).

All this is relevant from the perspective of the producer who plans and organizes themed areas; yet visitors are a crucial part of theming, too, just as their interpretations, desires, reactions, and feedback—most visibly through fandom and cosplaying—are also important facets of theming (see VISITORS). Indeed, visitors' reactions are crucial: in order to be successful, the themed space needs to be characterized by a certain "duplicity," in the sense that visitors must perceive it as "something that is entirely new but that still seems oddly familiar" (Wright 2009, 23). To capture the former point, scholars such as Bryman, Kolb, Younger, and Steinkrüger have identified the principle of thematic "externality" or "opposition." In *Sprawling Places*, Kolb notes that "the reference to what is outside the themed place is a constitutive element in the place. Though they pretend to offer total absorption in their themes, themed places rely on a continual awareness of the outside and their difference from it" (2008, 112). From this, Steinkrüger concludes that the source of an immersive space's theme is not just "typically [...] external to the institution or object to which it is applied," as Bryman had argued in *The Disneyization of Society* (2004, 15), but that "spatial, temporal or worldly externality is a necessary condition for a space to be perceived as a themed space"; "To be recognizable as a theme the represented times, spaces, and worlds have to become 'other' times, spaces, and worlds" (2017, 88).

Similarly, Younger defines “opposition” as a design principle that is not only used within theme parks to emphasize specific details, but also to heighten the contrast between the theme park and its surroundings (2016, 165).

Diametrically opposed to the principle of “externality” is the principle of “recognizability,” which ensures that visitors are able to cognitively and emotionally connect to the theme, despite its strangeness. For instance, while he insists that in a themed place “the social norms are self-consciously presented *as other* and *as different from the local everyday* expectations,” Kolb also stresses that the “definition of a themed place requires that the theme present itself as referring to an established unity of meaning or identity” (2008, 112–113; emphases original). Similarly, Lukas has argued that theming “relies on guests’ previous associations with certain themes” (2013, 68). The following two sections will thus explore the specific strategies and techniques used by theme park designers to produce both strangeness and familiarity, externality and recognizability.

3 Externality

Thanks to theming, visitors entering a theme park come to a place that is ontologically different from the place they come from. Within the park, visitors are not in the here and now, in the contemporary space of their everyday life; they travel virtually through time (to the past or the future) and space (to a foreign and/or imaginary country) to visit another world, designed to fulfill their fantasies. Hence, theme parks are vehicles transporting visitors to and immersing them within an attractive distant country, provided they agree to suspend their disbelief (see IMMERSION). This virtual or metatourism (Wöhler 2011; Freitag 2018) is not a poor substitute for tourism itself, it is much “better”: the journey is shorter, cheaper, and safer, and the destination is less disappointing, as it more closely corresponds to the visitors’ expectations (see below; see also AUTHENTICITY). As with tourism, the theme park is worth the trip because the place where visitors go to differs from the place they come from, and because it provides them with the liminal experience of a place where the usual rules and habits can and even should be transgressed. Even natural or physical forces such as gravity do not apply in theme parks, where visitors can enjoy the experience of flying or being a superhero. One could thus say that theme parks are designed for escapism (Tuan 1998): they give visitors a break from their daily life, its boring routines, and frustrating constraints. The path to escapism may take three directions: nostalgia, exoticism, or utopia.

Nostalgia takes visitors back to an idealized past, a time when life was (supposedly) easier and simpler. This may work on an individual level, where visitors travel to the happy moments of their childhood, or on a more collective level, where they visit a historical period which according to the regional or national(istic) narratives of the area where the theme park is located was “the good old times” (see WORLDVIEWS). The little town of Mount Airy, North Carolina, has been themed to look like

Mayberry, the quiet and peaceful town from *The Andy Griffith Show* (CBS, 1960–1968), but African-American tourists may not feel any more welcome in Mount Airy than they were in Mayberry, as the show casts only Caucasians and never mentions the Civil Rights movement (see Alderman et al. 2012). Likewise, it is not so sure that Disneyland's "Frontierland" or "Main Street, U.S.A." were happy times and places for every American, but they surely are for Walt Disney and the white middle-class family audience he had in mind (see also Lonsway 2009, 174–175; INCLUSION AND EXCLUSION).

Exoticism, in turn, takes visitors to a faraway country, where the people, the natural and cultural landscapes, the food, the music, etc. are attractive because they are different from those they experience at home. Exoticism is thus based on geographical imagination and is culturally specific; as such, it can have very different geographical referents, from individual cities (such as Venice, synecdochically used to signify the entire country of Italy in the "World Showcase" at Epcot in Orlando, Florida), countries (stereotypically perceived from the distance as homogeneous, regardless of their dimensions and in spite of internal differences, as in the case of China or Mexico in Phantasialand, Brühl, Germany, or most of the themed areas in Europa-Park), or even entire continents (which may likewise be perceived as largely homogeneous by the target public, as in the case of "Deep in Africa" in Phantasialand; see Steinkrüger 2013, 266–276).

Utopia differs from nostalgia and exoticism insofar as it is defined as a place that is not supposed to exist or has existed anywhere. Utopias are fictional places and assumed as such, often conceived, following Thomas More's eponymous book (1516), to describe a perfect society, and are thus opposed to "dystopias," nightmarish representations of fictional negative societies of violence, abuse, and despair. Etymologically, however, a utopia is simply a "no-place." Fictional places could just as well belong to the past, the present, or the future, and be located close by or far away. In this sense, Peter Pan's Neverland (in Disneyland's "Fantasyland" in Anaheim, California), Harry Potter's Hogwarts (in Universal Studios' "The Wizarding World of Harry Potter" in Orlando, Florida), or Avatar's Pandora (in "Pandora: The World of Avatar" at Disney's Animal Kingdom in Orlando, Florida) are utopias. Theming a land or an entire park as a utopia thus allows visitors to escape to a place and a time that is only limited by their creators' imagination.

To be sure, the three categories of nostalgia, exoticism, and utopia should be understood as ideal types, which often overlap and intersect. Exoticism, for example, is never only spatial, but also strongly temporal in nature. Each representation of an "exotic" culture corresponds to a precise time layer, chosen to best represent the stereotypes of the target public and be easily recognizable. This time layer is, more often than not, the past: with the Global North serving as the "standard" of modernity and civilization, other parts of the world are usually conceived as somehow backward and primitive, and are, therefore, represented as traditional societies, "older" places that lag behind Europe and America (see WORLDVIEWS). "China Town" in Phantasialand, for instance, is not a representation of the thriving contemporary megalopolis of Shanghai or the financial district of Hong Kong, but a conflation of the time of the Qing and Ming dynasties. However, from a primitivist perspective, such places might as well be considered more authentic, closer to God or to Nature. Such myths as the Noble Savage or the Garden of Eden refer to a former and better humankind, unspoiled by so-called civilization: exoticism is thus closely linked to nostalgia (see TIME). Conversely, nostalgia is also never completely devoid of aspects of exoticism: David Lowenthal has shown how *The Past is a Foreign Country* (2015), a different place that people long to visit.

Both exoticism and nostalgia are, in turn, related to utopia. Although it is true that themed areas based on exoticism and nostalgia refer to existing places and historical cultures, they are by no means limited to realism. The representation of myths and legends from more or less exotic cultures may also include monsters and ghosts, as well as the violations of the laws of physics, that one may expect from an imaginary, utopian world. In Europa-Park's representation of Greece, for instance, the Fluch der Kassandra madhouse ride makes visitors spin in a room full of objects referring to different phases of Greek history, after restoration works in a Mykonos church have supposedly freed from an amphora the soul of the prophetess Cassandra, who is now looking for revenge (Carlà-Uhink 2020, 58–61). At the same time, utopian worlds, which must be reified in architectural and decorative forms as a themed area in the theme park, inevitably find inspiration and draw on references from historical images. Following the film adaptations that set the standard for the visualization of the literary source texts, the "Wizarding World of Harry Potter" at Universal Studios is inspired by old British town centers, and more generally by medieval architecture—the historical referent for most representations connected to the world of magic and the genre of fantasy (Carlà-Uhink 2020, 6–7).

In its three ideal forms of nostalgia, exoticism, and utopia, externality thus allows theme park visitors to travel to a strange time, place, and world, to escape their daily life for a limited period of time. The most common way for theme park designers to achieve externality, strangeness, and escapism is through a careful selection of themes, as well as the deliberate exclusion of aspects that could remind visitors of the outside world or otherwise alienate them. Hence, choosing a nostalgic, exotic, or utopian theme—a theme that is temporally, geographically, and/or ontologically distant from its surroundings—constitutes the first step of theming. Designers may enhance the distance or contrast between the themed space and the outside world through visual barriers and monumental entrances (portals, tunnels) that signal a passage into a different world—or through a direct collision of the two: at Disney's Animal Kingdom, for instance, "the decision was made to keep the parking lot visible from the entrance, so that the barren, dry asphalt would make the lush, dense foliage inside the park that much more of an oasis" (Younger 2016, 165). Likewise, at Tokyo Disneyland, the neighboring Tokyo Bay is visible from inside the park, but the view is carefully framed so as to make the Pacific appear like the Atlantic.

Choosing a theme is a matter of *selection*, which is a crucial strategy within the principle of externality, because not everything from a chosen theme finds its way into the themed area: theming an area to China within a German theme park also implies choosing specific aspects from Chinese culture, geography, and history; this further selection is absolutely necessary, if only because of the limited dimensions of any themed environment. Just as the general selection of a theme depends on the location of the themed space, the line between what may be included in the theme park and what needs to be excluded from it is both culturally specific and historically shifting. Allusions to slavery in Disneyland's New Orleans-themed spaces may have been permissible in the 1950s and early 1960s, but not after the Civil Rights movement (see Freitag 2021, 131–136; see also INCLUSION AND EXCLUSION). Indeed, the various changes that themed spaces have undergone in culturally changing climates, as well as some of the cases of failed theming discussed below—notably, that of Disney California Adventure in Anaheim, California—stress the relevance of the strategy of selection and the principle of externality.

For Carlà and Freitag, selection thus not only entails the general selection of a theme, but, even more importantly, "the selection of specific aspects of the theme [as dictated] by the nature of theme parks as leisure spaces for all age groups: problematic and/or contested issues (such as war or slavery) are regularly omitted from the themed world" (2015a, 244). Themed spaces act here "as boundaries against innumerable forces—of evil, contagion and fear" (Tuan qtd. in Lukas 2007c, 276). The exclusion of aspects perceived as problematic and troubling, such as death, poverty, famine, etc. is not a fixed rule, however, and there are two main areas in which such topics can nevertheless be included in the theme park; in both cases, they are still the product of a reasoned selection, although one follows different criteria and rules. Firstly, "utopic" themes, defined as "unreal" spaces (see above), can include aspects that could be considered unsettling when inserted into a more "realistic" context. As part of (Classical) mythology, for example, war and its victims appear quite frequently in theme parks. Shows and attractions dedicated to the Trojan War such as Troja, la Conquista in Terra Mítica (Benidorm, Spain), or Flying over the Aegean Sea in Happy Valley also represent deaths, plundering, fires, and violence (Carlà and Freitag 2015b, 149; Carlà-Uhink 2020, 89; 144–147).

Secondly, planners and designers may intentionally decide to choose themes and aspects within those themes that are considered unsettling, uncomfortable, scary, or disturbing. This is known as "dark theming," "a subset of dark tourism which describes the ways in which disturbing, controversial, political, or other hidden topics and concerns are narrativized and performed" and which has been interpreted as "a meaningful way of expressing existential, nihilistic, and postmodern interpretations, experiences, and ideas in the contemporary world" (Lukas 2016, 226). Dark-themed spaces are, therefore, the product of a coherent and planned strategy of selection, albeit one that follows very different criteria to those adopted, for instance, in the theming of a family-oriented theme park. A successful case in point is Dismaland, Banksy's 2015 "bemusement park" in Weston-super-Mare (UK): focusing on politics, conflict, war, violence, death, and destruction, Dismaland included what other theme parks excluded and vice versa, parodying through inversion these processes of selection, as in the representation of Cinderella's pumpkin carriage being surrounded by paparazzi after a tragic accident, in a clear reference to the car accident in which Lady Diana died (see Freitag 2017; see also PARATEXTS AND RECEPTION).

Theming requires both a "suspension of disbelief" and a "residual awareness" of the outside world and how the latter differs from the themed space (see IMMERSION). At the same time, however, visitors must perceive the themed space not only as strange or “entirely new,” but also and simultaneously as “oddly familiar” (see above). In order to be able to navigate and cognitively and emotionally respond to the themed space, visitors must be able to recognize it as a themed space, a strange space that is different from the outside world, but also a space that does not confront them with unprocessable, profoundly unsettling, and potentially dangerous experiences. Even in the case of dark theming, visitors must not feel that they are actually in danger. To achieve this, and lest visitors should feel disoriented, disappointed, or confused, theme parks must combine externality and recognizability.

4 Recognizability

Firat and Ulusoy have defined theming as “the patterning of space, activity, or event to symbolize experiences and/or senses from a special or a specific past, present, or future place, activity, or event *as currently imagined*” (2011, 195; our emphasis). Indeed, in spite of their necessary externality, themes must simultaneously be highly recognizable to visitors—the visual referents, music, smells, colors, etc. deployed must be unambiguously and immediately connected to the chosen theme by the patrons. To ensure recognizability for the broadest audience possible, theming thus refers to (clusters of) images, representations, and symbols that are already known and widespread among the target public and that are situated within what can be defined as their “horizon of expectations.” This means, therefore, that theme parks routinely adopt, adapt, and perpetuate already available stereotypes. Scholars agree that theme parks and other touristic spaces do so by extensively drawing on previous depictions of the selected theme in popular media. In *Fairground Attractions: A Genealogy of the Pleasure Ground* (2012), for instance, Deborah Philips has argued that each of the eight themes or genres she identifies as regularly appearing in theme parks—the chivalric, the fairy tale, the Gothic, Egyptomania, explorers, science fiction, the Western, and Treasure Islands (Philips 2012, 2–3; see also below)—has “a basis in folk history and in an oral tradition” and has been reproduced in literature, illustrations, dioramas, panoramas, opera, ballet, drama, fairgrounds, film, and television, before eventually finding its way to the theme park (4–5).

Similarly, Karlheinz Wöhler maintains that tourism imaginaries are based on texts and images distributed through print media, radio, movies, television, and the internet (2011, 73). These texts and images, according to Wöhler, aggregate to form a virtual image of a tourism space—a process he refers to as “virtualization” or “cognitive mapping” (72)—that is then “realized” by the actors involved in the touristic space. To be sure, Wöhler's notion of the “realization of the virtual” could also be applied to the theme park itself and how theme park employees and visitors “realize” the virtual image of theme parks created by their paratexts (see AUTHENTICITY and PARATEXTS AND RECEPTION). Here, however, this model shall be used to illustrate the principle of the “recognizability” of theming, which uses virtual images of particular themes popularized through media and “realizes” them through theming in three-dimensional, immersive spaces.

In contrast to what Philips and Wöhler may suggest, however, virtual images of a particular theme do not necessarily have to be based on depictions in multiple media. Occasionally, and especially in the case of theming based on intellectual properties, one media source may suffice: "The Wizarding World of Harry Potter" may ultimately refer to the novels by J. K. Rowling, yet many of its visual and audio details are taken directly from the movie adaptations. As a popular medium that in itself already addresses multiple senses, the film provides theme park designers with valuable sources of easily and widely recognizable elements (see also MEDIA). Moreover, in the case of imitations such as Happy Valley Beijing's Happy World ride—a localized "lift" of Disney's It's a Small World attraction (see Carlà-Uhink and Freitag 2022)—it is the theme park and their paratexts themselves that provide the virtual image, which is then realized in a different park.

Just like "externality," virtual images and "recognizability" are culturally specific. For example, and as already mentioned above, the representation of China in Phantasialand must correspond to what the mostly German target audience of this park understands, knows, and recognizes as Chinese—the Tatar Mosque in Ürümqi is located in the People's Republic of China, but unlike, e.g. the Lama Temple in Beijing, it would not be understood by a broad German public as "Chinese." With its minarets and golden domes, a replica of Ürümqi's Tatar Mosque would rather be associated with the Near East, as it would fit into the Orientalist stereotypes widespread among this target public. Especially for a Western target audience, Orientalism has indeed proven a powerful discourse and image that has been employed for multiple different themes. As analyzed by Edward Said (1978), Orientalism is a further form of exoticism that does not differentiate between Morocco, Turkey, Arabia, and the Near East in general: the same desert, palm trees, camels, minarets, domes, magic carpets, belly dancers, snake charmers, etc. are all signs that evoke an undifferentiated Orient, in novels, movies, paintings, as well as in theme parks. The religious theme park Tierra Santa in Buenos Aires, Argentina, for example, seeks to provide an immersive experience of a "Holy Land" that is filtered through these Orientalist stereotypes—therefore, the restaurants at Tierra Santa offer kebab, shawarma, and falafel while dancers perform what the park calls "Arabic dances" in front of Pontius Pilatus' *praetorium*. This exoticism relies on a highly ambiguous fascination with the Other, whose difference is stereotyped, essentialized, fetishized, and commodified in an often racist and (neo)colonial way and is deeply intertwined with stigmatization and discrimination. As mentioned above, this also implies the representation of specific "time layers," generally reifying the idea that non-Western societies are more "primitive" (see WORLDVIEWS).

The success of "recognizability" is particularly relevant for those theme parks and themed environments that have a primarily commercial aim: they cannot afford to "irritate" the public with unrecognizable theming that would reduce the enjoyment and hinder visitors' immersion (see IMMERSION). However, even theme parks and themed environments with a primarily educational aim, such as religious theme parks, must follow the same rules; in this case, theming that makes immersion difficult or impossible and that irritates the public's expectations would not make visitors "understand" more philologically correct reconstructions of other places or cultures; on the contrary, it would "block" the learning processes the park aimed for. Regardless of the particular goals of the themed space, theming must thus take into account existing stereotypes or virtual images in the development of a representational strategy.

This is not only specific to theme parks, but also to their antecedents (see ANTECEDENTS): in a very famous incident, the German pioneer of zoos and organizer of ethnographical shows Carl Hagenbeck incurred a huge financial loss when he organized a spectacle involving the Native American group of the Bella Coola. Highly praised by scholars, this exhibition was a huge failure with the public, as the Bella Coola, with their masks and traditional costumes, did not correspond to what the German public expected from a display of Native Americans—their imagination was mostly shaped by the Native tribes from the Prairies (Steinkrüger 2013, 222; Carlà-Uhink 2020, 9). Yet, successful theming is not, or rather should not be, a mere repetition of existing stereotypes, but instead

> must go beyond the surface stereotypes and instead focus on a use of key symbols, materials, and other forms of material culture that will create a balance between the familiarity of the theme that will strike instant chords with the guest and the unique elements that will cause the guest to linger in the space and reflect upon the thematic design. (Lukas 2013, 79)

In this respect, Lukas formulates three principles of theming: "Theming is highly distinctive because it uses recognizable symbols; Theming relies on guests' previous associations with certain themes; Theming uses 'shorthand' or abbreviation to communicate with guests" (2013, 68).

The latter point is indicative of the process of *abstraction* that necessarily takes place in theming. Abstraction implies that "typical and characteristic features of the theme are reduced to iconic and evocative signs or symbols" (Carlà and Freitag 2015a, 245) and leads to the creation of what could be called "pictograms" that immediately evoke a theme—be it a period, a culture, an exotic place, or an unreal world. To evoke and represent the ancient Greek world, for example, theming generally adopts the architectural form of the white temple, Doric or Ionic, with six or eight columns on the front—in spite of the fact that in ancient Greece, temples were not white and could assume very diverse architectural forms. The "Greek temple" in the theme park is also, of course, not a temple: it can house a ride, a loading station, a restaurant, a shop, or simply function as a backdrop—independent of its specific function, its role is to unambiguously communicate to visitors the "ancient Greekness" of the themed area (Carlà-Uhink 2020, 18–19). Abstraction may also work through layout: Disneyland's "New Orleans Square," for instance, neither features the grid layout of the Crescent City's French Quarter nor constitutes a proper square. Instead, designer Herbert Ryman distributed the land's different attractions among several smaller, irregularly shaped structures with "winding streets curving out of view" to "arouse people's curiosity and invite them in to explore" (Ryman qtd. in Gordon and Mumford 2000, 180). Thus, Ryman managed to create an atmosphere of urbanity in spite of the comparatively small acreage of the land.

Incidentally, theme parks may also arrange for recognizability to take place at different levels, integrating multivalent elements that may be recognized differently by different target audiences. Entering the queue of Mystic Manor at Hong Kong

Disneyland, for instance, visitors discover an oil painting that shows the fictional hosts of the attraction, Lord Henry Mystic and his aide-de-camp Albert, together with other members of the "Society of Explorers and Adventurers." Most visitors to Mystic Manor may accept the painting as a suitable piece of wall adornment for a late nineteenth century colonial mansion or perhaps, given the clothes of the individual members, recognize the virtual image of the "archaeologist as adventurer" (Holtorf 2007, 63–75). Others, however—particularly Disney fans who have completed the Grand Tour of all the Disney parks worldwide—may recognize that the person standing on the far left in the painting is none other than Harrison Hightower, the fictional owner of the Hotel Hightower (the Tower of Terror attraction) at Tokyo DisneySea. Through such transmedia storytelling (see MEDIA) and autotheming (see below), the painting becomes multiply recognizable.

5 A Wide Range of Themes

Following the criteria of externality and recognizability, one can find a relatively limited series of recurrent themes or theme archetypes in theme parks or other themed environments. Several scholars have tried to establish comprehensive lists and classifications of the most popular themes. For example, in 1987, Fichtner and Michna (26–27) proposed five categories of themes. While the fifth and final one, "play worlds," includes non-themed water parks, circuses, etc. and can thus be left aside, the other four are arranged according to different kinds of externalities (even if the authors do not use this word): time (past, nostalgia, future); space (wild nature, foreign people); social dimension (folklore); and reality (unreal worlds). The limits inherent to all forms of classification of this kind are obvious: they are ideal descriptions, and most themes fall into more than one of these categories. A different strategy is followed by Lukas (2013, 69–70), who identifies place and culture, brand, interest and lifestyle, and mood and association as different forms of theming. Here too, however, most themes will cover more than one form—the world of Parc Astérix in Plailly, France, for example, represents both a "place and culture" as well as a globally known "brand."

Table 1 compares the lists of recurring themes, motifs, or theme archetypes compiled by five theme park scholars, with the addition of examples and case studies (exclusively from theme parks) for each of them. While the scope of these lists is not exactly the same (whereas four of the critics focus on theme parks, Gottdiener discusses themed environments in general) and while all of these scholars seem to focus on the west (as suggested by the examples they use in their texts), the comparison nevertheless allows for some interesting observations: exoticism or adventure, the Wild West, and the future are mentioned by all five scholars; fairy tales and ancient Mediterranean civilizations, in turn, appear in three out of five lists. With the exception of the latter, these are of course the themes featured in Disneyland (Anaheim, California)—which may suggest Disney's farsightedness and vision, or the extraordinary impact of Disneyland on later parks also in terms of choice of

Table 1 List of recurring themes, motifs, and theme archetypes in theme parks and themed environments

Gottdiener (1997)	Anton Clavé (2007)	Steinecke (2009)	Philips (2012)	Younger (2016)	Examples
Status					Superstar Limo, Disney California Adventure, California
Tropical paradise	Adventure	Faraway exotic places	Treasure islands/Explorers	Exotica	Tropical Islands Resort, Halbe, Germany
Wild West	Wild West	Wild West	Wild West	Wild West	"Frontierland," Disney Parks
Classical civilization		Classical civilization	(Egyptomania)		Terra Mítica, Benidorm, Spain
Nostalgia and retro fashions				Victoriana	"Sunshine Harbour," Happy Valley Shanghai, China
Arabian fantasy					Holy Land Experience, Orlando, Florida
Urban motif		Urban culture			"À Travers le temps," Parc Astérix, Plailly, France
Fortress architecture and surveillance					Fort Comstock, Parc Disneyland, Paris, France
Modernism and progress	Future and science/Futurism	Future of mankind	Science and technology	Sci-fi	Futuroscope, Poitiers, France

(continued)

Table 1 (continued)

Gottdiener (1997)	Anton Clavé (2007)	Steinecke (2009)	Philips (2012)	Younger (2016)	Examples
Representing the unrepresentable			Gothic tradition	Dark theming	"Mystery," Phantasialand, Brühl, Germany
	Travel/International	Tourism destinations			Europa-Park, Rust, Germany
	Comics				Parc Astérix, Plailly, France
	History				Hansa-Park, Sierksdorf, Germany
	Children's tales/fantasy		Fairy tale	Fairy tales	"Fantasy," Phantasialand, Brühl, Germany
	World of toys				Legoland parks
	Nature				Sea World parks
	Local mythological culture				Dragon World, Singapore
	Music				Dollywood, Pigeon Forge, Tennessee
	Movies				Movie Park Germany, Bottrop, Germany
			Chivalric romance		Enchanted Storybook Castle, Shanghai Disneyland, Shanghai, China

(continued)

Table 1 (continued)

Gottdiener (1997)	Anton Clavé (2007)	Steinecke (2009)	Philips (2012)	Younger (2016)	Examples
				Otherworlds	"Pandora: The World of Avatar," Disney's Animal Kingdom, Orlando, Florida
				Autotheming	"The Boardwalk," Knott's Berry Farm, Buena Park, California
				Self-referential theming	Europa-Park Historama, Europa-Park, Rust, Germany

Sources Gottdiener (1997), 145–151; Anton Clavé (2007), 35–36; Steinecke (2009), 8–11; Philips (2012), 2–3; and Younger (2016), 50–55

themes, or perhaps the dominant position that Disney parks occupy in theme park criticism (see INTRODUCTION).

At the same time, it is also interesting to observe that with the exception of Younger's, these lists do not yet reflect two recent trends in theming: self-reflexive or autotheming, and IP-based theming (theming based on intellectual properties or IPs). In "Late Modernity and the Dynamics of Quasification: The Case of the Themed Restaurant" (1999), Beardsworth and Bryman have argued that a self-reflexively themed space exclusively draws on internal sources for its theming (243). In the case of theme parks, self-reflexively themed attractions can mostly be found in successful and well-established parks, where they reflect on the history of the sites and seek to generate nostalgia by referring to their own previous forms and formats. Whereas theme parks usually seek to hide the passage of time and the changes they have undergone (see METHODS), self-reflexive theming, via nostalgia, lets "external time" flow into the themed representation. Self-reflexive theming thus frequently appears in the context of anniversaries: on the occasion of its 35th anniversary in 2010, for example, Europa-Park opened the Europa-Park Historama, a revolving theater show that employed animatronics, projections, as well as music and sound effects to give an overview of the park's history (Act 1), to tell the history of the park's parent company and offer glimpses behind the stage (Act 2), and to present the park's main rides (Act 3) and shows (Act 4; see Freitag 2021, 132–135). For its 50th anniversary in 2005, meanwhile, Disneyland not only created the Parade of Dreams, whose floats showcased such iconic Disneyland buildings as Sleeping Beauty Castle or Main Street Station, but also the Dream Suite, an apartment whose individual rooms are themed to the park's various lands (see Freitag 2021, 178).

Similarly, autothemed places have been defined as "places whose theming reflects their function, such as a shop themed to a shop, a restaurant themed to a restaurant, a hotel themed to a hotel, and so forth" (Freitag 2016b, 141). Prominent examples from Parc Disneyland in Paris include the Emporium, a souvenir shop themed to a turn-of-the-century department store; Parkside Diner, a restaurant themed to a New York diner; and the Hotel Sequoia Lodge, a hotel themed to an American National Park hotel. In the specific context of theme parks, however, one can also find entire themed areas that are themed to theme parks or their historical antecedents (see ANTECEDENTS): thus the "Boardwalk" at Knott's Berry Farm (Buena Park, California) and "Pixar Pier" at Disney California Adventure (Anaheim, California; see also below) are themed to seaside amusement piers, whereas the Toyville Trolley Park section of the "American Waterfront" area at Tokyo DisneySea and "Lost Kennywood" at Kennywood (West Mifflin, Pennsylvania) represent turn-of-the-century trolley and amusement parks. With its numerous references to Coney's turn-of-the-century amusement parks, including Steeplechase, Dreamland, and the original Luna Park, Luna Park on Coney Island (New York) constitutes an entire park themed to one of the theme park's predecessors (see Freitag 2020/2021). Hence, both self-reflexively and autothemed spaces and elements dramatize and celebrate the parks' histories and the history of the theme park medium as a whole.

The growing popularity of self-reflexive and autotheming may not only be related to the increasing relevance of branding for theme parks, as well as budgetary concerns (see Freitag 2016b, 143–147), but also to the phenomenon of "emergent authenticity," according to which over time, some theme parks are no longer perceived as mere copies of places but acquire a certain sense of place or authenticity of their own (see AUTHENTICITY)—which can itself then be exploited as a source of theming. The recent surge of IP-based theming, meanwhile, generally has not only to do with theme parks' increased emphasis on stories (see ANTECEDENTS), but also with the fact that virtually all destination parks are owned by transmedia conglomerates such as Disney or Universal, which seek to profit from synergy effects through transmedia storytelling.

Certainly, theme parks have participated in the practice of transmedia storytelling—the telling of stories across different media platforms—since the opening of Disneyland in Anaheim in 1955, though mostly in the shape of individual IP-based rides (see MEDIA), then related to Disney's popular animated pictures. Especially since the 2010s, however, many destination parks have opened entire themed lands based on transmedia franchises, including the "Wizarding World of Harry Potter" at Universal parks (since 2010), "Cars Land" at Disney California Adventure (2012), "Pandora: The World of Avatar" at Disney's Animal Kingdom (2017), and "Star Wars: Galaxy's Edge" at Disneyland in Anaheim and Magic Kingdom in Orlando, Florida (2019). Some smaller parks have joined this trend via licensed IPs: both Europa-Park (1990–2013) and Holiday Park (Haßloch, Germany; since 2013), for example, have featured lands based on the animated TV series *Vicky the Viking*. As the last example also shows, however, the longevity of IP-based theming not only depends on the general popularity of an IP (see Younger 2016, 80–81), but sometimes also on such legal issues as the limited duration of a license agreement (see also below).

Returning to the themes listed in Table 1, one may notice that most of them belong to what Younger has referred to as "manifestation themes," a term that identifies themes according to their setting ("where it is located in space and time") and subject ("what it is about"; 2016, 47). By contrast, "adventure" or "fantasy" belong to the "dramatic themes," which describe "how the guest should feel about their experience. At its simplest, this might [be] an emotion such as 'scary' or 'thrilling,' but might progress to abstract concepts such as 'fantasy' or 'adventure,' or complex philosophical notions such as the 'intrinsic value of nature'" (2016, 47). Rather than being two separate and mutually exclusive categories, Younger's differentiation between "manifestation" and "dramatic" themes actually establishes two different viewpoints on themes, and thus two different possible approaches to planning. Just as a specific dramatic theme can be realized using a variety of manifestation themes—Younger discusses the example of the "Fantasyland" dark rides planned for Magic Kingdom, which would keep the dramatic themes of their California counterparts, but use different manifestation themes (2016, 65)—a specific manifestation theme may evoke different associations and emotional responses, e.g. in different cultural contexts (see, for example, the discussion of the manifestation theme of "ancient Greece" in theme parks worldwide in Carlà-Uhink 2020).

Beyond the selected themes, theming practices can also be identified and classified by their "style," or what in literary studies would be called their "mode."

David Younger (2017) has identified four key styles (the fifth, called the "Themed Amusement Park Style Design," merely identifies the application of themed design techniques to attractions without the aim of providing visitors with an immersive experience). The four key styles are:

(1) Traditional Style Design, characterized by "immersive design […], Amalgamative Lands [i.e. lands that combine different "manifestation themes"], and experiential story," as opposed to an explicit storytelling (Younger 2017, 64);
(2) Presentational Style Design, which "relies heavily on screen-based attractions that can show images of the real thing, typically designed with updateability in mind to keep things fresh, and then making use of abstraction and symbolism as the means of connecting the various elements together, thus complementing the theme rather than copying it" (Younger 2017, 69);
(3) Postmodern Style Design, breaking the fourth wall and thus the suspension of disbelief, often in an ironic way, and offering glimpses "behind the scenes";
(4) New Traditional Style Design, an evolution of the Traditional Style employing more explicit narratives.

According to Younger, these four styles should also be broadly understood as a chronological sequence, moving from the Traditional Style of Disneyland in Anaheim in the 1950s and 1960s to the New Traditional Style of many a "remake" and "revision" of those first-generation rides—even if there are no clear boundaries, as the choice of style can be typical of the corporate identity of a specific park or chain. Here too, these styles should thus be understood as ideal types and some attractions may even combine elements from different styles: the Traditional Style of the water coaster Chiapas at Phantasialand becomes Postmodern when riders travel back in time to a Mayan "ghost disco" (see TIME).

Both the choice of theme as well as the manner of its execution are subject to changing trends and fashions. Theming trends are one of the reasons why theme parks frequently "re-theme" attractions and even entire areas: rather than creating new rides, restaurants, shops, etc. from scratch, designers simply apply a new theme to existing buildings, ride systems, and infrastructure (see Younger 2016, 391). Budget concerns, legal issues (i.e. the temporary nature of licensing and sponsorship agreements), lack of space, promotional campaigns, changes in the general cultural climate, and, relatedly, brand concerns are some of the other reasons why theme parks engage in re-theming (see ATTRACTIONS). In most cases, theme parks try not only to carefully hide any traces of re-theming—because of legal concerns in case of expired commercial agreements and copyright, but also so as not to break the immersion and suspension of disbelief by revealing the "historicity" and "changeability" of the themed environment. Nonetheless, many traces remain and are visible to the attentive visitor and researcher (see METHODS)—either because they have been deliberately left as "Easter eggs" (see, for example, the references to Disneyland's Country Bear Jamboree in The Many Adventures of Winnie the Pooh dark ride at Disneyland), or simply because they have not been removed: in Terra Mítica, for example, the decorations of some shops still reveal that in 2010 the park received the franchise, quickly lost again, for the popular cartoon characters Heidi, Vicky the Viking, and Maya the Bee (Carlà-Uhink 2020, 81; see also MEDIA).

The "land" successively referred to as "Erlebnisspielplatz" ("Adventure Playground"), "Wikingerland," "Welt der Kinder" ("Children's World"), and, most recently, "Irland" at Europa-Park provides an excellent case study, as most of the factors listed above have played a role in the various re-themings that make up the complex history of this small area. Separated from the rest of the park by the river Elz, this section has always featured multistory slides, which can still be found there today. In 1990, however, the simple red-and-yellow color scheme that visually tied the area together was replaced by an elaborate Viking theme when Europa-Park obtained the rights to the animated series *Vicky the Viking,* in an early example of IP-based theming (see above). For a brief period during the 1990s—around the time when Europa-Park also opened the Dinosaur-themed Universum der Energie dark ride, loosely based on the eponymous Epcot pavilion—the area also featured two dinosaur-themed spinning rides, no doubt in response to the dinosaur fad that was sweeping the world after the 1993 release of the blockbuster *Jurassic Park.* When, due to the construction of the Poseidon water coaster Poseidon and the surrounding "Griechenland" area, the Nivea-sponsored "Nivea Kinderland" playground was forced to move in 2000, its nautically themed swings and kiddie rides were installed next to the *Vicky*-themed elements, with the result that the area now had no fewer than two themes. These had to be slightly altered when the sponsorship deal with Nivea and the license agreement concerning *Vicky* ended in 2008 and 2013, respectively. In 2016, finally, the playground reopened with an Irish theme, which helped to integrate this area into Europa-Park's principal metatouristic theme of a journey through Europe, thus strengthening the brand while still making use of the existing infrastructure and at least some of the playground installations.

Interesting examples of re-themings due to promotional campaigns and changing cultural climates (as well as brand concerns) can also be found at Disneyland, among many other parks. For instance, in order to promote the *Gummi Bears* animated television series, the Motor Boat Cruise in Tomorrowland was briefly re-themed to Gummi Glen during the summer of 1990, with the painted plywood images of Gummi Bear characters already indicating the temporary nature of this re-theming, and of promotional re-themings in general. By contrast, the permanent re-themings of Aunt Jemima's Kitchen to Magnolia Tree Terrace in 1970 and, 50 years later, of Splash Mountain's *Song of the South* theme to *The Princess and the Frog* illustrate Disneyland's continuing attempts to remove allusions to slavery from its New Orleans-themed areas in response to changing cultural climates, as well as concerns about the family-friendly image of the brand (see above). Of course, re-themings can also simply be a result of the original theme being a failure with the audience (see below).

6 Theming Beyond the Parks

As Lukas's comment on the "growing ubiquity" (see above) of theming illustrates, theming is in no way limited to theme parks, and has increasingly been expanding to the most diverse areas and settings (Zukin 1991; Gottdiener 1997; Lukas 2007a; Mitrasinovic 2006; Lonsway 2009). Movie theaters, bars and restaurants, hotels, casinos and tourist resorts, shops and shopping malls, outlet villages, private residences and apartment buildings, hospitals and retirement homes, and even entire boroughs can be themed, as is famously the case in many new districts of Chinese cities: Thames Town in Shanghai, for example, reproduces an ideal "British" town (Piazzoni 2018). Built in 2006 over one square kilometer with the aim of hosting 10,000 inhabitants, it is only one of the newly developed, Western-themed districts of the Chinese megalopolis. Yet, this is not completely new: the European Chinoiserie fashion during the seventeenth century or the French Egyptomania at the beginning of the nineteenth century led to the production of pseudo-Chinese and pseudo-Egyptian costumes, furniture, rooms, and buildings, most of which were for private use. In international exhibitions, themed exotic villages or streets, like the strongly Orientalistic (see above) Streets of Cairo, where the first belly dancer appeared in Paris in 1889, used theming to allow a broad audience to travel virtually. Themed streets or villages brought theming to another scale and dimension, as it was applied not to a single building but to several, and even a whole urban space, including shops, restaurants and tea houses, theaters, etc. The street itself was also a themed space, where shows and sometimes parades were performed; Disney's "Main Street, U.S.A." and Epcot's "World Showcase" belong to this tradition (see ANTECEDENTS).

The reproduction of a Swiss village or a Paris Street in large exhibitions demonstrates that theming is a possible and attractive way to produce urban space. Similarly, because of their large-scale and holistic projects, theme parks offered more than a mere juxtaposition of themed attractions. They produced not only themed urban spaces dedicated to leisure and entertainment, but also transportation, education, accommodation, etc. Walt Disney's original "Main Street, U.S.A." project at Disneyland included habitations. His urban vision eventually became real with the Experimental Project Community of Tomorrow (Epcot, Walt Disney World, 1982) and the nearby development of Celebration by the Walt Disney Company (1996). Epcot includes the "World Showcase" and "Future World," and originally planned a community of 20,000, a project abandoned after Walt Disney's death in 1966. Celebration, a master-planned Florida community built in the 1990s and for the most part still operated by the Walt Disney Company, became famous as an iconic production of New Urbanism, an antimodernist urban design movement born during the 1980s in the U.S. that had a major impact on town planning and development all over the world (see Roost 2005).

Disney's "Main Street, U.S.A." has been identified (or rather denounced by its opponents) as a main source of inspiration for New Urbanism. New Urbanism shares with Disney the same utopian and nostalgic dream of a happy community in a small town, the same taste for pastiche and illusion, and produces the same depoliticization, sanitization, privatization, and commodification—some would say annihilation—of urban space. New Urbanism, also applied to the rehabilitation of the city center and to new developments in the suburbs, would then result in a "Disneyfication" (meant in a negative way) of contemporary cities, which would lose their public space and urban dimension, and become nothing but simulacra (Young and Riley 2002). Many opponents of New Urbanism criticize outdoor-themed retail and entertainment centers, with their Main street and fountain (such as The Grove, Los Angeles, 2002), and neo-traditionalist neighborhoods (such as in Seaside, Florida, 1981, where *The Truman Show* was shot) for the same reasons that they criticize theme parks: their lack of authenticity. But, just as in a theme park, the authenticity which matters is not that of the place, but of the experience (see AUTHENTICITY), and visitors enjoy it because they are not merely naïve victims of an illusion, but rather appreciative active participants willingly suspending their disbelief (see VISITORS).

Undoubtedly, by themselves and through New Urbanism, theme parks have played a large role in the recent development of more or less themed areas and buildings in contemporary cities, perhaps most spectacularly in Val d'Europe, France, the "ville nouvelle" developed around Disneyland Paris (see Chabard 2012; d'Hauteserre 2020). Theming was not invented in theme parks, but they are the place where it has been experimented with and developed as an art and a technique to produce urban space. However, successful urban (re)development, especially that influenced by New Urbanism, might in turn inspire designers to return to parks, in a "feedback loop of disneyfication" (see Freitag 2021, 178–185) that is highly typical of the postmodern urban fabric. The "Pacific Wharf" area at Disney California Adventure may thus be argued to represent not the historic cannery row area in Monterey, California, but rather its contemporary evolution as a James Rouse-style "festival marketplace" (see Bryman 2004, 36; Bloom 2004; Lukas 2019).

7 The Limits of Theming

As its expansion beyond theme parks demonstrates, theming is a highly successful endeavor and an expanding business—and has been so for some decades now. And yet, theming can also result in failure, sometimes leading to bankruptcy and legal problems. This can happen first of all when theming is badly implemented—either too poorly or, perhaps surprisingly, too richly, going way beyond the levels of public recognizability and costing excessive amounts. This subchapter will provide examples for both cases, before then moving on to another reason for the failure of theming: bad choices of themes, which, as we will see, can also have different facets, from choosing a theme with a low level of recognizability to one that evokes extremely negative associations, which can even lead to public outrage.

As the strategies discussed above illustrate, theming is a rather expensive practice that only makes sense in privately owned and commercially operated spaces (as most theme parks are) if the Return on Investment (ROI) is positive. A study conducted in 2012 and 2013 on 20 new attractions in European theme parks has shown that attractions featuring a higher degree of immersivity—which is determined to a huge extent, even if not exclusively, of course, by more detailed theming (see IMMERSION)—enjoyed greater commercial success, resulting in higher ROIs (Cornelis 2017). While this clearly explains why the theme park industry is successful and expanding—and why theming has also been adopted in many other sectors—there is an important consequence: namely, that the investments in theming are very high, also because of growing competition, and that the theme park is thus a high-risk industry, resulting in high volatility (Cornelis 2017, 225; see ECONOMIC STRATEGY). This causes, among the different possible responses by the parks' managements, a tendency to invest less than one might expect in better and more detailed theming, as "when it comes to the bottom line most people are uncertain whether the extra revenues outweigh the extra costs" (Cornelis 2017, 237).

This leads us to investigate what we might define as the "limits of theming"—cases in which wrong decisions connected to theming and its detail had very impactful and even disastrous consequences for theme parks or other themed environments. The first case of such failures derives, rather obviously, from the observations above: if management decides to cut theming costs in order to reduce expenses, the resulting theming may lead to disappointed visitors, fewer return visits, bad publicity, and in the end even to closing the park. One extreme example was Lapland New Forest, a Christmas-themed theme park that opened in Matshams, Hampshire, UK, in time for the Christmas season at the end of November 2008 (see Lukas 2013, 104–105). Thanks to an advertising campaign that promised, among others, "Hollywood special effects" and "a magical tunnel of light," around 40,000 tickets had been sold before the opening (Williams 2011). When the park opened, however, the "magical tunnel of light" turned out to be a few strings of lights hanging from trees, the Nativity scene was nothing but a painted billboard, it became clear that the animals were mishandled, the ice-skating rink melted, and the Christmas market, for which an additional entry fee had to be paid, consisted of just four booths (BBC News 2008). Children cried, especially after seeing Santa Claus smoking, and one angry parent hit an elf. The park closed after only a week; in 2011 the park's owners, two brothers, were sentenced to 13 months in prison for "misleading the public." As the judge who delivered the sentence put it,

> You promised customers in your advertising an amazing snow-covered Lapland village which was—in your own wonderful words—"Where dreams really do come true" and "Where we have prided ourselves on attention to detail." You told consumers that it would light up those who most loved Christmas. You said you would go through the magical tunnel of light coming out in a winter wonderland. What you actually provided was something that looked like an averagely-managed summer car boot sale. The failure of Lapland New Forest was caused by the unrelenting greed shown by you and your desire to squeeze every drop of profit rather than build and create the winter wonderland you promised thousands of consumers. (BBC News 2011)

In this case, even fraudulent activity and a court case developed around a broken promise of well-detailed theming, which then corresponded to a level of investment kept so cynically low as to result in what visitors experienced as a real dystopia.

On the other hand, excessively detailed theming may cost so much that it makes a positive ROI virtually impossible, as the necessary number of visitors per year to reach the break-even point becomes unrealistic. A case in point is Terra Mítica, a theme park in Benidorm, Spain. Planning for the park started in 1996 and the project soon developed into one of the most ambitious of its kind in Europe. During its opening year in 2000, the park welcomed an impressive two million visitors—to break even, however, the park needed an almost impossible three million visitors per year (Carlà-Uhink 2020, 80–82). Among the reasons for this—in addition to, e.g. an excessive amount of shops and restaurants—was the excessive theming, which included precise reconstructions of archaeological ruins. This corresponded with the aim, formulated at the very outset of the project, of also using the park as an educational site—as evidenced by the presence of numerous explanatory tables. Theming is, as we have seen, all about recognition—more detailed theming facilitates immersion, and yet, beyond a certain level of detail, "the process of recognition stops. [...] In this sense, the extremely high level of theming of Terra Mítica may have even increased the costs of the park and contributed to its financial unsustainability" (Carlà-Uhink 2020, 88). One of the park's most immediate responses to the disappointing visitor numbers was, indeed, to reduce the level of theming: during its most difficult years, Terra Mítica instead invested in new thrill rides that were barely themed beyond their name and their location (which implied indirectly taking advantage of the already existing, highly detailed theming; see Carlà-Uhink 2020, 92).

Theming failures are not only limited to planning and investment; the very choice of theme is, quite obviously, a delicate aspect that can lead to unfortunate results, for instance, if a theme is selected that the target public does not recognize or identify with. Thus, Europa-Park dedicated an area to Luxembourg, a state that, even within Europe, does not have any prominent symbols or emblems that may activate a successful process of recognition. Even the page dedicated to the themed area on Europa-Park's website does not contain any information about Luxembourg or its representation, but simply shows a giant pyramidal spinning theater (in fact, the site of the former Europa-Park Historama; see above) with a performance dedicated to the water park Rulantica (outside the theme park), and points to the area's location at the center of the park (perhaps alluding to the geographic position of the country within Europe, although this is not made explicit; see Europa-Park 2020b). Beyond the name of the central part of the area (Luxembourg Square) and the presence of Luxembourg flags, the area is thus substantially unthemed, because the choice of theme made a consistent implementation impossible.

A different problem was met by Disney's California Adventure, which opened in Anaheim, California, in 2001 and did not at all meet the expectations of management in terms of visitor numbers and earnings. Among the reasons for the very strong criticism that the park encountered was what we might call a lack of externality: as a theme park about California in California, the site was not even separated by a clear boundary from the surrounding areas (the fact that nearby buildings lying outside the park were visible at all times was negatively commented on). This led to a lack of immersion and eventually resulted in commercial failure (see IMMERSION). David Younger has described the design style of the park as "Postmodern" (2017, 75), a direction taken under then-CEO of the Walt Disney Company Michael Eisner to save money after the commercial woes that surrounded Euro Disney, and based on a desire "to create a project markedly different from Disneyland" (Younger 2017, 75). Eventually, this mistake was rectified by a major redesign process between 2007 and 2012, and again more recently between 2017 and 2021. The park has now largely moved away from its original theme of California toward IPs, including various animated Pixar films in "Pixar Pier" and "Cars Land," old cartoon characters such as Oswald the Lucky Rabbit in "Buena Vista Street," and the Marvel franchise in "Avengers Campus."

More sensational are the various scandals around themes evoking negative associations that make them unpleasant or even repulsive (without it being possible, however, to transform them into "dark themes"). This can lead to protests and demonstrations, especially when political issues are involved, or even to acts of vandalism against single attractions or the entire park. At Mini Israel, for example, "the group of mini Muslim prayers at the Dome of the Rock" (Feige 2017, 165) has been vandalized in order to protest how the park represents the State of Israel. While this cannot be considered a mistake on the part of the park, it must be stressed that such forms of contestation and refusal of theming can have large consequences for commercial enterprises such as theme parks.

One famous example, which involves a themed restaurant rather than a theme park, is Hitler's Cross, which opened in Mumbai, India, in 2006. The owner asserted that he simply wanted to capitalize on a name that everybody knew—and thus to develop a theme that would be recognizable to a large number of patrons. The mistake was, of course—if we believe his statements that he had no political affiliation whatsoever with the Nazi or Neo-Nazi movements—that this theme has extremely negative and uncanny associations that nobody would connect to an environment in which to enjoy a meal. As expected, the uproar was enormous, and the case became internationally known and discussed. Within a few weeks, the restaurant changed its name to Cross Cafe and the Third Reich theming was removed (Lukas 2008, 212–214).

A similar case, although much less known, took place in Busan, South Korea, where an Adolf Hitler-themed Techno Bar and Cocktail Show met with public outrage and uproar, was renamed "Ditler Techno Bar and Cocktail Show," but eventually failed and closed. According to its owner, "I am not a Nazi, the bar name came out at a meeting with [an] interior designer. We just wanted a name that can be related to or represent Germany, easy to remember and easy to design the interior with" (Lebow 2000). Of course, the Third Reich is a highly recognizable aspect of German history and meets the criteria of selection and abstraction for the representation of Germany that most people all over the world would recognize. Yet, like the Cross Café, the example illustrates that to consider theming only in terms of recognizability constitutes an unforgivable shortcut—above all from an ethical perspective, as trying to commercially exploit the Holocaust and World War II, or any other historical experience implying the genocide of entire peoples and cultures, cannot be justified by any means. This also leads—fortunately, as this implies a healthy outrage by potential customers—to the financial failure of such premises.

On the other hand, one must also be aware that the opposite is true—a "dark theme" or even a theme that, without being dark, is connected to adventure and to images of violence can be implemented in ways that are perceived by the visitors as too "sweetened" and unsatisfactory. For instance, classical myths are often chosen to represent war, violence, and death because the very nature of myth "makes them appear less 'real' and their reception less delicate" (Carlà and Freitag 2015b, 149), a representation of the Trojan War in which nobody dies and no violence takes place would hardly be successful. In Terra Mítica, the El Rescate de Ulises ride, built—like the rest of the park (see above)—with a huge financial investment and greeted as the biggest ride in Spain in 2001, was not well received by the public and was closed between 2005 and 2013. In this dark ride, visitors would embark on boats and accompany Ulysses' son Telemachus on the search for his father, who was kept prisoner by the god Poseidon. Yet, the ride consisted of nine quite static scenes, and although it was made livelier by an actor on each boat who embodied Telemachus and provided narrative details contributing to the immersion, the resulting experience was quite boring—especially when compared to the adventures of Ulysses, which are such a relevant part of Western cultural memory and are thus so popular, also among children. Consequently, the ride failed to become the "anchor" attraction it was expected to be (Carlà-Uhink 2020, 99–102). Similarly, when in 1997 Disney parks modified selected scenes of the Pirates of the Caribbean rides to make them less sexist and more politically correct, not everyone was enthusiastic, and designer X Atencio commented: "The show's called Pirates of the Caribbean, not Boy Scouts of the Caribbean" (see INCLUSION AND EXCLUSION).

Hence, theming is a much more complex endeavor than a mere commercial or industrial strategy. As a strategy of representation, theming is deeply embedded in culture, politics, society, historical cultures—and more generally in discourse, in the Foucauldian sense of a system producing meaning. Alongside many stories of successful theming, which have inspired numerous themed businesses and spaces all over the world, there are probably just as many—and much less well-known—stories of failure, which demonstrate that theming is not something that can be improvised in a meeting with an interior designer, but requires a great deal of awareness of the dynamics discussed in this chapter.

8 A Glimpse into the Future

One of the most recent trends in theming concerns neither the choice of theme nor the style or mode of theming, but simply the medium of execution: virtual theming. Of course, as Baker has pointed out, hybrid or mixed reality theming—which blends "physical/practical effects with virtual elements like screens, projections without screens, holograms, or other effects"—"has been around for decades" (2018, 258). For example, the integration of screens into three-dimensional, physical sets can be considered one of the key design strategies of Presentational Style Design, which from the 1980s onward has eschewed representation, preferring to "show images of the real thing" (Younger 2017, 69; see also above). A case in point is El Rio del Tiempo, a dark ride operating from 1982 to 2007 at the "Mexico" pavilion at Epcot—often considered to be "the theme park that had defined the [Presentational] style" (72): here, visitors passed screens that were framed by three-dimensional rocks and architectural elements and featured shots of Mexican beaches, festivals, and street markets—sometimes, but not always, to the same scale as the surrounding physical parts of the set (see Sheppard 2016).

While the Presentational Style Design may have gone out of fashion somewhat (Younger 2017, 72), hybrid or mixed reality theming and especially screen-based theming has continued to be employed, particularly in complex IP-based dark rides such as Harry Potter and the Forbidden Journey at Islands of Adventure (Orlando, Florida), *Star Wars*: Rise of the Resistance at Disneyland, or *Arthur*: The Ride at Europa-Park. The connection to IP-based theming is no coincidence: as Lauren Rabinovitz has stated with respect to simulator ride adaptations of movies that use screens, "it is much easier to simulate reality when the reality being depicted is already a movie. It's not difficult to make a movie that looks like another movie" (2012, 167).

In the mid-2010s, however, several theme parks also started introducing "full" virtual theming, mostly by retrofitting existing roller coasters with VR systems: while the coasters and their physical theming are left unchanged, riders wear head-mounted gear that displays a computer-generated phantom shot synchronized with the movements of the coaster as well as the movements of the rider's head, resulting in a new (visual) experience. In 2015/2016, for example, The New Revolution at Six Flags Magic Mountain (Valencia, California), Space Fantasy–The Ride at Universal Studios Japan (Osaka), Galactica at Alton Towers (Alton, UK), and Alpenexpress Enzian at Europa-Park all offered VR experiences, mostly as a temporary upcharge option (see Burt and Zika 2018; Younger 2016, 428).

Whereas scholars have noted the technological and operational challenges in connection with VR coasters ("synchronizing the video with the roller coaster dynamics," see Younger 2016, 428; keeping headgear hygienic, see Baker 2018, 263), their implications for theming have yet to be explored in greater detail. In most cases so far, the visual theme of the VR experience has been utterly disconnected from the surrounding physical, multisensorial theming. At Europa-Park, for instance, the temporary VR experience offered in 2016/2017 at Pegasus (a coaster located in the Greek area and themed to an archaeological excavation) revolved around the characters of the computer-animated horror comedy *Happy Family* (2017; co-produced by the park; see MEDIA), while that offered since 2015 at Alpenexpress Enzian (located in the Austrian area and themed to a mine train) has featured the "Ottifanten" cartoon characters (note the persisting connection to movies and/or animation). Moreover, since in both cases the headgear has been left unthemed except for a generic logo (allowing the park to use the same headgear for all of its VR coasters and thus save on operational costs), the experience is also thematically jarring for non-riders.

One of Europa-Park's more recent VR offerings, Eurosat Coastiality, has sought to mitigate these issues by featuring separate waiting and loading areas for riders who wish to experience the coaster "virtually," and by connecting the theme of the VR experience to the history of the coaster. In 2018, the enclosed space-themed coaster originally known as Eurosat that is located in the French section of Europa-Park reopened with a French theme that takes riders to the Paris of the Belle Epoque. Despite the retheme, the coaster's old name was incorporated into its new name (Eurosat CanCan Coaster), perhaps for marketing reasons, but perhaps also to nostalgically evoke a past chapter in the history of the park and the attraction itself (see AUTHENTICITY and TIME). However, the old space theme is also referred to in Eurosat CanCan Coaster's new VR experience, which was added during the retheme: based on French filmmaker Luc Besson's 2017 space opera *Valérien et la Cité des Mille Planètes*, Eurosat Coastiality ensures, according to the park's website, "that the Eurosat continues with its popular [space] theme" (Europa-Park 2020a). While selecting a movie as the source material for the VR experience may not seem especially innovative, the particular choice of the movie has allowed the park to connect the virtual coaster to the themed area in which it is located (via Besson) and to the history of the ride infrastructure that it uses (via the space theme). Eurosat Coastiality has thus been integrated much more organically into its surroundings than previous VR coasters.

This also applies to the logistics of the virtual coaster, which make the transition from "analog" to digital/virtual remarkably smooth: rather than sharing a loading area with its "analog" sister ride (as previous VR coasters have done), Eurosat Coastiality features a separate queue line and load/unload station. Hence, visitors who choose to ride Eurosat CanCan Coaster and who supposedly travel back in time to the Belle Epoque are not exposed to the riders of Eurosat Coastiality with their futuristic VR headgear. Riders of Eurosat Coastiality are in turn invited to put on their headgear as soon as they enter the waiting line: the latter has been left (physically) unthemed (see Fig. 1), and visitors experience the entire waiting and loading process virtually.

This ensures a thematically coherent visitor experience that can be easily and instantaneously rethemed in the future. Somewhat ironically, Eurosat Coastiality thus points the way toward a future of theming where physical theming no longer plays much of a role.

Fig. 1 Europa-Park's (unthemed) Coastiality headgear. *Photograph* Torsten Widmann

References

Alderman, Derek H., Stefanie K. Benjamin, and Paige P. Schneider. 2012. Transforming Mount Airy into Mayberry: Film-Induced Tourism as Place-Making. *Southeastern Geographer* 52(2): 212–239.

Anton Clavé, Salvador. 2007. *The Global Theme Park Industry*. Wallingford: CABI.

Baker, Carissa Ann. 2018. *Exploring a Three-Dimensional Narrative Medium: The Theme Park De Sprookjessprokkelaar, the Gatherer and Teller of Stories*. Diss., University of Central Florida.

BBC News. 2008. In Pictures: Lapland New Forest. *BBC News Online*. http://news.bbc.co.uk/2/shared/spl/hi/pop_ups/08/uk_lapland_new_forest/html/1.stm. Accessed 28 Aug 2022.

BBC News. 2011. Brothers Jailed over Lapland New Forest Park. *BBC News Online*, March 18. https://www.bbc.com/news/uk-england-12783389. Accessed 28 Aug 2022.
Beardsworth, Alan, and Alan Bryman. 1999. Late Modernity and the Dynamics of Quasification: The Case of the Themed Restaurant. *Sociological Review* 47: 228–257.
Bloom, Nicholas Dagen. 2004. *Merchant of Illusion: James Rouse. America's Salesman of the Businessman's Utopia*. Columbus: The Ohio State University Press.
Bolter, Jay, and Richard Grusin. 2000. *Remediation: Understanding New Media*. Cambridge: The MIT Press.
Bryman, Alan. 2004. *The Disneyization of Society*. London: Sage.
Burt, Malcolm, and Joel Zika. 2018. A Century of Virtual Amusement. *Refractory: A Journal of Entertainment Media* 30. https://refractory-journal.com/30-Burt-Zika/. Accessed 28 Aug 2022.
Carlà, Filippo, and Florian Freitag. 2015a. Ancient Greek Culture and Myth in the Terra Mítica Theme Park. *Classical Receptions Journal* 7 (2): 242–259.
Carlà, Filippo, and Florian Freitag. 2015b. The Labyrinthine Ways of Myth Reception: Cretan Myths in Theme Parks. *Journal of European Popular Culture* 6 (2): 145–159.
Carlà-Uhink, Filippo. 2020. *Representations of Classical Greece in Theme Parks*. London: Bloomsbury.
Carlà-Uhink, Filippo, and Florian Freitag. 2022. Theme Park Imitations: The Case of Happy World (Happy Valley Beijing). *Cultural History* 11 (2): 181–198.
Chabard, Pierre. 2012. Val d'Europe: De la Ville Nouvelle au Nouvel Urbanisme. In *De la Ville Nouvelle à la Ville Durable: Marne-la-Vallée*, ed. Clément Orillard and Antoine Picon, 167–179. Marseille: Parenthèse.
Cornelis, Pieter C.M. 2017. Time and Temporality in Theme Parks: The Economic Impact of Immersion. In *Time and Temporality in Theme Parks*, ed. Filippo Carlà-Uhink, Florian Freitag, Sabrina Mittermeier, and Ariane Schwarz, 223–239. Hanover: Wehrhahn.
D'Hauteserre, Anne-Marie. 2020. *Disneyfying Ile de France?* Newcastle upon Tyne: Cambridge Scholars.
Europa-Park. 2020a. Eurosat: CanCan Coaster. *Eurosat: CanCan Coaster*. https://eurosat-cancan-coaster.de/en. Accessed 24 December 2020.
Europa-Park. 2020b. Luxembourg. *Europa-Park Erlebnis-Resort*. https://www.europapark.de/en/attractions/themed-areas/luxembourg. Accessed 24 December 2020.
Feige, Michael. 2017. Mini Israel and the Subversive Present. In *Time and Temporality in Theme Parks*, ed. Filippo Carlà-Uhink, Florian Freitag, Sabrina Mittermeier, and Ariane Schwarz, 155–168. Hanover: Wehrhahn.
Fichtner, Uwe, and Rudolf Michna. 1987. *Freizeitparks: Allgemeine Züge eines modernen Freizeitangebotes, vertieft am Beispiel des Europa-Park in Rust/Baden*. Freiburg: N.P.
Firat, A. Fuat., and Ebru Ulusoy. 2011. Living a Theme. *Consumption Markets & Culture* 14 (2): 193–202.
Freitag, Florian. 2016a. Movies, Rides, Immersion. In *A Reader in Themed and Immersive Spaces*, ed. Scott A. Lukas, 125–130. Pittsburgh: ETC.
Freitag, Florian. 2016b. Autotheming: Themed and Immersive Spaces in Self-Dialogue. In *A Reader in Themed and Immersive Spaces*, ed. Scott A. Lukas, 141–149. Pittsburgh: ETC.
Freitag, Florian. 2017. Critical Theme Parks: Dismaland, Disney, and the Politics of Theming. *Continuum* 31 (6): 923–932.
Freitag, Florian. 2018. "Who Really Lives There?": (Meta-)Tourism and the Canada Pavilion at Epcot. In *Gained Ground: Perspectives on Canadian and Comparative North American Studies*, ed. Eva Gruber and Caroline Rosenthal, 161–178. Rochester: Camden House.
Freitag, Florian. 2020/2021. "Share a Luna Park Memory … and Make a New One!": Memorializing Coney Island. *Andererseits* 9/10: 247–263.
Freitag, Florian. 2021. *Popular New Orleans: The Crescent City in Periodicals, Theme Parks, and Opera, 1875–2015*. New York: Routledge.
Gordon, Bruce, and David Mumford. 2000. *A Brush with Disney: An Artist's Journey, Told through the Words and Works of Herbert Dickens Ryman*. Santa Clarita: Camphor Tree.

Gottdiener, Mark. 1997. *The Theming of America: Dreams, Visions, and Commercial Spaces.* Boulder: Westview.
Holtorf, Cornelius. 2007. *Archaeology Is a Brand! The Meaning of Archaeology in Contemporary Popular Culture.* Walnut Creek: Left Coast Press.
Kolb, David. 2008. *Sprawling Places.* Athens: University of Georgia Press.
Lebow, Jeff. 2000. A Visit to the Hitler Bar. *Pusanweb*, December 16. http://www.pusanweb.com/feature/hitlerbar/hitlermain.htm. Accessed 28 Aug 2022.
Lonsway, Brian. 2009. *Making Leisure Work: Architecture and the Experience Economy.* New York: Routledge.
Lowenthal, David. 2015. *The Past is a Foreign Country Revisited.* Cambridge: Cambridge University Press.
Lukas, Scott A. 2007a. The Themed Space. Locating Culture, Nation, and Self. In *The Themed Space: Locating Culture, Nation, and Self*, ed. Scott A. Lukas, 1–22. Lanham: Lexington.
Lukas, Scott A. 2007b. How the Theme Park Gets Its Power: Lived Theming, Social Control, and the Themed Worker Self. In *The Themed Space: Locating Culture, Nation, and Self*, ed. Scott A. Lukas, 183–206. Lanham: Lexington.
Lukas, Scott A. 2007c. A Politics of Reverence and Irreverence: Social Discourse on Theming Controversies. In *The Themed Space: Locating Culture, Nation, and Self*, ed. Scott A. Lukas, 271–293. Lanham: Lexington.
Lukas, Scott A. 2008. *Theme Park.* London: Reaktion.
Lukas, Scott A. 2013. *The Immersive Worlds Handbook: Designing Theme Parks and Consumer Spaces.* New York: Focal.
Lukas, Scott A. 2016. Dark Theming Reconsidered. In *A Reader in Themed and Immersive Spaces*, ed. Scott A. Lukas, 225–235. Pittsburgh: ETC.
Lukas, Scott A. 2019. Between Simulation and Authenticity: The Question of Urban Remaking. In *The New Companion to Urban Design*, ed. Tridib Banerjee and Anastasia Loukaitou-Sideris, 327–336. London: Routledge.
Mitrasinovic, Miodrag. 2006. *Total Landscape, Theme Parks, Public Space.* Burlington: Ashgate.
Philips, Deborah. 2012. *Fairground Attractions: A Genealogy of the Pleasure Ground.* London: Bloomsbury.
Piazzoni, Maria Francesca. 2018. *The Real Fake: Authenticity and the Production of Space.* New York: Fordham University Press.
Pine II, B. Joseph, and James H. Gilmore. 2011. *The Experience Economy.* Updated ed. Boston: Harvard Business Review Press.
Rabinovitz, Lauren. 2012. *Electric Dreamland: Amusement Parks, Movies, and American Modernity.* New York: Columbia University Press.
Roost, Frank. 2005. Synergy City: How Times Square and Celebration Are Integrated into Disney's Marketing Cycle. In *Rethinking Disney: Private Control, Public Dimensions*, ed. Mike Budd and Max H. Kirsch, 261–298. Middletown: Wesleyan University Press.
Said, Edward. 1978. *Orientalism.* New York: Pantheon.
Schulze, Gerhard. 2005 [1992]. *The Experience Society.* London: Sage.
Sheppard, Randal. 2016. Mexico Goes to Disney World: Recognizing and Representing Mexico at EPCOT Center's Mexico Pavilion. *Latin American Research Review* 51 (3): 64–84.
Sorkin, Michael, ed. 1992. *Variations on a Theme Park: The New American City and the End of Public Space.* New York: Hill and Wang.
Steinecke, Albrecht. 2009. *Themenwelten im Tourismus: Marktstrukturen—Marketing—Management—Trends.* Munich: Oldenbourg.
Steinkrüger, Jan-Erik. 2013. *Thematisierte Welten: Über Darstellungspraxen in Zoologischen Gärten und Vergnügungsparks.* Bielefeld: Transcript.
Steinkrüger, Jan-Erik. 2017. Other Times and Other Spaces: Themed Places and the Doubling of Landscapes. In *Time and Temporality in Theme Parks*, ed. Filippo Carlà-Uhink, Florian Freitag, Sabrina Mittermeier, and Ariane Schwarz, 83–95. Hanover: Wehrhahn.
Tuan, Yi-Fu. 1998. *Escapism.* Baltimore: Johns Hopkins University Press.

Williams, Eleanor. 2011. How the Crowds Were Lured to Lapland New Forest. *BBC News Online*, February 18. https://www.bbc.com/news/uk-england-dorset-11842396. Accessed 28 Aug 2022.

Wöhler, Karlheinz. 2011. *Touristifizierung von Räumen: Kulturwissenschaftliche und soziologische Studien zur Konstruktion von Räumen*. Wiesbaden: VS.

Wright, Alex. 2009. *The Imagineering Field Guide to Magic Kingdom at Walt Disney World*. New York: Disney Editions.

Young, Terence, and Robert Riley, eds. 2002. *Theme Park Landscapes: Antecedents and Variations*. Washington, D.C.: Dumbarton Oaks Research Library and Collection.

Younger, David. 2016. *Theme Park Design & the Art of Themed Entertainment*. N.P.: Inklingwood.

Younger, David. 2017. Traditionally Postmodern: The Changing Styles of Theme Park Design. In *Time and Temporality in Theme Parks*, ed. Filippo Carlà-Uhink, Florian Freitag, Sabrina Mittermeier, and Ariane Schwarz, 63–82. Hanover: Wehrhahn.

Zukin, Sharon. 1991. *Landscapes of Power: From Detroit to Disney World*. Berkeley: University of California Press.

Time: Represented, Experienced, and Managed Temporalities in Theme Parks

Abstract This chapter discusses the complex temporality of theme parks, in which several distinct temporal layers continuously interact and manifest themselves in specific material objects. All the themes depicted in the parks are necessarily situated at a specific point in time ("represented time"); however, different theme park actors—from managers to front-line employees to visitors—need to manage (their) time in the theme park in multifarious ways. In addition to the layer of "represented time," "experienced time," as in visitors' feelings of acceleration and deceleration caused by rides and queues, and "managed time," through which the park organizes visitor flows, are also central to the theme park. Alongside these temporalities, which are intrinsic to the theme park, "external times" also have an impact on and are frequently utilized by the theme park, as is the case with holiday seasons and their special events, or the changing length of daylight throughout the year. Anchored by case studies, including waiting time boards and soundtracks (two examples where the complex temporality of theme parks manifests itself particularly well), this chapter discusses the ways in which theme parks construct or react to different layers of time.

1 Introduction

At the end of "Main Street" in Disneyland Paris, visitors find an information board which displays, among other things, the current waiting times for individual attractions, as well as the park's entertainment schedule. Similar waiting time boards can be found in many theme parks, but the one at Disneyland Paris is particularly intricate. With its detailed Victorian look, the board is perfectly integrated into the theme of "Main Street, U.S.A."; the board is further connected to the theme of the area via its middle section, where the front page of the fictive *Main Street Gazette* shows a map of the park, as well as a short illustrated article that links "Main Street, U.S.A." to

---

This work is contributed by Filippo Carlà-Uhink, Florian Freitag, Salvador Anton Clavé, Astrid Böger, Thibaut Clément, Scott A. Lukas, Sabrina Mittermeier, Céline Molter, Crispin Paine, Ariane Schwarz, Jean-François Staszak, Jan-Erik Steinkrüger, Torsten Widmann. The corresponding author is Filippo Carlà-Uhink, Historisches Institut, Universität Potsdam, Potsdam, Germany.

F. Freitag et al., *Key Concepts in Theme Park Studies*,
https://doi.org/10.1007/978-3-031-11132-7_15

Fig. 1 The waiting times board at Disneyland Paris (France). *Photograph* Florian Freitag

Walt Disney's hometown (Marceline, Missouri), which according to Disney's paratexts (see PARATEXTS AND RECEPTION), served as the basis for the design of the area (Mittermeier 2021). Directly above this, a clock tells the current time, and small exchangeable signs indicate the opening hours of the park. On each side of this center piece, small screens inform visitors about the current wait time at the park's major attractions and the hours of the most important parades and shows (see Fig. 1).

The waiting times board embodies the complex temporality of theme parks. The board's decorative elements point to the temporal setting of the themed area (in this case, the late Victorian era that is also evoked in the rest of "Main Street"). In addition, though, there are other layers of temporality that shape the temporal dimensions of theme parks, and most of these have also left material traces on the information board: for instance, the board itself may evoke the past, but the information it displays (opening hours, wait times, etc.) refers to the present of the park's visit. Since they depend on the season of the year, as well as the day of the week and holidays, all these pieces of information further evoke the lapse of time outside the park, and the ways these external times are productively integrated into the park's operational rhythms. The relationship between these various temporal layers and the board are complex and not entirely clear, however: for instance, the scratches on the board's surface might be part of the theming and thus related to the time represented in the theme park, or they might simply be natural signs of wear and tear, and thus the park's history and the deterioration of its elements (see METHODS). Each element in the theme park supports different temporal layers, and every temporal layer of the theme park permeates each of the park's elements in various ways. The theme park thus constitutes a "time park," as King and O'Boyle have suggested, not just with respect to how "the mechanism of theming makes it possible to transmute time into physical space [and] the theme park can transport us into a hyperreality of past or future" (King and O'Boyle 2011, 11), but in a much more general way.

2 Represented Time

Surely, represented time is the most obvious and perhaps most important temporal layer of a theme park: from its opening in 1955, Disneyland has greeted visitors with a sign informing them that they were stepping not into a different place, but into a different time: "Here You Leave Today and Enter the World of Yesterday, Tomorrow and Fantasy." All themes employed in theme parks, and not only those inspired by historical periods and cultures, are situated at a specific time that is extraneous to the temporality of daily life outside the park (on externality, see THEMING). Exotic worlds, for instance, are frequently represented in a pristine, and therefore past, ideal condition, and even fictional worlds necessarily require a temporal situatedness in order to be visualized and thus representable within an immersive environment. In Western culture, magic and witchcraft are connected with the Middle Ages, as are the world of fairy tales with their kings and queens, princes and princesses. "Fantasyland" at Disneyland is thus architectonically represented through Middle-European "antiquizing" buildings (for more on Disney's take on the middle ages, see Pugh and Aronstein 2012), and the "Wizarding World of Harry Potter" at Universal Studios, Orlando, takes its visual cues from the Harry Potter movies which, in turn, have adopted an architectonically medieval setting for Hogwarts and the world connected to it (Carlà-Uhink 2020, 5–6).

While consistency in the time represented through theming may be crucial to creating an effective immersive experience (see IMMERSION), it does not always imply uniformity. For one thing, the temporal layer represented in a themed area can have its own past, which is sometimes represented within the park to add more depth to the immersive space. In Phantasialand (Brühl, Germany), for instance, the "Mexico" area is themed to the central American country around the time of the Mexican Revolution. Within this area, however, the water coaster Chiapas gives visitors the opportunity to travel back even further into the past: the ruins of a Mayan temple, represented as an archaeological excavation (the waiting area features boxes containing antiquities to be sent to the archaeological museum), allows for the visualization of yet another time layer, consistently represented as anterior to the layer of the rest of the themed area.

The latter point may also take precedence over chronological logic and lead to combinations or juxtapositions of temporal layers that may appear inconsistent at first sight. A very clear example comes from the German park Belantis, located close to the city of Leipzig. Here, visitors find an area themed to ancient, pharaonic

Egypt, called "Tal der Pharaonen" ("Valley of the Pharaohs"). As may be expected, the main architectural element in the area is a pyramid. And yet Egypt, as experienced by contemporary European tourists, comprises not only the pyramids and the archaeological excavations but also the modern cities and the Islamic culture of the country. In contrast to e.g. "Greece" in Europa-Park (Rust, Germany), however, there are no neatly separated subareas dedicated to specific times—instead, the temporal layers are mixed: the area is decorated with loudspeakers through which the prayers of the muezzin are regularly played to create an immersive "Egyptian" environment (Schröder 2012, 219).

Anachronisms of this kind can, of course, have an additional purpose: pure fun and silly humor. Despite the general temporal consistency of its representation of Mexico, for instance, Phantasialand does not shy away from breaking the carefully designed temporal logic by adding contemporary dance music and a disco ball to the interior of the Mayan temple, inviting the riders of Chiapas to a Mayan-themed dance party. At Parc Astérix (Plailly, France), the use of such anachronistic constructions as Ulysses' travel agency derives directly from the specific forms of humor used in the comics on which the park is based and can thus be considered a facet of its remediation (see MEDIA).

The representation of the future is even more complex: following the steampunk and retro-future crazes beginning in the late 1970s, for example, the different "Tomorrowlands" at Disneyland Paris, Walt Disney World's Magic Kingdom, and Disneyland in Anaheim, California, were all inspired by past visions of the future from various eras, including late nineteenth century science fiction novels, 1930s' space travel comic books, and the design of Disneyland's original Tomorrowland, respectively. The representation of the future in these lands thus combines past and future, but also the present: their nostalgic takes on past visions are the product of a dissatisfaction with the present (or, more precisely, with present visions of the future; see Mittermeier 2017). Hence, whether it is about the past or the future, even within the layer of represented time, the present is a crucial component. Similarly, when historical periods or cultures are represented, the forms of their visualization are always historically contingent and rely on popular ideas and stereotypes about past societies which are a product of the specific time and place of the park's target public (Carlà 2016).

At the level of represented time, therefore, theme parks clearly show a collapse of the neat separation between past, present, and future as three clearly identifiable time layers. This is by no means specific to theme parks: it has in general been observed that the specifically modern, tripartite time structure, which had dominated the Western world since the eighteenth century, has started to collapse by the 1970s, generating a new "time regime," which can be considered typically postmodern. The reasons for this change are still heavily discussed in scholarship, with processes of digitalization, anxiety about technological progress, the development of pessimistic or catastrophic views of the future, or the development of post-anthropocentric and post-human configurations of knowledge among the theories debated. Nevertheless, it is possible to identify a number of characteristics that define the new, postmodern temporality, most prominently the conflation of different time levels within a "broad present" (Gumbrecht 2010).

In theme parks, this new, postmodern temporal regime finds expression in the three-dimensional themed worlds and their constitutive elements. Nostalgia, for example, which has been identified as the central politics of Walt Disney's art in general and his theme parks in particular (Schickel 1997, 157), is a typical product of this new temporality. A loss of trust in the future generates a re-orientation, which often moves, as Svetlana Boym has highlighted, "sideways. The nostalgic feels stifled within the conventional confines of time and space" (2001, xiv). The retro-future and steampunk themes described above (see also Guffey 2006) are a clear example of this, and their temporality embodies the collapse of the earlier tripartite time structure. Moreover, it has been noted that in postmodernism history has become more and more popular, resulting in a real "history boom." This also impacted theme parks and themed environments, as "a redirected focus on the past [...] remains a central component of postmodern theme development" (Paradis 2007, 60). This is also a consequence of the "nostalgic approach" just described: in connection with the end of utopia, or "death of the future" (Hölscher 2002), representations of the past are configurated as a "transhistorical reconstruction of the lost home," or "restorative nostalgia" (Boym 2001, xviii).

A significant example is provided by the enclosed roller coaster Eurosat, located in the French area of Europa-Park, Germany. In its original version, inaugurated in 1989, the ride had a space travel theme, certainly one of the most canonical topics among representations of the future. The ride was then closed in 2017 and reopened a year later with a completely different theme, nostalgically re-oriented from the future to the past. With its new name of Eurosat CanCan Coaster, the ride now takes the visitors back in time, to the Paris of the Belle Epoque. After a waiting area themed to the universal exhibitions, ride vehicles climb the Eiffel Tower and then descend, accompanied by the music of Offenbach's "Galop infernal" from *Orphée aux Enfers*, in a speedy ride through Parisian landmarks such as the Moulin Rouge and Notre Dame. The futuristic motives of the early ride had gone out of fashion: "revamping" Eurosat therefore necessarily meant re-theming it to history and to a nostalgic taste. Nevertheless, the old, futuristic name "Eurosat" was kept, perhaps for marketing reasons, but perhaps also to nostalgically evoke a past chapter in the history of the park and the attraction itself (see also THEMING).

Indeed, in the context of theme parks, it is frequently the theme park itself (or parts of it) that is presented as a "lost home." Perhaps the most famous example is "Main Street, U.S.A.": in Disneyland in Anaheim, this themed area was created in 1955 with the specific aim of recreating the idealized image of the American small town at the beginning of the twentieth century. Older visitors would have had a chance to relive their childhood, while children would gain a better understanding of the lives of their parents. The significance of the area is stressed by the fact that this is the only section of the park which all visitors must cross, on their way from the entrance to the central plaza; in this sense, "Main Street" "brings together" all the visitors of the park, as well as serving as a mindsetter, and creates a sense of belonging to a "people," especially during the huge gatherings connected to the parades and fireworks shows. The lack of attractions in this area is also functional to this image, as they would contradict the nostalgic and utopian image of the "American way of life" that is constructed here (see Francaviglia 1981). Over the years, the unifying function of "Main Street" has expanded to the entire park, as Disneyland has evolved into a modern "pilgrimage center" (Moore 1980) whose visit creates and reinforces a sense of community (see VISITORS).

In this sense, the theme park is not only a product of nostalgia but also an object of nostalgia. A visit to the park—especially to iconic parks such as Disneyland—is constructed in advance through the images and narratives of other people's visits; the visit itself is thus turned into an instantly nostalgic event, which can be memorialized through pictures, videos, and the purchase of souvenirs in the park's shops (see Disneyland's tagline "Let the Memories Begin"). These mediated memories, when seen at home or shown to friends, in turn reinforce the nostalgic picture of the park and help to construct the expectations of future visitors (see PARATEXTS AND RECEPTION).

Finally, theme parks also illustrate the characteristic new approach to history which has developed alongside and within the new postmodern time regime—the journey through time. The possibility of presentifying the past, experiencing it, and living it in one's own skin, has resulted in the current success of reenactment, living history, and theme parks with historical themes: all of these allow people to "affectively turn" toward history (Agnew 2007) by travelling through time, that is by making "an experience and *social practice* in the present that evokes a past [...] reality" (Holtorf 2017, 7). In fact, the very structure of the theme park underlines the idea of travelling: the "rite of passage" implicit in crossing the park's entrance represents a departure, detaching the visitor from his daily life; and the passage from one themed area to the next is generally marked by a gate, which again allows visitors to ritually cross a boundary, moving from one (temporal) destination to the next. This can be built in a sequential and logic way, as in Belantis, where the park's map mimics a map of the world, with themed areas (such as Egypt, Greece, Germany, Great Britain, North and South America) located around a body of water that references the Mediterranean and the Atlantic Ocean.

It can also be constructed, however, as a more random set of destinations, as in Happy Valley Beijing (China), where visitors move from the Viking area to ancient Greece, and from here to the Maya world, Shangri-La, and Atlantis. The same also applies within themed lands: within the Greek area at Terra Mítica (Benidorm, Spain), for instance, the passage from the lower, Mycenaean and Minoan part (featuring a reproduction of the Lions Gate at Mycenae and a replica of the Palace of Knossos) to the area's "acropolis" (with its reproductions of Classical structures) follows a chronological pattern; by contrast, Disneyland Paris's "Discoveryland" moves from the 1950s retro-future of Autopia to the Leonardo da Vinci-themed Orbitron to the steampunk-inspired exterior of Space Mountain (and its now Star Wars-inspired interior).

There is yet another sense in which the represented time of theme parks is anchored in the present: no matter how intricate the temporal layers of the represented times and no matter how detailed the historical displays, theme park visitors are, of course, at all times aware of the fact that such immersion is temporary, and that they are not "really" moving through time and space (Kolb 2008, 123; see IMMERSION).

3 Experienced Time

Theme parks' focus on time travel notwithstanding, what visitors ultimately experience there is an imminent present, "the time in which the visitor moves through the environments given to him by the designers and operators of the park. [...] Similar to the time envisioned by casino managers, the time of the theme parks could be said to be that of the moment—of living, experiencing, and flowing with the course of events and enjoying it as fully as possible at that moment" (Lukas 2017, 27). Much like the represented times in theme parks, however, this imminent present is not uniform: moments of extreme acceleration alternate with moments of extreme deceleration. These changes may be of two different natures. On the one hand, visitors are exposed to physical movements and thus to bodily experiences of acceleration and deceleration; on the other hand, the rhythms of the park, the presence of other visitors, and organizational aspects can generate a sense of time passing quickly or slowly. Theme parks generally offer a wide spectrum of mechanically induced speeds, from which visitors may choose whatever they feel comfortable with. Indeed, while each theme park has its own attractions mix, and consequently, according to their target public, a stronger focus on thrill rides or family rides (see ATTRACTIONS), visitors have a certain liberty to configure their experience of mechanical speed variations. Visitors who only browse the shops and restaurants, for instance, will encounter speed variations only according to the amount of people occupying their path and slowing them down; on the contrary, visitors opting for all the thrill rides will have a much broader amplitude of far more brutal speed variations during their visit.

The experience of mechanical speed may be further manipulated by the theme park through design: "Kinetics, haptic or other sensory technologies, lighting and scenography, and other design approaches may be used to create corporeal and psychological alterations in visitors which result in temporal illusions" (Lukas 2017, 35). An early example of this kind of technique was the so-called "Speed Room" in If You Had Wings, a dark ride at Magic Kingdom (Orlando, Florida) which operated under different titles from 1972 to 1998. While the ride vehicles themselves did not speed up in the eponymous room, the illusion was created by projecting point-of-view phantom rides shot on accelerating trains, motorcycles, etc. Similarly, during the relatively slow lift hill section of the park's Space Mountain ride, visitors are surrounded by pulsating lights that make the ascent appear much faster.

The experience of speed can be further enhanced through a maximum amount of contrast. Indeed, theme parks tend to set contrastive speeds side by side: for instance, classic roller coasters with lift hills or rides using enhanced motion techniques do not always travel at the same speed, but accelerate and decelerate at specific moments to allow riders to experience sometimes quite extreme speed variations. At Tokyo DisneySea's Journey to the Center of the Earth, ride vehicles first slow down and then dramatically speed up following their pivotal encounter with the "lava monster." At Rocket Rods (Magic Kingdom), ride vehicles constantly accelerated and decelerated, to the point where the wear and tear on the ride's infrastructure led to its closure. Speed contrasts are often narratively motivated by the ride's theming, as is the case with Journey to the Center of the Earth or Europa-Park's Poseidon, where the switch from the "coaster" to the "leisurely cruise" portion of the ride signals the end of the storm created by the titular deity.

At Europa-Park's Blue Fire coaster, the experience of contrasting speeds begins even before the riders get on the vehicles: the last section of the waiting line runs parallel and right next to the coaster's catapult launch section, and visitors moving slowly forward are thus located next to the rapidly accelerating trains, contributing to the expectation of what is to come and thus the excitement of the ride. Indeed, visitors do not even need to bodily experience such speed changes to be aware of them—theme parks also construct systems of crossed gazes, through which even visitors avoiding the thrill rides can see (and hear) the consequences of the accelerations and decelerations experienced by fellow visitors. Talocan at Phantasialand, for instance, offers special "spectator" sections that allow visitors to observe riders while remaining perfectly still (see Schwarz 2017, analyzing the water coaster Poseidon in Europa-Park from this point of view; see also ATTRACTIONS).

Nonetheless, mechanical speed, however scenographically altered, is not necessarily correlated with time as perceived by the visitors. Sitting while watching a show is very different, for instance, from slowly moving in a queue—in the first case, one has the impression that time is running much faster than in the second. When Pinzer argues that theme park visitors need to experience a "loss of the sense of time" (2012, 108), he is referring to exactly this form of temporal perception: whether using mechanical speed or other means, theme parks seek to make visitors feel they are making the best out of the time they are spending in the park. In this sense, they seek to avoid the sense that time spent waiting in line, for instance, is not worth the thrill experienced on the rides.

Indeed, one of the most important trends in theme park architecture in recent times has been the optimizing of queuing systems. Since the very beginning of theme parks, waiting areas have been planned so as to hide the actual length of the queues, lest visitors should be confronted with the sight of a huge mass of people in front of them, and so as to break up their waiting time into smaller, "less boring" chunks. Strategies for this can be to divide the waiting area into different rooms which are hidden from each other, or to organize serpentine queues whose length is difficult to estimate. Yet while theme parks would previously just announce waiting times at the entrances of rides and attractions, thus discouraging visitors from immediately moving toward overcrowded attractions and therefore redistributing their fluxes, the subsequent introduction of physical and virtual waiting time boards, themed waiting areas, interactive queues, or virtual queues and ride reservation systems such as FastPass, has practically done away with the classical "waiting in line." Whereas waiting time boards simply allow visitors to reduce their waiting time, themed and interactive waiting areas seek to make the time actually spent in line more enjoyable, for instance by "extend(ing) the narrative or storyline of a ride or attraction beyond the immediate experience itself" (Lukas 2017, 32). Virtual queues convert waiting time into experiencing time: in a simple form, like Disney's FastPass, visitors "book" an allotted time to come back to the ride and directly move to the front of the queue; meanwhile, they can visit other attractions, a restaurant, or a shop. Yet virtual queues can also come in much more elaborate forms: at Dumbo the Flying Elephant at Magic Kingdom (Florida), for instance, visitors receive ticket-themed pagers and spend their waiting time in a themed play area until prompted to go on the ride—rather than wait for a ride, visitors experience two attractions in one.

One of the problems of these newer types of waiting areas is, of course, that they make classical waiting lines that have not yet been "converted" into interactive and/or virtual seem even more bland and boring. Since the advent of smartphones in particular, visitors frequently use their mobile phones while standing in line, which may cause an intrusion of the outside world and, hence, a breach of the immersion. Theme parks have reacted by offering apps with social media features (see PARATEXTS AND RECEPTION), but the fact is that classic waiting lines remain weak spots in the parks' time management: the boredom they may generate could impact overall visitor satisfaction (and thus the motivation for repeat visits; see ECONOMIC STRATEGY). Perhaps even more importantly, visitors also have less time to visit the restaurants and shops. Souvenir and ice cream vending points located within waiting lines—popular, for instance, at Europa-Park—are not just indicative of the park's consideration for its visitors' needs, but also of the parks' need to make up for lost time.

4 Managed Time

As these examples of queues illustrate, time can be "equally varied and thrilling for the guest as it is a source of possible efficiency and control for park management" (Lukas 2017, 35). Consider, for instance, Terra Mítica, which appears to be organized simply according to the geography of the Mediterranean, with themed areas referring to (counter-clockwise) Egypt, Greece, the Aegean islands, Rome, and Spain grouped around a central body of water. Yet a more detailed look reveals the pivotal role time plays in the organization of this park. This applies not only to represented time, which underpins the ideological message of the park (see WORLDVIEWS), but also to managed time, which impacts the precise location of themed areas, rides, restaurants, etc., the operational practices of the park, and, relatedly, how it is experienced by visitors.

The layout of Terra Mítica follows a loop pattern: the entrance is located in Egypt, and in theory, visitors could follow the path both clockwise and counterclockwise. Yet the park has clearly been planned with a preference for the latter (which is also connected to the park's implicit political agenda): to achieve this there is, for example, a much higher concentration of attractions and visual magnets on the right side of the entrance, leading visitors to almost naturally start moving toward them. Considering this distribution, it is no surprise that most of the restaurants in the park are located in Rome, as this is the themed area which, according to this model, would be visited around lunch time. Ice cream parlors, present in other areas of the park, are absent here, while multiple seated restaurants are available. The distribution of the shows stresses in an analogous way the "correct" way of visiting the park, as they are scheduled to follow the visitors' progression through the themed areas in a model visit taking the entire day. The main show, Barbaroja, takes place during the afternoon in the Iberian area, while the latest shows, just before the park's closure, are located in Egypt, directly in front of the entrance, to lure visitors back to the park exit at closing time. Show personnel as well as operational staff are implicated in this temporal organization of the park: rides, restaurants, and shops are staffed according to the progression of the visitors, with rides and shops in the back areas—which visitors only reach later during the day—even being closed during the morning hours.

Terra Mítica constitutes an extreme example of the temporal organization of a theme park, yet other parks at least partially employ this model as well. Ren, for instance, notes that in the Chinese Ethnic Culture Park (Beijing), "daily performance is scheduled carefully to direct the flow of visitors. Managers use time management to manipulate visitors to spend money at appropriate places at appropriate time[s]" (1998, 85). Similarly, around the turn of the millennium Disneyland Paris introduced an "Enter Stage Right" operational schedule that guided visitors from one themed land to the next to save on staff, heating, and electricity costs. Disney parks in general use nighttime parades and fireworks and projection shows to gently guide visitors to the park's exits—as well as to "Main Street"'s shops, which remain open longer than the rest of the park.

Right next to the alleged "loss of a sense of time" (Pinzer 2012, 108), we thus also find an acute awareness of the preciousness of theme park time on the part of both visitors and park managers. A good example are single rider lines, which allow visitors to reduce waiting times by splitting up and thus freeing up potentially boring waiting time for other attractions and/or shops and restaurants. The park landscape itself points to the importance of time and its management. In spite of the separation of the park from its external reality to achieve immersion, indeed, clocks are not extraneous to the theme park and are, in fact, a very common decorative element inserted into the theming of every single area.

As theme parks know very well, in the end, time is a commodity—time is money. Visitors pay for entrance tickets to "buy" time in the park and strive to exploit this time as much as possible; time allows them to spend additional money (on food, drinks, and souvenirs; see ECONOMIC STRATEGY). It is therefore not surprising that theme parks often also offer opportunities to purchase additional time in the park—in some parks, for instance, virtual queue services are not included in the entrance ticket, but are sold as an optional add-on; similarly, parks with hotels and resorts offer special packages that guarantee visitors who spend the night there a different "time structure": they have separate entrances and sometimes different opening times. In Parc Astérix, for instance, the entrance for hotel guests opens half an hour before the general entrance and is located on the opposite side of the park, thus allowing hotel guests quicker access to the most popular attractions—and therefore shorter queues.

5 External Time

The times and temporalities examined thus far—Represented Time, Experienced Time, and Managed Time—originate within the parks and may thus be considered unique to them. In addition, the complex temporality of theme parks also includes times or temporal layers that lie beyond the influence of the parks and may therefore be categorized as "external." This is not surprising: even if theme parks are characterized by their "externality" (see THEMING) and represented time, for example, is thus always different from and exotic to the temporal layer outside the park, visitors never completely lose their awareness of the outside world to which they will return at the end of their visit; this residual awareness of the outside world also includes the latter's time and temporalities (see IMMERSION). Theme parks cannot ignore these "external" temporalities but must acknowledge and react to them in one way or another, while also making sure that they do not dominate the theme park experience, resulting in a difficult and unpleasant visit.

Perhaps the simplest and most basic example of "external time" is connected to the biological rhythms of human beings. Obviously, a place where people spend several hours—whether as visitors or as employees—needs to provide them with food, drink, and bathrooms. While the latter are generally distributed evenly throughout the park, the location of facilities such as restaurants, snack bars, ice cream and popcorn carts, etc. is determined in accordance with Managed Time, and therefore with the forms in which the park expects visitors to move at different moments of the day. As shown above, at Chinese Ethnic Culture Park, for instance, the various shows are coordinated to "naturally" lead visitors to specific restaurants around lunch time (Ren 1998, 85–86).

In addition, theme parks need to react to the natural rhythms of daytime and nighttime and the cycle of the seasons. The decision to include nighttime operation, for instance, implies the need to provide illumination (both for safety and for theming), while at the same time allowing the park to offer visitors additional attractions, such as fireworks shows or illuminated parades (see ECONOMIC STRATEGY). The seasons and the natural cycles of plants also impose themselves on the theme park's landscaping operations: not only must plants be watered when it rains less, but they are arranged so as to provide color patterns through their blossoms at specific moments. The decision to include all-year-round operation in turn implies the need to provide protection from colder or rainier seasons. The southern part of the island of Taiwan, for example, is characterized by a tropical monsoon climate, with heavy rains throughout the months of June, July, and August. The theme park E-Da, in Kaohsiung, is therefore composed of three themed areas, all subdivided into an indoor and an outdoor part. In case of bad weather, visitors can move between the three indoor sections with a monorail and thus enjoy many of the park's attractions without having to be outside.

The intersection of the cycles of day and night and the seasons of the year frequently manifests itself in longer opening hours during the long summer days and shorter opening hours during the rest of the year. This once again interacts with

Managed Time, as it can lead the park to assume particular operational strategies to reduce costs. Disneyland Paris's "Enter Stage Right" program (see above), for instance, was restricted to the winter season, when the park expected much fewer visitors due to the colder and shorter days of the French winter. The theme park is, of course, by no means the only destination affected by such astronomical and weather cycles: in fact, theme parks are deeply embedded in the tourism seasons of their geographical location, and this influences their operational strategies. One of the reasons why theme parks are often located in proximity to large metropolitan areas, in addition to the large local audience and the established infrastructure, is that city tourism works year-round. Perhaps the most extreme cases of external times that theme parks need to address are completely unforeseeable, short, and localized events such as fires, natural catastrophes, terrorist attacks, pandemics, etc. Like any other place, theme parks need to provide protocols for the safety of the people on site. Such protocols generally override other layers of the parks' complex temporality: for example, during the invasion of Disneyland by members of the "Youth International Party" on August 6, 1970, or during the fire that destroyed Europa-Park's Piraten in Batavia ride on May 26, 2018, the parks were evacuated and closed early; during the COVID-19 pandemic from 2020–2022, theme parks remained closed for entire seasons.

Some external temporalities, however, can be productively integrated into the theme parks' represented, perceived, and managed times so that they appear as internal. This applies to cultural festivities such as civil or religious holidays, as well as extraordinary events such as the Olympic Games, the FIFA World Cup, etc. While this can mean nothing more than some minor adjustments, such as theming the daily parade to the current event, the impact of the "intromission" of these temporalities on the theme park is often much larger and more visible. Parks regularly embrace these events and stretch them out, for instance with holidays being celebrated over a period of several weeks and a park anniversary spreading over several months or even years. Many parks that are usually closed for the winter nevertheless open their doors for special Halloween and Christmas seasons. In others, there is barely enough time between the Halloween and Christmas celebrations to change the respective decorations. Sometimes, nearby festivities are subsumed into one longer, unified celebration: one example is the "Miao Sisters Festival" in the Chinese Ethnic Culture Park in Beijing, which arranged a display of the Miao ethnic culture around International Women's Day on 8 March and Chinese Tree-Planting Day four days later (see Ren 1998, 82). In fact, these "internalized external temporalities" have become so successful that parks celebrate as many of them as possible, regardless of whether they fit within their themes or whether the festivity has traditionally been celebrated in the park's location. The yearly calendar of theme park festivities has thus become ever more globalized, with Chinese New Year celebrated at Disneyland Paris and Happy Valley Shenzhen hosting Christmas celebrations.

Bridging the gap between internal and external temporalities and their effects on the theme park are park or company anniversaries. While these derive from internal circumstances, such as the year of the park's opening or the creation of famous cartoon characters, they also depend on the progression of linear time that visitors are supposed to leave behind at the entrance. Indeed, in most cases theme parks tend to hide the signs of their own history—or better still, display them only at specific moments and in a self-celebratory strategy (see THEMING).

6 In the Rhythm of the Theme Park

If there is one theme park element in which the parks' complex temporality manifests itself perhaps even more clearly than in the waiting times board discussed at the beginning of this chapter, it is the park soundtrack. Along with the noises made by visitors and employees and the sounds emanating from the park's attractions, the soundtrack constitutes one of the central elements of the park's soundscape and comprises live music, but especially recorded music (instrumental and vocal), announcements and narrations, and sound effects that are broadcast over the PA system either in the entire park or in selected sections. While studying the waiting time board at the end of "Main Street" at Disneyland Paris, for instance, visitors may hear a collection of ragtime pieces that firmly situate the area in the early 1900s (Represented Time). During the morning hours, the music is lively and energetic, as if to urge visitors to move on and thus make way for the other visitors who enter the park at this point. In the afternoon and evening, by contrast, the music sounds more relaxed, inviting visitors to maybe linger a bit longer, perhaps browse the shops on "Main Street" or have a coffee in one of the cafés before leaving the park (Managed Time). Depending on the time of day, visitors may thus perceive the atmosphere on "Main Street" as bustling and full of life, or charmingly quaint and cozy (Perceived Time). In very rare cases, however, visitors may hear no music at all on "Main Street." In May and June of 2002, for instance, the volume of the music was frequently lowered for the announcement of yet another goal at the FIFA World Cup in South Korea and Japan. And a few months before, the music had even been entirely stopped during a minute of silence for the victims of the terror attacks on September 11, 2001 (External Time).

During the Holiday season, however, it does not matter what time of day visitors study the waiting time board—they will always hear the Christmas songs that have temporarily replaced the two ragtime soundtracks. Accordingly, the waiting time board is also decked out for the holidays, suggesting further subtle shifts towards a different timelessness through minor retheming. Hence, both together and individually, the music played at the park and the waiting time board suggest the multi-layered, complex temporality of Disneyland Paris in particular, and the theme park in general.

References

Agnew, Vanessa. 2007. History's Affective Turn: Historical Reenactment and Its Work in the Present. *Rethinking History* 11 (3): 299–312.
Boym, Svetlana. 2001. *The Future of Nostalgia.* New York: Basic Books.

Carlà, Filippo. 2016. The Uses of History in Themed Environments. In *A Reader in Themed and Immersive Spaces*, ed. Scott A. Lukas, 19–29. Pittsburgh: ETC.
Carlà-Uhink, Filippo. 2020. *Representations of Classical Greece in Theme Parks*. London: Bloomsbury.
Francaviglia, Richard V. 1981. Main Street U.S.A.: A Comparison/Contrast of Streetscapes in Disneyland and Walt Disney World. *The Journal of Popular Culture* 15 (1): 141–156.
Guffey, Elizabeth. 2006. *Retro: The Culture of Revival*. London: Reaktion.
Gumbrecht, Hans Ulrich. 2010. *Unsere Breite Gegenwart*. Frankfurt: Suhrkamp.
Hölscher, Lucian. 2002. Die Zukunft: Ein auslaufendes Modell neuzeitlichen Geschichtsbewußtseins? In *Die Zeit im Wandel der Zeit*, ed. Hans-Joachim Bieber, Hans Ottomeyer, and Georg Christoph Tholen, 129–149. Kassel: Kassel University Press.
Holtorf, Cornelius. 2017. Introduction: The Meaning of Time Travel. In *The Archaeology of Time Travel: Experiencing the Past in the 21st Century*, ed. Bodil Petersson and Cornelius Holtorf, 1–22. Oxford: Archaeopress.
King, Margaret J., and J.G. O'Boyle. 2011. The Theme Park: The Art of Time and Space. In *Disneyland and Culture: Essays on the Parks and Their Influence*, ed. Kathy Merlock Jackson and Mark I. West, 5–18. Jefferson: McFarland.
Kolb, David. 2008. *Sprawling Places*. Athens: University of Georgia Press.
Lukas, Scott A. 2017. Time and Temporality in the World of Theme Parks. In *Time and Temporality in Theme Parks*, ed. Filippo Carlà-Uhink, Florian Freitag, Sabrina Mittermeier, and Ariane Schwarz, 19–41. Hanover: Wehrhahn.
Mittermeier, Sabrina. 2017. Utopia, Nostalgia, and Our Struggle with the Present: Time Travelling through Discovery Bay. In *Time and Temporality in Theme Parks*, ed. Filippo Carlà-Uhink, Florian Freitag, Sabrina Mittermeier, and Ariane Schwarz, 171–187. Hanover: Wehrhahn.
Mittermeier, Sabrina. 2021. *A Cultural History of the Disneyland Theme Parks: Middle Class Kingdoms*. Chicago: Intellect.
Moore, Alexander. 1980. Walt Disney World: Bounded Ritual Space and the Playful Pilgrimage Center. *Anthropological Quarterly* 53 (4): 207–218.
Paradis, Thomas W. 2007. From Downtown to Theme Town: Reinventing America's Smaller Historic Retail District. In *The Themed Space. Locating Culture, Nation, and the Self*, ed. Scott A. Lukas, 57–74. Lanham: Lexington.
Pinzer, David. 2012. Erlebniswelten und Technikphilosophie. In *Erlebnislandschaft—Erlebnis Landschaft? Atmosphären im architektonischen Entwurf*, ed. Achim Hahn, 97–120. Bielefeld: Transcript.
Pugh, Tison, and Susan Aronstein, eds. 2012. *The Disney Middle Ages: A Fairy-Tale and Fantasy Past*. New York: Palgrave Macmillan.
Ren, Hai. 1998. *Economies of Culture: Theme Parks, Museums and Capital Accumulation in China, Hong Kong and Taiwan*. Diss., The University of Washington.
Schickel, Richard. 1997 [1968]. *The Disney Version: The Life, Times, Art and Commerce of Walt Disney*, 3rd ed. Chicago: Ivan R. Dee.
Schröder, Jörg. 2012. Belantis: Erlebnisgestaltung zwischen Funktion und Emotion. Interpretation des Interviews mit dem Architekten Herrn Rudolf. In *Erlebnislandschaft—Erlebnis Landschaft? Atmosphären im architektonischen Entwurf*, ed. Achim Hahn, 201–224. Bielefeld: Transcript.
Schwarz, Ariane. 2017. Staging the Gaze: The Water Coaster Poseidon as an Example of Staging Strategies in Theme Parks. In *Time and Temporality in Theme Parks*, ed. Filippo Carlà-Uhink, Florian Freitag, Sabrina Mittermeier, and Ariane Schwarz, 97–112. Hanover: Wehrhahn.

Visitors: The Roles of Guests as Customers, Pilgrims, Fans, Performers, and Bodies in the Theme Park

Abstract This chapter discusses the significance of visitors and customers in the theme park industry. The chapter begins with a discussion of the concept of "guest" in the theme park industry—including its significance in theme park training programs and curricula. Exploring different heuristic approximations to the complex experience of the theme park visitor, the chapter then discusses the visitor as a customer, pilgrim, fan, performer, and experiencing body. First, the focus is on demographics, the economics of customer spending, and the catchment area in terms of customer base. This is followed by a discussion of visitors as pilgrims who visit and interact with theme parks in ways that replicate religion, ritual, and other sacred forms of culture. The chapter then shifts first to a subgroup of visitors—namely, fans—and then to visitors as performers in order to analyze their manifold ways of interacting with the theme park. Finally, the chapter considers the visitor experience on rides and why this experience is significant for the continuing popularity of theme parks, notably as theme parks and other "brick and mortar" or physical spaces (like shopping malls and retail stores) compete with virtual entertainment and lifestyle retail for customer attention.

1 Introduction

Alongside those who are professionally engaged with theme parks (designers, employees, and researchers; see LABOR and METHODS), theme parks are also lived and experienced by their visitors. Within internal communication and paratexts (see PARATEXTS AND RECEPTION), theme park visitors have often been referred to as "guests," but depending on one's conceptualization of the theme park, they may also be thought of and researched as customers, pilgrims, performers, fans, and experiencing bodies. This chapter will engage with all of these different conceptions.

This work is contributed by Salvador Anton Clavé, Ariane Schwarz, Thibaut Clément, Scott A. Lukas, Crispin Paine, Sabrina Mittermeier, Florian Freitag, Astrid Böger, Filippo Carlà-Uhink, Céline Molter, Jean-François Staszak, Jan-Erik Steinkrüger, Torsten Widmann. The corresponding author is Salvador Anton Clavé, Departament de Geografia, Universitat Rovira i Virgili, Vila-seca, Spain.

F. Freitag et al., *Key Concepts in Theme Park Studies*,
https://doi.org/10.1007/978-3-031-11132-7_16

The label "guests" was assigned by the Walt Disney Company, and due to the company's long history and dominant position in the industry, most of the academic inquiry into theme park visitors has been into visitors to Disney parks. In reference to Disney, parks have been framed as "landscapes of power" (Zukin 1991), most prominently by European scholars such as Louis Marin (1973), Umberto Eco (1986), Jean Baudrillard (1983), and their American continuators, including Stephen Fjellman (1992) or, to some extent, the authors behind *The Project on Disney* (Kuenz et al. 1995). Drawing from structuralism, semiotics, and postmodernist theory, such analyses essentially rest on two premises: first, the idea that representations—especially those found in Disney parks—are fundamentally deceptive; and second, the notion that theme parks operate with signs and symbols that form an ideological discourse in the service of capitalist hegemony (Clément 2012). As those critics argue, Disneyland's evocative power is such that visitors have come to accept copies as originals, and the fake as real: this is what makes theme parks prime examples of "hyperreality," where "[t]he 'completely real' becomes identified with the 'completely fake'" (Eco 1986, 7), or of "simulation," where "the radical negation of the sign as value" leads to a "reversion and death sentence of every reference" (Baudrillard 1983, 6; see AUTHENTICITY).

The resulting "blunting of visitors' powers of discrimination" (Fjellman 1992, 256) between fantasy and reality serves as a ploy to manipulate the visitors' states of mind and profit from them. Not only is an immersion into fantasy conditioned to real-life purchases (see ATTRACTIONS), but parks also tap directly into visitors' dreams and aspirations, to the effect that "[w]hat is falsified is our will to buy, which we take as real" (Eco 1986, 43) and that "the last remaining vestige of uncommodified life—the unconscious—is brought into the market system" (Fjellman 1992, 300). As a result of the disrupted relation between "signifier" and "signified" (to use the language of semiotics), the true meaning of the parks' potent cultural symbols is perverted in the service of ideology—in most cases that of capitalist hegemony (see WORLDVIEWS). Disneyland is thus defined as "the quintessence of consumer ideology" (Eco 1986, 43), and "Disneyworld" as "the most ideologically important piece of land in the United States" in that it exposes "[t]he hegemonic meta-message of our time," namely that "the commodity form is natural and inescapable" (Fjellman 1992, 9–10).

These early analyses fall short in several respects. Firstly, for all their efforts at laying bare Disney's control over visitors, they generally fail to acknowledge that they are themselves expressions of power: the critics' aesthetic judgments (some attractions are rated as "unbearably corny," Fjellman 1992, 72) or their distancing themselves from popular cultural practices, express an elitism that relies purely on cultural capital. Secondly, and just as questionably, these accounts usually deprive visitors of any agency and present park-goers as cultural dupes: Eco, for instance, speaks of Disneyland as "a place of total passivity" (1986, 48), while Baudrillard argues that visitors to Disneyland play at being children in order to sustain the illusion that, outside its bounds, they are really adults (1983, 25). In their apparent adherence to the "hypodermic needle model," according to which mass media messages are directly "injected" into the brains of passive receivers (see Danesi 2013), the critiques by Eco, Baudrillard, etc. generally disregard the visitors' (and the workers') interpretive activity and ignore deviations from the parks' prescriptive standards of behavior (for examples of creative or disruptive behavior by visitors or employees; see Van Maanen 1991; Kuenz 1995b; Choi 2007; Clément 2016; see also below and LABOR).

While giving nods to earlier critiques in the postmodernist tradition—one of them pronounces "Disneyworld [...] this planet's most elaborate and sustained simulation" (Willis 1995, 48)—the writers involved in *The Project on Disney* already warn against "degrad[ing] visitors [...] as an unreflective band of consumers eager to enjoy the benefits of a simulated world" (Kuenz 1995a, 78) and instead maintain that the "willing suspension of disbelief by the park's guests is neither simple nor simple-minded" (Kuenz 1995b, 111). Yet it is primarily more recent work that has come to view theme park audiences in much more differentiated ways, granting them agency and categorizing them, e.g. according to their demographics, spending habits, as well as their level of engagement with the theme park, from dedicated theme park antagonists to equally dedicated theme park fans. Jennifer A. Kokai and Tom Robson's collection *Performance and the Disney Theme Park Experience* (2019), for instance, employs a multiplicity of perspectives to engage with visitors as actors in a performative space. With the publication of the revised edition of Janet Wasko's *Understanding Disney* (2020)—the first edition from 2001 had contained one of the earliest discussions of Disney (park) fandom—and the publication of Rebecca Williams's *Theme Park Fandom* (2020), theme park fans and their activities have further moved to the center of attention in current research on theme park visitors. Hence, although repeated calls for an "integrated" approach to theme parks that takes into account the production and reception of theme park spaces by employees and visitors (see Raz 1999, 6; Wasko 2001, 152; Lukas 2016, 168; and Clément 2016, 17–18) have so far remained largely unanswered, it is nevertheless safe to say that today, the traditional "scholarly and mainstream notion of theme-park visitors as naïve, controlled and duped into excessive consumption" (Williams 2020, 9) has been definitely overcome.

Of course, there are people who, for financial, socio-cultural, political, or other reasons can or will not visit a theme park—the Chairman of the Council of Ministers of the Soviet Union Nikita Khrushchev, for example, was famously denied a visit to Disneyland due to safety concerns in 1959; in 1992, French president François Mitterrand declared that Disneyland Paris was not his cup of tea, implying higher tastes. Theoretically, however, theme parks attract anyone: as commercial enterprises that depend on large audiences, theme parks have taken great care to be as popular—that is, as physically, culturally, and cognitively accessible—as possible (see Freitag 2021, 13–14). And while we only have data for individual parks and park chains, and therefore, can only discuss why people will or will not visit a specific park (see METHODS), we can at least state that the diversity of visitors that theme parks seek to attract is matched by the diversity of offerings theme parks make to their audience. As Goronzy (2006) and Steinkrüger (2013, 146) have noted, theme parks offer a blend of explorative (including cognitive), biotic (bodily and sensual), social, and emotional experiences, often within the same element: for example, riding the Poseidon water coaster not only allows visitors of Europa-Park (Rust, Germany) to make biotic, but also emotional (fear, excitement, immersion), cognitive (learning about Greek mythology), and social (interaction with fellow riders and visitors watching from below) experiences (see Schwarz 2017).

The following subchapters each highlight specific aspects of the visitor experience: the visitor as a customer, with a focus on the commercial role played by the "guests" within the theme park as a business venture; the visitor as a pilgrim, underlining the explorative and emotional aspects of the theme park visit; the visitor as a performer, foregrounding the social experience; the visitor as a fan, investigating the phenomenon of theme park fandom; and the visitor as a body, highlighting the "corporeal" aspects of a visit to the theme park. Nonetheless, all the subchapters merely constitute heuristic approximations of the complex experience of the theme park visitor, which results from a mixture of all these aspects. This chapter thus attempts to disentangle and identify the various ways in which we have thought and can continue to think about visitors, discussing visitors not only in their passive, but also—and especially—in their more active roles within the theme park space.

2 Visitors as Customers

Within the corporate world of theme parks, as well as the service industry in general, the term "visitor" has often been replaced with the term "guest." Sharing an etymological root with the word "host," the term "guest" is used to refer to "someone with whom one has reciprocal duties of hospitality" (Luke 1990, 17). This connects to how the naming of theme park visitors as "guests" has gained traction in the theme park industry. For instance, in the 1990s, Six Flags theme parks relied on the concept of guest to structure their employee training programs. Programs like "Guest First" stressed the significance of each theme park worker making a positive, intimate, and emotional connection with visitors (Lukas 2007, 190; see also LABOR). The 1994 *Six Flags AstroWorld Host and Hostess Handbook* offered the following welcome letter to all new employees of the theme park:

> You have been cast as a Host or Hostess. Each day you will create a very special occasion for the many Park Guests you will serve and entertain. Talk to the people you encounter. Smile and anticipate their needs. Offer suggestions, directions or a helping hand whenever possible. This attention to detail and the caring attitude we convey to our Guests creates a truly memorable experience. You must remember, even though it may be just another work day for us, it is a very special time for our Guests. (Six Flags AstroWorld 1994, 7)

Much like the Disney parks, Six Flags has relied on the connotation of the visitor as a guest to create a service industry culture that values personalized and intimate constructions of the visitor experience. Most large theme park corporations, in addition to avoiding the term "visitor," avoid representations and interpretations of visitor experiences that highlight corporate, economic, and work-based connotations; on the contrary, they tend to display and promote images of visitors having fun and enjoying themselves (see PARATEXTS AND RECEPTION).

However, theme park-owning companies or conglomerates still very much perceive visitors as paying customers and treat them accordingly, with a clear attention to the definition and pursuit of a specific target public and, over time, to develop new strategies to cope with socio-demographic change. The most important visitor groups for a majority of theme parks are families, teenagers, and young adults; however, specific aspects vary depending on the general attractions mix of the park (see ATTRACTIONS and ECONOMIC STRATEGY). For instance, unreleased data from a recently conducted study shows that the average age of adult theme park visitors in Germany is around 28 years, while thrill-oriented theme parks like Heide Park (Soltau, Germany) have a younger audience with an average of 22 years. In this case, young adult "thrill-seekers" aged 14–24 years and families with children between 12 and 16 years each make up a third of the total customers. Meanwhile, the share of visitors older than 50 years has been decreasing significantly in German themed environments (Stiftung für Zukunftsfragen 2013). Yet as demographic change makes this age group more and more important, theme parks have been keen to target older visitors, or to develop specific offers for them. Europa-Park, for instance, offers a circuit of attractions, shows, and other installations that seeks to attract this demographic by emphasizing aspects of enjoyment and recreation through a special program called "Europapark für Genießer" ("Europapark for connoisseurs"; Europa-Park 2019).

Equally important for theme park operators is to identify different visitor groups according to their staying and spending habits (see ECONOMIC STRATEGY). The core geographical target group for German (and many other) theme parks is usually same-day-visitors, that is, residents within a 1.5–2 h radius or the so-called "catchment area." While the "guest potential" for smaller parks declines significantly once driving time exceeds 1.5 h, well-established theme parks often have a catchment area of up to 4.5 h (Fichtner 2000, 82). As a rule of thumb, 2–5% of the residents within the 1.5 h-drive catchment area can be activated for at least one theme park visit per year. Unpublished research at a mid-sized theme park in southern Germany with 318,000 visitors in 2011 showed that up to 9% of the residents within this area had visited the park at least once in that year. Measuring the catchment area in driving time is important because many theme parks, particularly in Germany, are located in non-urban areas, and the main mode of transport to reach them is the car (89%), followed by the coach (6%), bike (2%), and only 1% by train (Steinecke 2011, 75). Yet even for theme parks in so-called "agglomeration areas," the car is still the most important mode of transport, as exemplified by Phantasialand in Brühl (Germany), where 70–80% of visitors arrive by car (Thuy and Wachowiak 2008, 11).

If the park is situated within the range of a larger tourist destination, overnight-staying guests are also an important target group. The longer tourists stay in a particular area, the more willing they are to take day trips, e.g. to a nearby theme park. Hence, tourists who stay for more than four nights and reside within a catchment area of 60 min can be considered an additional target group. Usually, 10–15% of these visitors can be activated for a theme park visit (Valdani and Guenzi 2001, 163). In the case of Disneyland Paris, the Walt Disney Company had banked on the central location of Paris, considering the French capital, with its public transport connections and accessibility by car, as superior to other locations such as Barcelona in Spain, despite the differences in weather (Mittermeier 2021, 106–107). While this proved a wise decision, Disney nevertheless originally massively misunderstood their potential customers, as they conceived of the theme park as a resort, similar to their U.S.-based ventures, rather than as a day trip destination for tourists to Paris (Mittermeier 2021, 142–144). It is thus vital for companies to understand and target customers based on their specific location and the market they operate in (see ECONOMIC STRATEGY).

This also plays a central role when it comes to customers' per capita spending, as this is a crucial figure when measuring the economic success of a theme park. The entrance fee for a family with two children in German theme parks ranged from €117 to €190 in 2018 (Statista 2019, 19). In order to increase per capita spending, theme parks offer various upselling opportunities like food and beverage outlets, merchandise, or other experiences that require extra payments (see also INCLUSION AND EXCLUSION). This so-called secondary spending, or rather the lack thereof, also hurt Disneyland Paris in its early days. Disney generally targets a middle-to-upper-class customer base with ample disposable income for secondary spending, but had originally failed to do so in the case of their Paris park (Mittermeier 2021, 143–145).

To stimulate overall per capita spending, theme park operators also try to increase the duration of stays in the parks. Europa-Park, for example, managed to increase the average length of stay from 5 to 6 h to more than 8 h by providing accommodation for overnight stays and staging shows and events in the evening (Steinecke 2011, 261). Another measure to motivate visitors to stay longer is the development of differently themed areas as a so-called "second gate," with examples including the Walt Disney Studios in Disneyland Paris (2001) or the water park Rulantica at Europa-Park (2020; see ECONOMIC STRATEGY). Yet despite such purely economic considerations, theme park operators also have to consider their visitors in other ways, as they ultimately have individual needs and bodily experiences based on their interactions with and inside the theme park space (see Birenboim et al. 2013).

3 Visitors as Pilgrims

One of these needs may be the ritualistic transformation of the self during a pilgrimage. In tourism studies, overlaps between religious pilgrimage and tourist travel have long been recognized. As theme parks are the inscription of idealized fantasies into space, they have sometimes been interpreted by scholarship as connected with the sacred, the imagined ideal "other" that humans establish through ritual practices in religious acts. Therefore, it is no surprise that theme park tourism—which includes, but definitely also goes beyond visits to religious theme parks—has been associated with pilgrimage, and visits to theme parks have been understood as ritual. Paine (2019, 91–92), for example, explains how theme parks are sacred places in the sense of being "set apart" from the ordinary by architectural and social boundaries. Moore (1980) reflects on the mutuality of ritual and play and concludes that in the age of postmodernity, (religious) ritual was losing ground and being replaced by its predecessor, play. According to Moore, Walt Disney World shares every formal aspect of "traditional" pilgrimage centers and constitutes

> a bounded liminal place that one visits on a playful meta-pilgrimage. At a time when some proclaim that God is dead, North Americans may take comfort in the truth that Mickey Mouse reigns at the baroque capital of the Magic Kingdom and that Walt Disney is his prophet. (1980, 216)

Disney might indeed be at the core of this association, since the company has assigned itself a socio-cultural function that goes well beyond mere amusement and into the sphere of the ideal:

> In marketing the brand, Disney has situated itself as a key actor in defining what North American (and increasingly worldwide) childhood fantasies are, or at least ought to be. In doing so, a visit to a Disney theme park has become a pilgrimage that many middle-class kids seem hard-pressed to live without. (Rutherford 2011, 46; see also Pettigrew 2011, 145)

But what is the revelation that Disney pilgrims might seek? In his essay on Walt Disney World, Schultz repeatedly refers to the patrons as "pilgrim visitors" (1988, 282) on a quest for an earthly paradise, a futurist social utopia that had originally been envisioned by the pilgrim fathers and finally realized by Walt Disney. "This is the mythical North American Garden of Social Innocence and Prosperity, sought ever westward by groups of immigrant pioneers, now found at last in the wilderness of Florida" (Schultz 1988, 282). This social utopia, as Schultz points out, is not just a place for fun and happiness, but a specifically white middle-class ideal of America, cleansed of its historical "blood on the tracks" (1988, 286; see INCLUSION AND EXCLUSION).

Schultz, therefore, links the sacrality of the theme park visit to a specific middle-class American worldview. In line with his work, other commentators have interpreted Disney theme parks as spatialized nostalgic utopias, aimed at grown-ups prospecting for the new millennium. Inspired by Jean Baudrillard's thoughts on the "principle of achieved utopia" as a distinctive feature of Americanness, Scheer-Schätzler (1993, 53) has even argued that the aspiration of turning utopia into reality, of testing theory in practice, had been rooted in American self-perception from the first day that settlers had set sail for the new continent. Consequently, she interprets Epcot (Disney's "Experimental Prototype Community of Tomorrow" at Walt Disney World in Florida) in line with other, often religious, utopian community projects like the Shakers (which, like the Shaker Village of Pleasant Hill, Kentucky, have nowadays also been turned into living history museums that feature elements inspired by theme parks and other themed environments; see THEMING).

Steiner, meanwhile, uses Disney's "Frontierland"—a cleansed and comforting version of the Western frontier, "America's most potent myth" (1998, 4)—as an example of Disney's ideological framing (see WORLDVIEWS). For Steiner, the key to Disney's success was the transformation of unsettling and contradictory historical truths into soothing, simple, and unifying myths for white middle-class people. The new theme park medium allowed visitors to relive these myths through architecture and theater and to more or less consciously overwrite the common past with a more pleasant narrative (1998, 4–5; for comparable information on Disney's representations of New Orleans, see Freitag 2021). Cher Krause Knight's in-depth study of Walt Disney World, in turn, also discusses the more individual feeling of "wonder" or "optimism" the site may bring to many of its visitors as they engage in this form of pilgrimage (2014, 43). What all these studies of Disney parks and their visitors thus suggest is that more than amusement, for theme park pilgrims a visit also means an engagement with a collective identity, a ritual transformation of the individual, be it intended or not. Giving agency to visitors, these readings of theme parks as sites of secular pilgrimages stress both the importance of *communitas* as well as the individually assigned meanings these spaces offer to visitors.

4 Visitors as Fans

A very special *communitas* is that of theme park fans. Frequently based on their longevity and, for many, rooted in their childhood nostalgia, certain theme parks and theme park attractions in particular—organized fandom has gathered mostly around specific types of attractions and notably, roller coasters—have seen an established and ever-growing fandom. Fans of Europa-Park, for instance, have founded an official fan club called "EP Fans," complete with a website and a messaging board; and the fan veneration of one of Disney's oldest attractions, the Haunted Mansion, has led to the availability of vast amounts of merchandise, including apparel to perform cosplay (see Williams 2020, 101–131). Unsurprisingly, the most visible fan communities are those connected to the parks owned and operated by the big conglomerates, namely the Walt Disney Company and Comcast/Universal. Universal Studios parks, for instance, have a dedicated, cross-generational group of fans, which connects through several websites.

The big conglomerates have become aware of such fandom and fannish activities, and have consequently bought into this demand by developing special offers, which, in turn, further increase the role of such activities in visitors' engagement with the park and its structures. While adults may not dress up as characters within Disney parks (see below), the company has encouraged a practice called Disneybound—dressing in certain colors and looks that correspond to a character (see Williams 2020, 194–205). With the Disney Dress Shop clothing line, they have begun to officially produce merchandise to encourage this activity while also financially benefiting from it, offering an alternative to the usually fan-made costumes.

As theme parks rely more and more on intellectual properties (IPs) as thematic sources (see ANTECEDENTS, MEDIA, and THEMING), media franchises have come to play a central role in theme park fandom. Indeed, one way to draw fans of specific media franchises to theme parks is by dedicating attractions or entire themed lands to these IPs. One of the most notable events over the last years in this regard was Universal/Comcast obtaining the theme park rights to the *Harry Potter* franchise. In 2010, the company opened "The Wizarding World of Harry Potter" at their Orlando Resort, where it now spans two different themed areas in both of their theme parks that are connected by the Hogwarts Express. They also opened smaller versions at Universal Studios Japan (in 2014) and Hollywood (in 2016). "The Wizarding World" is arguably the most immersive environment Universal has ever created for their theme parks; yet, as studies have shown, visitors' immersion into the story is largely dependent on their previous knowledge of the content it is based on (Hofer and Wirth 2008, 167). For those unfamiliar with the *Harry Potter* books or, even more importantly, their filmic adaptations, by contrast, the themed areas might fall flat. For fans of this IP, however, "The Wizarding World" constitutes an "authentic adaptation of the *Harry Potter* story-world, a place where *Harry Potter* fans can [...] experience the story-world in an embodied manner" (Waysdorf and Rejinders 2018, 174).

Functioning as a draw for established fanbases, IP-themed areas thus represent one of the most important sites for what has been described as "film-induced tourism" (Beeton 2016) or "mediatized tourism" (Månsson 2011), where traveling to a location one is familiar with from a film or other media product "represents a realization of an earlier imaginary journey" (Reijnders 2016, 673; see Fig. 1).

More recently, this phenomenon has been described as "fan tourism" (Mittermeier 2019) or "transmedia tourism" (Garner 2019). This also makes theme parks part of what Henry Jenkins has famously termed "convergence culture" or, to be more precise, of the underlying concept of "participatory culture" (Jenkins 2006). Participation here not only means emotional, affective engagement, e.g. through wearing cosplay to the parks (Hogwarts wizarding robes and other apparel can be bought at the shops in "The Wizarding World"), but also by literally "taking part" in such experiences as Ollivander's wand shop. Wandmaker Ollivander picks a visitor—or rather, lets the wand pick the visitor/wizard, as per Harry Potter lore; and by purchasing

Fig. 1 Fans of the British preschool animated television series *Peppa Pig* connecting with their idols at Paultons Park in New Forest (UK). *Photograph* Salvador Anton Clavé

this wand, visitors can quite literally buy their way into the story world (Mittermeier 2020). The interactive wands have not yet been completely integrated into the theme park experience, although they also function as cos- or even roleplay as they can be used to perform spells throughout "The Wizarding World" via infrared sensors (Mittermeier 2020; see ATTRACTIONS).

In such cases, the theme park space primarily becomes one of fannish activities, something Williams has termed "haptic fandom" (2020, 12) as it is fandom tied to a specific space and is "also concerned with the physical and sensory experiences of the theme park space and how fans themselves accord value and meaning to the immersion of 'being there'" (2020, 12–13). Theme parks may thus function in similar ways to fan conventions—something Universal has also made active use of with its A Celebration of Harry Potter fan event, organized yearly at the Orlando park from 2014 to 2018. At the same time, fans have also taken to the parks themselves, for instance, to honor the Harry Potter-films' actor Alan Rickman (who portrayed Severus Snape) after he passed away in 2016, raising their wands in front of Hogwarts castle at Universal's Islands of Adventure (Orlando, Florida), as well as leaving a lily in front of the door to his character's classroom (Williams 2020, 175). Such activities illustrate the importance of an emotional attachment to the theme park space many fans have built.

5 Visitors as Performers

It is not only fans who engage and interact with the theme park, but performers and visitors, too, engage in at least two forms of "audience labor": first, what has variously been referred to as "experience work, "fun work," "anticipatory labor," or "plandom" and second, performances for other visitors (see LABOR). Firstly, in theme parks, the visitors' experience and performance oscillate between two extremes: on the one hand, the freedom to choose what to do and which attraction to experience, and on the other, the specific offer of the commercial mixed-use center or the so-called attractions mix (see ATTRACTIONS). The visitors' role is thus generally situated between the pole of the purely passive consumer and that of the active designer of their own experience. These two opposing positions can also be found in academic literature: in addition to those already mentioned, Miodrag Mitrasinovic refers to the interplay of theming as a means of control (2006, 36), while Judith A. Adams notes: "Everything about the park, including the behaviour of the 'guests,' is engineered" (1991, 97). While acknowledging the theme park's "seductive, controlling character" (Legnaro 2000, 296; our translation), other scholars nevertheless maintain that visitors can and do play an active role. Aldo Legnaro and H. Jürgen Kagelmann have thus developed the concepts of "Erlebnisarbeit" ("experience work"; Legnaro 2000, 293) and "Spaß-Arbeit" ("fun work"; Kagelmann 2004, 175), respectively. Focusing on theme park trip preparations, Williams has coined the expressions "anticipatory labor" and "plandom" (2020, 67–99). All these concepts describe a more active part of the visitor in the theme park—a role through which visitors ultimately become "producers and consumers in one person" (Legnaro 2000, 293; our translation). Hence, the controlling character of the theme park and its pre-structuring of the visitor's experience on the one hand, and the activity of the visitor, "the visitor's own actions, willingness to experience and desires" (Legnaro 2000, 293; our translation) on the other, are interdependent and intertwined, forming extreme poles on a scale. Even though the theme park experience may be scripted and even though it may vary little between visitors, the experience still has to be performed by the visitors to become a lived experience.

Secondly, once visitors have decided on a specific experience, the park's paratexts, its spatial and thematic arrangement, and the parts played by other theme park actors (visitors as well as employees) all suggest or invite them to play specific roles, which they may accept or refuse (see PARATEXTS AND RECEPTION and SPACE). Perhaps one of the most interesting among these roles is that of the tourist. This particularly occurs in so-called "meta-touristic" theme parks such as Europa-Park, Erlebniszoo Hannover (Hanover, Germany), or Epcot's "World Showcase" (Orlando, Florida; see Freitag 2018, 167), whose individual areas are themed to tourist destinations and where tourists to the theme park thus play at being tourists to other places. Through the role assignment as a tourist, it becomes easier to accept the large crowds that usually populate themed environments, as this is precisely what one would expect at a popular tourist spot. Perfectly blending into the environment, visitors to the theme parks thus double as performers, playing visitors to the tourist spot, and thus reinforce the parks' theming. But even without this roleplay, it is important to mention that the presence of other people changes the perception of the surroundings and space for the individual visitor—just imagine a nearly empty theme park in the morning or during the extra hours for hotel guests and compare it with a crowded day during summer holidays (see also SPACE).

Other roles that visitors frequently play are those of photographer and model. At theme parks, visitors often take pictures of each other and strike poses for the photographer as well as for the other people in the park. Taking a picture of someone changes the attitude and appearance of the person whose picture is being taken, for both the photographer and bystanders (Schwarz 2017, 108), but the act of photographing may also change visitors' perception of the theme park itself, making certain spots more interesting than others, and fixing them on material support—previously film, now digital supports—for future memories and display to friends and family as well as followers on social media. Of course, people taking pictures—of each other as well as the landscape—particularly fit the environment in "meta-touristic" theme parks. At Europa-Park, for instance, the "Griechenland" ("Greece") area is themed to the popular tourist destination of Mykonos (Carlà-Uhink 2020, 47–50), and visitors taking pictures of their friends and the surroundings may appear to be tourists on the island rather than visitors to Europa-Park.

Visitors also pose and perform for so-called on-ride pictures, often placed at certain key points of attractions where visitors scream and raise their arms. Here, the people on the ride not only express their joy and excitement to themselves, but also perform for the camera and the other visitors. Raising one's arms on a roller coaster is, on the one hand, something of a cultural trope that visitors may have adopted from other visitors or from visual representations of theme parks and their visitors, e.g. in advertisements (see PARATEXTS AND RECEPTION). And since visitors usually respond to the same parts of the ride, their reactions may be said to follow a certain dramaturgy that is synched to the ride's loops, curves, and screws. The screams of the visitors are sometimes even used as soundtracks for paratexts (Freitag and Schwarz 2015/16). On the other hand, the act of raising one's arms and screaming constitutes the result of an individual's decision to display their bravery, joy, or excitement. Hence, the visitor's "performance" on the roller coaster oscillates between a cultural, scripted path and an individual expression; it can thus be seen as one example of a performance that may be suggested by the layout of the attraction but only comes to life when visitors actively play their part.

Finally, pictures are also taken during encounters between visitors and the so-called characters (see again Fig. 1). These "meets and greets," which are particularly popular at Disney parks, constitute one type of encounter between professional employees and theme park visitors in which both take part in a form of roleplay or performance. A case in point is the Princess Pavilion at Disneyland Paris, where visitors meet a Disney princess from an animated film for a short chat, get her autograph, and take a picture with her in front of a special backdrop. During this encounter, visitors have to react to the performative offer made by the princess, which leads to different interactions, especially as most of the visitors are little girls. The princess sometimes suggests a certain pose for them and uses special phrases, facial expressions, and gestures that correspond to her character. Rebecca Williams has fittingly discussed these character meet and greets at Disney or Universal parks as encounters of "metonymic celebrity" (2020, 133–152).

Hence, in these situations—as well as during encounters between theme park visitors and employees in general—employees whose job it is to perform the generic role of the ever smiling, ever courteous, and helpful theme park employee, meet visitors who are encouraged to play the equally generic part of the "perfect" theme park guest. In practice, of course, the performances of both visitors and employees are frequently impacted by the extreme conditions of the theme park visit: the pressure generated by "fun work" and the stress engendered by "performative labor" and "deep acting" under the eyes of masses of fellow visitors and employees, the constant sensory overload, etc. (see LABOR). Children, in particular, will become easily tired, feel shy in front of characters, and may thus "fail" to perform their part in the theme park spectacle. Even in the case of such "glitches"—especially in these cases, in fact—employees are urged to stay in character.

This also applies to visitors who deliberately "fail" to play their part, e.g. those who show disturbing behavior or ask annoying questions. In fact, even if visitors' actions and reactions to the theme park tend to be foreseen and scripted by its layout, there is always the possibility of visitors evading these guidelines on purpose. For instance, a visitor may spend their entire visit just sitting on a bench and reading a book, thus subversively undermining the park's performative offers, or they may openly break the park's rules. At Disney parks, for example, adults may not enter the park in costume. Performance artist Pilvi Takala has based an artistic intervention exactly on these regulations. In her “The Real Snow White,” Takala tried to enter Disneyland Paris dressed as Snow White. When security stopped her at the entrance, she engaged the employee in a dadaistic dialog about the existence of the real Snow White, thus revealing the complex structure of real and fake in the context of theme parks (Schwarz 2016; see AUTHENTICITY). Transcending the visitor roles pre-scripted by paratexts, spatial layouts, and thematic choices, as well as encounters with other theme park actors, visitors have thus found multiple different ways of engaging with the theme park space.

6 Visitors as Bodies

Performance always and necessarily involves the body. What distinguishes out-of-home entertainment attractions such as theme parks from their in-home competitors are the manifold and unique bodily sensations that they offer. Of course, theme park visitors create bodily experiences everywhere in the park, with the precise nature of these experiences depending on the individual conditions of the site (see SPACE). Yet it is probably in such comparatively “extreme” situations as riding a roller coaster or some other thrill ride (see ATTRACTIONS and IMMERSION) that theme park visitors most conspicuously experience themselves as bodies. In such rides, the visitors’ bodies are confronted with forces up to 6.3 G—a force bigger than that which astronauts have to deal with during the launch of a spaceship (even though roller coasters expose visitors to these forces for a much shorter amount of time). In general, the layout of a roller coaster can be described as a sequence of plays with the forces—particularly G-Forces—that affect the body.

Silver Star and Blue Fire, to name two examples from Europa-Park, stage the forces working on the body in different ways: whereas Silver Star has a classic “up-and-down” ride dramaturgy in which visitors’ bodies alternately experience positive and negative G-forces, Blue Fire also includes loops and screws. And while Silver Star starts with a 73-m-high lift hill, Blue Fire begins with a catapult launch from 0 to 100 km/h, which is directly followed by the first loop. On both rides, visitors experience moments of weightlessness, but on Blue Fire, these are combined with a sort of spatial disorientation due to the loops and screws. These special types of inversion elements are designed so that the tracks wind around an imaginary “heart line,” and abrupt lateral movements are thus reduced to a minimum and visitors’ heads move along a clearly defined line. The rapid changes in direction put a lot of strain on visitors’ sense of balance and are also difficult to reconcile with the movement patterns learned by the body. Centrifugal force acts on the body, pushing visitors outwards in the curves. Due to the rapid succession of movements, this force has a double effect by pushing the visitor first in one direction and then in the other.

For safety reasons, and specifically due to Blue Fire’s overhead elements, visitors are secured by over-the-shoulder safety bars, and people with physical disabilities, as well as children under the age of seven and shorter than 1.30 m, are not allowed to ride. At the entrance of a roller coaster ride, visitors usually find a benchmark and sometimes even a test seat to find out if they physically fit into the coaster's ride vehicles. At the same time, most theme parks also feature rides for small children from which adults are excluded, for example, the Bumper Klumpen bumper car ride in Phantasialand, which is only accessible to people between 1 and 1.45 m tall. Likewise, most rides exclude pregnant women, visitors with certain illnesses, be they psychological or physical, and people under the influence of drugs or alcohol. While theme parks may advertise themselves as places for everyone, these examples of exclusion due to safety reasons show that not all experiences are available to everybody (see INCLUSION AND EXCLUSION). Theme parks try to compensate for this by offering a variety of different rides and attractions (an "attractions mix") that allows visitors to choose from different rides according to their age, their bodies, their physical conditions, and of course their personal preferences (see ATTRACTIONS).

Due to these choices and exclusions, the experience of a theme park, including the bodily experience, may strongly differ from visitor to visitor. For instance, riding a fast roller coaster like Silver Star or Blue Fire will result in an experience that is quite different from that of going on the smaller, calmer rides, or indeed that of completely ignoring the rides. In addition, thrill rides will make even slightly calmer rides seem much more relaxed as, similar to hot and cold water, speed accelerations work by contrast. Finally, even once visitors leave a particular ride, the latter's effects on the body do not immediately cease, as riding a roller coaster might lead to vertigo or motion sickness for a long time after the ride itself is over. Consequently, thrill seekers, the so-called "screamers," may experience the theme park as a place of excitement, whereas for "dreamers" it might seem like a relatively relaxed place (see ATTRACTIONS).

But there are also facets of the theme park experience that are involuntary, or rather independent from visitors' planning and desires. Although theme parks are supposed to be (and advertise themselves as) fun places, negative feelings like frustration or boredom can, in fact, often be found in the "happiest places on earth." The concept of "fun work," according to which visitors play an active part in creating their theme park experience (see above), already evokes negative feelings like stress or exhaustion: plans may not work out, something may go wrong, or visitors may simply need to go to the bathroom. As a trip to a theme park usually lasts several hours and visitors are exposed to different speed variations or body sensations, it is not surprising that they get tired or stressed—and yet, theme park marketing usually excludes these factors and instead seeks to evoke images of endless fun. But of course, anyone who has visited a theme park has seen (or heard) children crying, perhaps because they have been excluded from a ride due to safety rules. Note that these obviously unhappy children merely openly express, or perform, what many adult visitors around them may also experience. Paradoxically, however, since theme parks try to ban such negative feelings from their premises, it is the crying children who appear as out of place. Visiting a theme park not only entails exposing your body to different experiences, but it often also means deliberately controlling and censoring your body, or playing a role.

7 Conclusion

As noted in the introduction to this chapter, visitors have become the focus of recent academic inquiry into theme parks—and rightly so. Visitors' relationships with these spaces, with their employees, and with other visitors are multi-layered and complicated, and all of these relationships deserve more detailed attention. The bodily experience of visitors riding attractions and otherwise interacting and existing in these spaces is often seen in terms of categories that are too broad or simplistic, as visitors are mainly perceived as demographics to be targeted; however, the perspectives on visitors offered by sociology, theater and performance studies, fan and audience studies, or even biology may prove to be as fruitful as purely economic perspectives. Moreover, not only have studies of visitors so far mostly focused on Disney and Comcast/Universal (or more generally on Western theme parks), but they have also far too often presupposed white, able-bodied, heterosexual, and cisgender people of certain body types as visitors (see INCLUSION AND EXCLUSION). Visitors deserve more in-depth and nuanced studies, however, as in many ways they constitute the central group of theme park actors and thus the most important part of theme parks.

References

Adams, Judith A. 1991. *The American Amusement Park Industry: A History of Technology and Thrills*. Boston: Twayne.

Baudrillard, Jean. 1983. *Simulations*. Los Angeles: Semiotext(e).

Beeton, Sue. 2016. *Film Induced Tourism*, 2nd ed. Bristol: Channel View.

Birenboim, Amit, Salvador Anton Clavé, Antonio Paolo Russo, and Noam Shoval. 2013. Temporal Activity Patterns of Theme Park Visitors. *Tourism Geographies* 15 (4): 601–619.

Carlà-Uhink, Filippo. 2020. *Representations of Classical Greece in Theme Parks*. London: Bloomsbury.

Choi, Kimburley. 2007. *Remade in Hong Kong: How Hong Kong People Use Hong Kong Disneyland*. Diss., Lingnan University.

Clément, Thibaut. 2012. "Locus of Control": A Selective Review of Disney Theme Parks. *In Media: The French Journal of Media and Media Representations in the English-Speaking World*. https://inmedia.revues.org/463. Accessed 28 Aug 2022.

Clément, Thibaut. 2016. *Plus vrais que nature: Les parcs Disney, ou l'usage de la fiction de l'espace et le paysage*. Paris: Presses de la Sorbonne Nouvelle.

Danesi, Marcel. 2013. Hypodermic Needle Theory. In *Encyclopedia of Media and Communication*, ed. Marcel Danesi, 343–344. Toronto: University of Toronto Press.

Eco, Umberto. 1986 [1975]. Travels in Hyperreality. In *Travels in Hyperreality: Essays*, trans. William Weaver, 1–58. San Diego: Harcourt Brace Janovich.

Europa-Park. 2019. *Europapark für Genießer: Sommersaison 2019. Mit tollen Angeboten für alle ab 60 Jahren*. Rust: Europa-Park.

Fichtner, Uwe. 2000. Freizeit- und Erlebnisparks. In *Nationalatlas Bundesrepublik Deutschland. Vol. 10: Tourismus und Freizeit*, ed. Christoph Becker and Hubert Job, 80–83. Leipzig: Institut für Länderkunde Leipzig.

Fjellman, Stephen M. 1992. *Vinyl Leaves: Walt Disney World and America*. Boulder: Westview.

Freitag, Florian. 2018. "Who Really Lives There?": (Meta-)Tourism and the Canada Pavilion at Epcot. In *Gained Ground: Perspectives on Canadian and Comparative North American Studies*, ed. Eva Gruber und Caroline Rosenthal, 161–178. Rochester: Camden House.

Freitag, Florian. 2021. *Popular New Orleans: The Crescent City in Periodicals, Theme Parks, and Opera, 1875–2015*. New York: Routledge.

Freitag, Florian, and Ariane Schwarz. 2015/16. Thresholds of Fun and Fear: Borders and Liminal Experiences in Theme Parks. *OAA Perspectives* 23(4): 22–23.

Garner, Ross. 2019. Transmedia Tourism. *JOMEC Journal* 14: 1–10.

Goronzy, Frederic. 2006. *Spiel und Geschichten in Erlebniswelten: Ein theoriegeleiteter Ansatz und eine empirische Untersuchung zur Angebotsgestaltung von Freizeitparks*. Berlin: LIT.

Hofer, Matthias, and Werner Wirth. 2008. Präsenzerleben: Eine medienpsychologische Modellierung. *Montage A/V* 17 (2): 159–175.

Jenkins, Henry. 2006. *Convergence Culture: Where Old and New Media Collide*. New York: New York University Press.

Kagelmann, H. Jürgen. 2004. Themenparks. In *ErlebnisWelten: Zum Erlebnisboom in der Postmoderne*, ed. H. Jürgen Kagelmann, Reinhard Bachleitner, and Max Rieder, 160–180. Munich: Profil.

Knight, Cher Krause. 2014. *Power and Paradise in Walt Disney's World*. Gainesville: University Press of Florida.

Kokai, Jennifer A., and Tom Robson, eds. 2019. *Performance and the Disney Theme Park Experience: The Tourist as Actor*. Cham: Palgrave Macmillan.

Kuenz, Jane. 1995a. It's a Small World After All. In *Inside the Mouse: Work and Play at Disney World*, ed. Jane Kuenz, Susan Willis, Shelton Waldrep, and Stanley Fish, 54–78. Durham: Duke University Press.

Kuenz, Jane. 1995b. Working at the Rat. In *Inside the Mouse: Work and Play at Disney World*, ed. Jane Kuenz, Susan Willis, Shelton Waldrep, and Stanley Fish, 111–162. Durham: Duke University Press.

Kuenz, Jane, Susan Willis, Shelton Waldrep, and Stanley Fish, eds. 1995. *Inside the Mouse: Work and Play at Disney World*. Duke University Press.

Legnaro, Aldo. 2000. Subjektivität im Zeitalter ihrer simulativen Reproduzierbarkeit: Das Beispiel des Disney-Kontinents. In *Gouvernementalität der Gegenwart*, ed. Ulrich Bröckling, Susanne Krasman, and Thomas Lemke, 286–314. Frankfurt: Suhrkamp.

Lukas, Scott A. 2007. How the Theme Park Gets Its Power: Lived Theming, Social Control, and the Themed Worker Self. In *The Themed Space: Locating Culture, Nation, and Self*, ed. Scott A. Lukas, 183–206. Lanham: Lexington.

Lukas, Scott A. 2016. Research in Themed and Immersive Spaces: At the Threshold of Identity. In *A Reader in Themed and Immersive Spaces*, ed. Scott A. Lukas, 159–169. Pittsburgh: ETC.

Luke, Helen M. 1990. The Stranger Within. *Parabola* 15 (4): 17–23.

Månsson, Maria. 2011. Mediatized Tourism. *Annals of Tourism Research* 38: 1634–1652.

Marin, Louis. 1973. Dégénérescence Utopique: Disneyland. In *Utopiques: Jeux d'espace*, 297–324. Paris: Editions de Minuit.

Mitrasinovic, Miodrag. 2006. *Total Landscape, Theme Parks, Public Space*. Burlington: Ashgate.

Mittermeier, Sabrina. 2019. (Un)Conventional Voyages? Star Trek: The Cruise and the Themed Cruise Experience. *The Journal of Popular Culture* 52 (6): 1372–1386.

Mittermeier, Sabrina. 2020. Theme Parks: Where Media and Tourism Converge. In *The Routledge Companion to Media and Tourism*, ed. Maria Månsson, Annæ Buchmann, Cecilia Cassinger, and Lena Eskilsson, 27–34. London: Routledge.

Mittermeier, Sabrina. 2021. *A Cultural History of the Disneyland Theme Parks: Middle Class Kingdoms*. Chicago: Intellect.

Moore, Alexander. 1980. Walt Disney World: Bounded Ritual Space and the Playful Pilgrimage Center. *Anthropological Quarterly* 53 (4): 207–218.

Paine, Crispin. 2019. *Gods and Rollercoasters: Religion in Theme Parks Worldwide*. London: Bloomsbury.

Pettigrew, Simone. 2011. Hearts and Minds: Children's Experiences of Disney World. *Consumption Markets & Culture* 14 (2): 145–161.

Raz, Aviad. 1999. *Riding the Black Ship: Japan and Tokyo Disneyland*. Cambridge: Harvard University Press.

Reijnders, Stijn. 2016. Stories That Move: Fiction, Imagination, Tourism. *European Journal of Cultural Studies* 19: 672–689.

Rutherford, Stephanie. 2011. *Governing the Wild: Ecotours of Power*. Minneapolis: University of Minnesota Press.

Scheer-Schätzler, Brigitte. 1993. Die realisierte Utopie: Disneys EPCOT als "Arcadia in Florida." Versuch einer ideengeschichtlichen Interpretation. *AAA: Arbeiten aus Anglistik und Amerikanistik* 18 (1): 53–74.

Schultz, John. 1988. The Fabulous Presumption of Disney World: Magic Kingdom in the Wilderness. *The Georgia Review* 42 (2): 275–312.

Schwarz, Ariane. 2016. Spieglein, Spieglein an der Wand: Wer ist das echte Schneewittchen im Land? In *Where the Magic Happens: Bildung nach der Entgrenzung der Künste*, ed. Torsten Meyer, Julia Dick, Peter Moormann, and Julia Ziegenbein, 149–152. Munich: Kopaed.

Schwarz, Ariane. 2017. Staging the Gaze: The Water Coaster Poseidon as an Example of Staging Strategies in Theme Parks. In *Time and Temporality in Theme Parks*, ed. Filippo Carlà-Uhink, Florian Freitag, Sabrina Mittermeier, and Ariane Schwarz, 97–112. Hanover: Werhahn.

Six Flags AstroWorld. 1994. *Host and Hostess Handbook 1994*. Houston: Six Flags AstroWorld.

Statista. 2019. *Statista Dossier Vergnügungs-, Freizeit- und Naturparks*. Hamburg: Statista.

Steinecke, Albrecht. 2011. *Tourismus*. Braunschweig: Westermann.

Steiner, Michael. 1998. Frontierland as Tomorrowland: Walt Disney and the Architectural Packaging of the Mythic West. *Montana: The Magazine of Western History* 48 (1): 2–17.

Steinkrüger, Jan-Erik. 2013. *Thematisierte Welten: Über Darstellungspraxen in Zoologischen Gärten und Vergnügungsparks*. Bielefeld: Transcript.

Stiftung für Zukunftsfragen. 2013. *Besucherstruktur in Freizeitparks in Deutschland 2013*. https://de.statista.com/statistik/daten/studie/261498/umfrage/besucherstruktur-von-freizeitparks-in-deutschland/. Accessed 28 Aug 2022.

Thuy, Peter, and Helmut Wachowiak. 2008. *Gutachten über die Beschäftigungseffekte einer Erweiterung des Phantasialands*. Bad Honnef: Schmidt-Löffelhardt GmbH & Co. KG.

Valdani, Enrico, and Paolo Guenzi. 2001. Marketing von Brand Parks. In *IndustrieErlebnisWelten: Vom Standort zur Destination*, ed. Hans Hinterhuber, Harald Pechlaner, and Kurt Matzler, 153–196. Berlin: Erich Schmidt.

Van Maanen, John. 1991. The Smile Factory: Working at Disneyland. In *Reframing Organizational Culture*, ed. Peter J. Frost, Larry F. Moore, Meryl Reis Louis, Craig C. Lundberg, and Joanne Martin, 58–77. Newbury Park: Sage.

Wasko, Janet. 2001. *Understanding Disney: The Manufacture of Fantasy*. Cambridge: Polity.

Wasko, Janet. 2020. *Understanding Disney: The Manufacture of Fantasy*, 2nd ed. Cambridge: Polity.

Waysdorf, Abby, and Stijn Reijnders. 2018. Immersion, Authenticity and the Theme Park as Social Space: Experiencing the Wizarding World of Harry Potter. *International Journal of Cultural Studies* 21 (2): 173–188.

Williams, Rebecca. 2020. *Theme Park Fandom: Spatial Transmedia, Materiality and Participatory Culture*. Amsterdam: Amsterdam University Press.

Willis, Susan. 1995. The Problem with Pleasure. In *Inside the Mouse: Work and Play at Disney World*, ed. Jane Kuenz, Susan Willis, Shelton Waldrep, and Stanley Fish, 1–11. Durham: Duke University Press.

Zukin, Sharon. 1991. *Landscapes of Power: From Detroit to Disney World*. Berkeley: University of California Press.

Worldviews and Ideologies: Nationalism, Regionalism, Capitalism, Religion, and Other *Weltanschauungen* in Theme Parks

Abstract The premise of this chapter is that theme parks always reflect, represent, and even implement a certain series of values and worldviews, both on the representational level (reflected in the choice of themes and how they are depicted) as well as on the practical level (through rules of behavior inside the park). A wide range of overlapping and intersecting worldviews and ideologies can be found in theme parks, but while some are consciously implemented as the backbone of the theming, others are merely accidental or a matter of perspective, deriving from individual park politics: consumer demands, company branding, funding, legal requirements, or state politics. In many cases, the dissemination of a certain worldview goes beyond the immersive experience of the park, through cooperation with schools, the publication of learning material, merchandise, and cross-references in other media. This chapter offers a discussion of the most common worldviews in theme park theming and conduct: cultural imperialism, nationalism, regionalism, capitalism, religion, and environmentalism. Finally, theme parks are shaped not only by certain perspectives, but as media they also create a unique frame for viewing the world. The final section of this chapter consequently deals with theme park worldviews in the outside world.

1 Introduction

In 2017, the German late-night show *Neo Magazin Royale* published a 45-min investigative report on a new theme park project in Germany that would allow visitors to experience life in the Third Reich. At the beginning of the broadcast, stunned reporters watch the TV trailer for the "Reichspark" project:

> "Reichspark: Living History" comprises four themed adventure areas, called "Countries," that visitors are free to roam and explore: Rememberland, Warland, Historyland, and Alt-Berlin. Don't fancy being burdened with the controversy of Germany's troubled past? No

This work is contributed by Céline Molter, Filippo Carlà-Uhink, Salvador Anton Clavé, Crispin Paine, Astrid Böger, Thibaut Clément, Florian Freitag, Scott A. Lukas, Sabrina Mittermeier, Ariane Schwarz, Jean-François Staszak, Jan-Erik Steinkrüger, Torsten Widmann. The corresponding author is Céline Molter, Institut für Ethnologie und Afrikastudien, Johannes Gutenberg-Universität Mainz, Mainz, Germany.

F. Freitag et al., *Key Concepts in Theme Park Studies*,
https://doi.org/10.1007/978-3-031-11132-7_17

> problem! Come to "Rememberland" instead! As well as a romantic Black Forest open-air museum and a life-sized replica of central Königsberg, Old Prussia's picturesque former capital destroyed in WWII, "Rememberland" lets you stroll through a historic Dresden, still completely unscathed by the Allied bombing terror of 1945, in a larger-than-life 2:1 scale! But watch out! If you don't want to be accused of ignorance later, then be sure to stay until dawn. When the sun sets, make sure you don't miss out on the daily fire spectacle of roaring Reichspark Reichskristallnacht, of course reenacted in a historically faithful, responsible, and family-friendly performance. (Arbeiter and Teitge 2017; our translation)

After a while, it becomes clear that the park project is a fake, invented by the TV show, but another twist in the story comes when the fake investor promotes his fake project at the real Europe Attractions Show. There, potential suppliers praise the idea and are only too willing to agree to produce "extremely realistic" toy weapons and explosives for the war- and concentration camp-themed attractions, e.g. "the battle of Stalingrad as a VR-enhanced dark ride with authentic temperatures of −20 °C" (Arbeiter and Teitge 2017; our translation).

Yet the goal of the report was not to criticize the theme park industry, but to raise questions about the ethics of theming and the politics of remembering, including: what happens if the past is rendered "family-friendly"? Are there moral limits to what can be turned into a theme? And are there mechanisms of regulation, or are investors free to turn their worldview into a consumer attraction? Some of these questions have been addressed in theme park, heritage, and tourism studies in the context of dark or thanatourism (Lennon and Foley 2000; Stone 2006) and dark theming (Lukas 2016; see THEMING). These studies mostly focus on enjoyable representations of death and catastrophe, or dark spots in the bright theme park world. Questions about processes of "whitewashing" history and deleting its more controversial aspects, such as war, poverty, slavery, and oppression, have, in turn, crystallized around the terms "Distory" (Fjellman 1992, 59–63) and "Mickey Mouse history" (Wallace 1985), and have been raised most prominently during the 1990s in connection with park projects such as Disney's America (Mittermeier 2017; Carlà-Uhink 2020, 14–15).

This chapter will take these considerations as a starting point but will go on to argue that even lighter and/or seemingly neutral themes are shaped by a more or less ideological framing. "Reichspark" is a satirical perversion of the fact that theme parks always reflect, represent, and even implement a certain set of values and worldviews as part of their creation, no matter whether they are state-owned or privately operated, or whether they are run by a single person, a family, or a religious or ideological group. The spectrum ranges from showing sections of the world from a certain perspective to being built with the intention to foster a certain ideology. Examples of the former include Phantasialand (Brühl, Germany), which has a themed area of "Old Berlin" next to a quite indifferent "China" and an even more unspecific "Deep in Africa" section, and Global Village Dubai (UAE), which has a pavilion for almost every country of the Arab world, whereas Europe is represented by pavilions for Turkey, Russia, western Europe, and the Balkans, and the Americas by only one pavilion (see Fig. 1). The intentional projection of a clearly identifiable ideology can be observed, for example, in Ark Encounter in Kentucky, which immerses visitors into the worldview of Young Earth Creationism.

Fig. 1 View from Global Village Dubai (UAE). *Photograph* Salvador Anton Clavé

Theme park scholars need to be aware of this kind of underpinnings, especially when they themselves find them highly familiar, as they can conceal a political agenda. For example, the great temple-cum-theme park Akshardham in New Delhi, India (on which more below) shows evidence of an underlying Hindutva worldview. The Cultural Boat Ride through dioramas of life in ancient Vedic India presents a blissful era in which scientific and technological advances such as space travel and modern medicine had already been achieved. The boat ride ends with a paean to the future of humankind, with Swaminarayan Hinduism subtly elided. The complex also includes a sculpture park of figures of Indian heroes, freedom fighters, and great rulers—all higher-caste Hindus. It seems unlikely that many of the park's five million visitors notice these things, yet the visit can reinforce their already existing prejudices.

In the following, we use the term "worldview" to refer to the (individual or collective) positioning of the self and others in the world. A wide range of overlapping worldview types can be found in theme parks—some of which are intentional and implemented as the backbone of the theming, while others are rather unintentional, deriving from individual park politics: consumer demand, company branding, funding, legal requirements, and state politics. In contrast, we use "ideology" in the sense of shared beliefs about the world but acknowledge that "ideology is perhaps best seen as a field of struggle and negotiation between various social groups and classes, not as some worldview intrinsic to each of them" (Eagleton 2013, 13). The boundaries between worldview and ideology are fluid. Theme parks are often initiated by individuals who may be representatives of an ideology (Walt Disney/Disneyland: mainstream conservatism, Ken Ham/Ark Encounter: Creationism, Philippe de Villiers/Puy du Fou: right-wing Catholic conservatism), and who integrate their individual ideas within this ideology into the park. The following subchapters give examples of the worldviews that we find most prevalent in theme park theming: cultural imperialism, nationalism, regionalism, capitalism, religion, and environmentalism—as well as their intersections and overlaps.

Since theme parks are representations of the world and often use "real" places as blueprints for their themed lands, visitors experience representations of real and fictitious places whose order is based on a structural division of the earth. This includes power relations, which inform the representations of self and others in the form of (1) cultural imperialism. Theme parks further condense (2) nations, (3) regions, political entities, entire continents, or individual cities, while the underlying mechanisms of inclusion and exclusion always turn theme parks into perspectives *on*, rather than images *of* the world. For operators, this opens up the possibility to influence collective perceptions, and sometimes the medium is used for precisely this purpose. (4) Capitalism is inherent in the medium and its organizational structure, yet is probably not perceived by most visitors as a worldview—which is precisely why it is worth discussing it here. Just as (5) religion is part of the world, it is also a design element, sometimes even a basic motif, in theme parks. Architecturally and culturally, temples, sacred buildings, iconographies, nativity scenes, and theaters can be considered religious forerunners of the theme park medium (see ANTECEDENTS and VISITORS). Today, religious theme parks have been established as a genre in their own right (Paine 2019). They share with their secular counterparts and predecessors a notion of paradise: theme parks are modeled as perfect world imaginations, utopias in which nature is tamed, safety and happiness reign, and death, fear, and destruction are banned into the realm of illusion and entertainment. (6) Environmentalism, in the form of celebrated nature, is another ideology that guides theme park theming, a remnant of their joint history with zoos, aquariums, and botanical gardens.

In many cases, the dissemination of theme parks' worldviews goes beyond the immersive experience in the park, through cooperation with schools, the publication of learning material, merchandise, cross-references in other media, and so on. For instance, the aforementioned discussion about representations of history in theme parks is not only about who history "belongs" to and who has the right to produce the "correct history," but also a debate on forms of representation and their deployability for different aims, including education. The final paragraph of this chapter will deal with such processes of dissemination.

The list of concepts presented here is far from complete, and within theme parks, there are various further types of ideologies that theme park researchers should be aware of. Park ethics and ideologies attract certain visitor target groups. From a global perspective, the biggest group of visitors are families, and whether a theme park is "family-friendly" (enough) can be an issue of heated debate. Consequently, theme parks are tailored to represent mainstream conceptions of gender roles and family (Griffin 2000, 121; see also INCLUSION AND EXCLUSION). A theme park visit is mostly affordable and considered culturally appropriate for middle-class citizens, so it is usually middle-class ideals that are represented (Mittermeier 2021). Upper class exists in the form of aspirations (feeling like a princess at Disney) or condemnation (in the themed coaster Baron 1898 at the Dutch theme park Efteling, where patrons become exploited miners who work for a greedy rich baron). Lower classes, meanwhile, are instead allocated into the realms of the romanticized exotic Other or the dark and spooky (see also INCLUSION AND EXCLUSION). This middle-class ideology usually includes the valorization of democracy and human rights as dominant theme park imperatives. So far, a German "Reichspark" celebrating Third Reich ideology, as presented in the introduction to this chapter, is clearly—and hopefully—not realizable.

Another issue is the field of human-animal relations: theme parks may feature animals as mascots, display them as decorations, pets, and dangerous threats (e.g. in the Black Mamba roller coaster at Phantasialand), or serve them in their restaurants—usually all of the above simultaneously. Celebrating nature and selling mass-produced meat is not a matter of controversy in most theme parks (as in the outside world). Dinosaurs assume a special role among the animals: since the dinosaur fad was started by the Jurassic Park movie franchise, dinosaurs have been a theme park classic, the perfect animal: based on reality but with room for imagination, necessarily dead, and therefore, not threatened by comparison with living counterparts, and thrillingly dangerous. They allow visitors to slip into the role of adventurers and explorers, without the imperialist aftertaste of cultural conquest. Yet their implementation can still be ideological: the Ark Encounter theme park in Kentucky displays ancient humans living in harmony with dinosaurs to support the Creationist Young Earth theory.

And finally, when writing about worldviews and theme parks, there is the need to critically reflect on our own perspective as researchers from mostly Western countries who tend to talk about theme parks as Western phenomena, while many other origins for this medium can be found throughout the world (see ANTECEDENTS and METHODS).

2 Types of Worldviews in Theme Parks

2.1 Cultural Imperialism

Upon entering Phantasialand, visitors walk onto an elevated platform and enjoy a first look at the park stretching out beneath their feet. "Old Berlin," a colorful 1920s imagination of Germany, presents itself below the elevated guest, and other themed worlds can be spotted in the distance. If a theme park is structured into different sections of the world, the visit often starts at home: in Disneyland (Anaheim, California), "Main Street, U.S.A." welcomes patrons before they head to more distant and exotic places. With the purchase of the entrance ticket, visitors "own" this entire world for the day and may now set out to explore it.

Geographical imagination inspires not only theme park landscapes, but also their spatial organization. Theme parks are microcosms in which the macrocosm is reproduced, sometimes explicitly as in Epcot's "World Showcase" (Orlando, Florida) or Disney's It's a Small World attraction (Carlà-Uhink and Freitag 2022). In this sense, the theme park is a map (see PARATEXTS AND RECEPTION and SPACE). Europa-Park (Rust, Germany) thus maps a continent, with 15 distinct themed areas dedicated to 15 European countries, but also with "Adventureland" (which is more or less African), "Grimm's Enchanted Forest," and the "Minimoys Kingdom." Many theme parks map the world itself and do so by referring to major geographical divisions, most often continents. Such a division is recent and arbitrary, and often reflects Western worldviews and colonial projects. There is a race for each continent and a continent for each race, as well as an obvious hierarchy between them. Although naturalized and taken for granted, continents result from an ideological construction: they show and reproduce a racist and Western view of the World.

This is often very obvious in the way theme parks represent Africa as a black and backward continent, a place mainly for nature and wild animals, but also a place to be looked at by visitors, who are invited to reenact the European exploration of the "dark continent." Indiana Jones-type attractions exploit the narrative of the heroic white explorer confronted with the mysterious and dangerous jungles and ruins of primitive Africa, South America, or Asia. Owing to one of their predecessors (world's fairs; see ANTECEDENTS), Western theme parks usually depict a teleological narrative of development in line with (Western) technological innovation: the future is usually themed as Western, whereas Africa, Asia, the Arab world, and South America are represented in a mythical, (pre-)colonial past: Harambe Market in "Africa" at Disney's Animal Kingdom (Orlando, Florida) is "constructed around an old colonial-era train depot" (Walt Disney World N.D.), while Phantasialand's "Deep in Africa" is supposedly based on a traditional Dogon village (Steinkrüger 2013, 272).

Another stigmatizing geographical worldview often reproduced in theme parks is Western Orientalism (Said 1978), which appears in Thousand and One Nights-themed attraction or areas that often play on generic Orientalist stereotypes and could thus equally recall Morocco, Egypt, Turkey, or Persia. The stories from these regions, which are familiar to a Western audience and which are, therefore, frequently taken up in theme parks, often take place in a distant past. Thus, the image of the "Orient" in form of romanticized backwardness is reinforced and solidified in theme parks. In Tierra Santa (Buenos Aires, Argentina), which depicts the Holy Land during Jesus' lifetime (see Sanchez 2012), visitors can simultaneously see the crucifixion scene at Calvary and, on the horizon beyond the park walls, the skyline of Buenos Aires—a view that seems to stretch from the past into the future (see Fig. 2). Such attractions do not differ much from the Streets of Cairo on Coney Island, and other ethnographic shows or exotic villages that have been so successful in large international exhibitions since the end of the nineteenth century (see also THEMING).

These forms of exoticism provide a very rich and powerful source of inspiration for park designers, who may not intend to reproduce a colonial and racist ideology

Fig. 2 View from Calvary at Tierra Santa in Buenos Aires (Argentina). *Photograph* Céline Molter

but nevertheless perpetuate a geographical worldview that contrasts the West with the Rest. This becomes particularly evident in some theme parks located in former imperial powers: Efteling (Kaatsheuvel, Netherlands), for instance, offers a Flying Dutchman ride whose theming is dominated by the logo of the *Vereenigde Oost-indische Compagnie*, the Dutch East India Company, which substantially contributed to the so-called "Dutch Golden Age," often seen today as an age of slavery and exploitation. Being more than just a collection of attractions, successful theme parks offer the satisfying experience (or rather, illusion) of imperialism. The world, in a thoroughly accessible form, exists for the visitors' enjoyment—they may marvel at its cultures and technologies and stroll through its tamed, yet "adventurous" wilderness. Nowadays, there is a growing awareness of the problems of racial and cultural stereotyping in theming, and therefore, a trend toward fictional story worlds, such as Harry Potter, Star Wars, or Avatar (see MEDIA), and autotheming (Freitag 2016).

Nonetheless, Western cultural imperialism is deeply rooted in the theme park's organizational structures. By linking theme park (more precisely Disney) representations to colonial cultural exhibitions, Hom (2013) stresses that producing simulacra is an act of colonizing the object (and sometimes even the subject), owning and dominating it, and condemning it to immobility in time and space. Her notion of "simulated imperialism" expresses the ideology inherent in every theme park representation. The other (and/or self) is adopted with an imperialist self-evidence and offered to paying visitors who move through it: "at the end of the day, exhausted, we felt like we had mastered the park, domesticating its characters and conquering its every land. We were imperialists in a Disney empire" (Hom 2013, 26).

2.2 *Nationalism*

As spaces of representation, theme parks reference localities: continents, nations, regions, or fictional places that allow visitors to travel in time and space just by walking through them. And since recognizability is the key to successful theming (see THEMING), the average geographical knowledge of the target audience determines the choice of representational icons for a certain region. Nationalism is one of the most prevalent ideologies in theme parks. Sometimes a theme park is established deliberately to promote the love of one's country, and at times patriotism is an underlying theme that is not immediately apparent to the visitor. Theme park scholars need to examine the motives of those who set them up, but also investigate how specific themes resonate with the visitors' assumptions, pre-knowledge, and worldviews.

Even if a park has no explicitly nationalist agenda, it often plays with ideas of the nation-state, presenting condensed versions of other countries or snippets from the past or future of the nation they are based in. An entire theme park genre linked to nationalism is that of miniature worlds, which often represent a nation as a condensate of popular monuments, together with national symbols like hymns and flags. At Miniatürk, which opened in Istanbul in 2003, for example, visitors find reproductions of famous monuments from the entire Republic of Turkey; yet there is also a selection of monuments from other countries. Significantly, all of these are from parts of the former Ottoman Empire, whose heritage is clearly also celebrated here: visitors are thus reminded of the fact that Hungary, Greece, Bosnia and Herzegovina, Syria, and Egypt were once part of the "Turkish Empire." The connection between such miniature parks and nationalistic feelings is particularly obvious in contested areas. Most notably, Mini Israel, which opened in 2002 in the Latrun area, has been the target of acts of subversion and contestation: extremist Jewish visitors, for instance, have thrown stones at the miniatures of Muslim prayers at the Dome of the Rock, disrupting the "national utopia" that the miniature park—shaped in the form of the star of David—seeks to represent (Feige 2017).

Perhaps more commonly, "patriotic" theme parks do not attempt to impose a unitary view of the country, but rather celebrate "unity in diversity." Taman Mini Indonesia Indah ("Beautiful Indonesia Miniature Park") in East Jakarta was founded by Siti Hartinah, the wife of Indonesian dictator Suharto, in 1970, soon after she had visited Disneyland. The park aimed to reinforce Suharto's "New Order," established after the army had consolidated its power by massacring over half a million communists and others. Taman Mini still promotes national unity and patriotism by celebrating Indonesia's cultural diversity; each of the country's 33 provinces has its own space there. Such an approach can be found in many theme parks, especially in Asia, e.g. Splendid China (Shenzhen, China). Sometimes, theme park nationalism stretches out into larger political entities: Europa-Park celebrates the European community (but without giving importance to completeness); Portugal dos Pequenitos in Coimbra (Portugal) features depictions of iconic Portuguese monuments and places, but also of the country's former colonies, now (questionably) labeled as "Portuguese-speaking countries."

Whenever the park's intention is overt, however, understanding the role of ideology in its operation becomes straightforward. Shortly after Russia's 2014 annexation of Crimea, plans were made for a theme park that was to be realized by the French company of Puy du Fou (see below) in a joint venture with the Russian banker K.V. Malofeev. According to the website of the project (since deleted), the park would be "based on familiarization with the history of the Crimean Peninsula as an integral part of Russia" (Zavadski 2014). The park—whose opening had been scheduled for 2017—has yet to appear, however.

A comparable example from western Europe is Terra Mítica in Benidorm (Spain). Opened in 2000, the park was realized by a society partially owned by the autonomous Valencian Community. Therefore, it is hard to argue that a nationalistic agenda (as, for instance, in Mini Indonesia) was explicitly intended as the main content of the park. Nonetheless, by drawing on well-known and very common historical narratives in Spain and their recognizability to the public, the park can be described as supporting a strongly nationalistic ideology. At its opening, Terra Mítica was subdivided into five themed areas: ancient "Egypt," ancient "Greece," the "Islands" (the Aegean), "Rome," and "Ibéria." The loop structure of the park and its strategy of differentiated opening times for the different themed areas (and the spatial and temporal distribution of shows) led visitors to move through the individual areas in a very specific order, in which only the "Islands," located at the center, could be visited at any time (see TIME). Starting from "Egypt," where the entrance was located, visitors thus moved in time and space through "Greece" and "Rome" toward Spain. "Ibéria," themed to medieval and early modern Spain and hinting at national myths (such as El Cid) as well as colonial expansion, thus constituted the final point of the park's visit and somehow the final point of a teleological narrative that brought the history of civilization to a culmination.

All this was made even clearer by the shows, as the most important show Barbarroja, performed every day in the late afternoon (and almost never changed throughout the long and complex history of the park), took place in the biggest theater in the "Ibéria" section of Terra Mítica. The plot of the stunt show is rather blunt: a small Spanish village, inhabited by an Edenic couple (a boy and a girl), is attacked by Ottoman pirates; through many vicissitudes and fights, the pirates are defeated so that the Spanish couple can marry and live happily ever after. The representation of the Ottoman East as absolute Alterity, as is common in European Orientalism, is thus accompanied here by the idea that this absolute Alterity is defeated by a representation *in nuce* of the Spanish people. Hence, the structure of the park and the show combine to create a teleological perspective in which Spain not only represents the product of centuries of civilization and progress, but also the most important bulwark for its defense (Carlà and Freitag 2015; Carlà-Uhink 2020, 80–85).

2.3 Regionalism

Theme parks and themed areas can also appeal to a different form of identification and to a different set of recognizable stereotypes, which are connected to the specific place where the park is built. This happens when the entire park, or parts of it, celebrate the history and specificity of the region of the park itself, appealing to the "geopiety" (Tuan 1976) of the local visitors. What is represented within the park refers to what is directly outside of it—and is thus also recognizable to visitors from other areas and countries. This is probably the point where the principle of "externality" (see THEMING) encounters its most serious contradictions, as what is represented here has the function of presenting an encounter with the Same—albeit the Same that appears in a distilled, idealized, and utopic form, and is often represented as "fixed" at a specific moment in time that is recognized as the "golden age" of the region itself (Carlà 2016). Accordingly, nostalgia is a strong driver of theme park regionalism.

It is not necessary for a park to develop entire themed areas to display a regional agenda and a strong sense of geopiety; nor is this in contradiction with any other worldview discussed here, from religion to globalism to nationalism itself. Terra Mítica is a case in point: at the entrance to the park, each themed area is represented by a statue that iconically identifies its corresponding civilization in an "abstract" way (see Carlà and Freitag 2015, 137 and THEMING on the concept of "abstraction"). "Ibéria" is represented by the Dama Oferente del Cerro de los Santos, a famous ancient Iberian sculpture dating to the third century BCE and found near Albacete. While Albacete does not belong to the Valencian Community, it is not far from it, and not far from Benidorm. Elche, where the famous Dama de Elche is from, is even closer and would have provided an even better-known symbol, but the Dama de Elche would not be suited for the task as it is a bust rather than a complete standing figure. The choice of the statue, therefore, seems to be motivated by the desire to provide a well-known symbol of Iberian culture, comparable to the other figures, and by a choice made from the images responding to these criteria, i.e. those found closest to the park.

Six Flags Over Texas (Arlington, Texas) is a theme park entirely dedicated to the representation of local history for local inhabitants. Its historical content may have become less visible over the years, but it was rather dominant during the park's early years and is still clearly noticeable today. The name of the park and its six themed areas evoke the six powers which, at different historical moments, ruled over Texas: Spain, Mexico, France, Texas, the Confederacy, and the U.S. In this sense, the park is "nationalist" as well as "regionalist": it promotes a Eurocentric view of history, not rendering native peoples as an autonomous chapter in the history of Texas, instead representing them in a highly stereotypical manner that places them on the margins of the European colonists. In this sense, "a Lone Star as well as an American patriotism is evident. The park stresses 'independence' and progress on the one hand but subconsciously perpetuates the ethnicities of those who had power," and thus remains clearly steeped in a U.S. nationalist vision (Francaviglia 1995, 35).

HansaPark was opened in 1987 in Sierksdorf, Germany. The name refers to the historical formation of the Hanse, or Hanseatic League, a confederation of guilds and towns which stretched from the Baltic to the North Sea and controlled the Baltic Sea commerce during the Middle Ages—in fact, the park is not far from Lübeck, which was one of the most important member cities of the League. Within the park, some themed areas pick up on this regional inspiration: "Wikingerland" and "Die Reiche des Nordens" ("the Empires of the North"), for instance, draw on the connection between the Baltic Sea and Scandinavia; in a much more direct and evident way, "Hanse in Europa" celebrates the historical experience of the Hanseatic League. As with Disneyland's "Main Street, U.S.A.," "Hanse in Europa" is located at the entrance of the park, and all visitors must cross it to get in and out—in this sense, it is a crucial vector of the regionalist worldview held by the park.

References to regional identity and history are not restricted to small or medium parks, as they can also be found in parks belonging to major chains. Happy Valley Shanghai, for instance, has a themed area called "Shanghai Beach" which is unique to this park within the entire chain, and thus invites the local public to identify with their "place." Disney is not exempt from this kind of approach, either: Disney California Adventure in Anaheim, California (opened in 2001) clearly insists on its Californian location and on Californian history as a theme. "Buena Vista Street" thus constitutes an immersive depiction of Los Angeles during the 1920s, "Paradise Pier" (since 2018 "Pixar Pier") refers to a Californian boardwalk, "Pacific Wharf" to San Francisco's Fisherman's Wharf, etc. Here, the park's regionalism is not only, and not even mostly, addressed at Californians living in the region and visiting the park, but at visitors from the rest of the U.S. and from other countries, who can "visit California" while "visiting a theme park in California" (in a way, the opposite of metatourism; see THEMING). In this case, the mechanisms of recognition are activated less by geopiety than by the identification of what is stereotypically known by tourists about California.

Such forms of regionalism do not automatically imply the choice of a theme directly inspired by the history of the region: the connection might be more "subtle," barely obvious to visitors from other areas and yet very clear to locals. An example is the Oasys MiniHollywood Theme Park in Almería (Spain). The theme of the park is the American Wild West—and yet, the reason for this choice is the fact that in the 1960s, European Westerns were shot on location here, as the deserts around Almería provide suitable backdrops for the genre. The choice of an apparently exotic theme thus has a very strong regional background, and a very strong connection to local identity, local history, and local economic structures.

The development of regionalism in theme parks can take two directions: parks can integrate special features of the region into their theming, as in all the examples discussed until now, or regional assets can develop into a theme park. An example of the latter is Le Puy du Fou in the Vendée region in France: in 1977, right-wing conservative politician Philippe de Villiers staged a spectacle ("Cinéscénie") that depicted the family history of a fictitious fighter of the counter-revolution of 1793, with references to the history of rural society in the region. The successful show has since developed into a theme park, with historically inspired spectacles, hotels, and immersive adventure locations. Still run by volunteers and professionals from the area, the park celebrates the region's way of life: "The whole production provides the conditions for the collective creation of the Puyfolais (as the actors call themselves), the perfect embodiment of the native of the Vendée, combining loyalty to the values of the past with at least formal acceptance of modern society" (Martin and Suaud 1992; see also Anton Clavé 2007, 268–269). The productions of the park convey the conservative, Catholic values of its founder—and the park is so successful (in 2012 and 2014, it was awarded best theme park in the world by IAAPA) that it has a decisive influence on the public image of the entire region, as well as on its self-image.

2.4 Capitalism

Theme parks might (re-)present different parts of the world and times in history, and usually, this happens through "culture," i.e. architecture, arts, crafts, music, food, and artistic performances—but the overarching theme park culture can be described as Western consumer capitalism. Theme parks are indeed businesses, and they operate in a global capitalist economy (see INDUSTRY). Moreover, their ubiquitous representative, Disney, has become one of the most notorious symbols of global capitalism (Ariès 2002, 24). The temporal *communitas* (in a Turnerian sense; see Turner 1969) among theme park visitors is permeated by capitalist structures: shopping is imperative, as there is no theme park without gift shops, and every themed area usually comes with its own themed shops, restaurants, and consumer products. Theme parks are places of hybrid consumption (Bryman 2004), which aim to fulfill every imaginable need and make visitors stay and play as long as possible. And while shopping malls have discovered the power of theming as an incentive (Gottdiener 1997), theme parks recreate nostalgic shopping venues like "Main Street, U.S.A.," African craft shops, or oriental bazaars, as well as fully themed shopping districts like Disney Springs or Universal Boardwalk directly outside the theme parks. Patrons are customers who expect value for money (see VISITORS).

This framing may lead to moments of friction where it does not match the theming, as is the case with the soft drink vending machines in Noah's Ark at Ark Encounter. But it is not always that subtle: the Mexican theme park/edutainment franchise KidZania is all about the capitalist experience. KidZania is a complete "grown-up world" for kids, in which they can earn money (referred to as "KidZos") by trying different jobs and careers, and spend it, too, although they may also save, invest, or donate their earnings. KidZania currently has facilities in 21 countries worldwide, with more opening soon. By cooperating with local brands, the model adapts to the cultural environment and presents an ever-fitting vision of the perfect capitalist society (see Montalvo and Daspro 2018). Like KidZania, theme park management video games involve customers in ludic capitalism (Makai 2019, 165): in the *RollerCoaster Tycoon* games series, *Theme Park World*, or *Parkitect*, for instance, players learn how to run a financially successful theme park by manipulating space, controlling the visitor flow, and pleasing patrons' needs to make them stay and spend money for as long as possible (see PARATEXTS AND RECEPTION). While these games reveal the capitalist nature of theme parks to future visitors, they transmediate the ideology, offering bonus packs with themed items and binding customers to theme park companies by allowing them to recreate certain rides or even entire parks.

2.5 Religion

A changing world brings changing patterns of leisure, and also changing patterns of religion. The new global middle class has been looking for new ways to enjoy itself, and new ways to express its religion. In some parts of the world, particularly in the U.S., the theme park has been adopted as a new place where religious people can celebrate and perform their faith—evangelical Christians are joining religion with fun as Chaucer's pilgrims did, while in Asia, traditional temples are expanding and adapting their attractions to respectable families, so that they too become theme parks.

Thus, the Akshardham in New Delhi (see above) was opened in 2005 by the Swaminarayan sect of Hindus, after a team of sadhus had visited Disneyland and Universal Studios. Visitors can walk through a series of dioramas of Swaminarayan's life, in which animatronic figures act out his teachings; they can take the Cultural Boat Ride through a series of dioramas showing the idyllic life of Vedic India, watch the large-format film about "Mystic India," which tells the story of Swaminarayan's spiritual journeying through late eighteenth century India, from the icy Himalayas to the beaches of Kerala, visit the sculpture garden, playpark, Food Court, and shop, and in the evening watch an elaborate water show presenting a moral tale from the Upanishads. The center of the complex is the golden idol of Swaminarayan himself, in his huge white marble and pink sandstone temple with its 20,000 sculpted figures, standing on a base carved with 148 half-life-size elephants.

Probably the best-known "religious" theme park, or religion park, in the world was Holy Land Experience in Orlando, Florida (2001–2020). Run and financed by the evangelical Trinity Broadcasting Network (TBN), this park allowed visitors to experience the stories of the Bible through ever-changing musical and theatrical performances like "The Wilderness Tabernacle" or "Jesus at the Temple." The park was themed as a bricolage of Holy Land imageries, including reconstructions of the Jerusalem Temple and Golgotha Mountain and a miniature version of Jerusalem in Jesus' days. "The Scriptorium" hosted a collection of biblical artifacts, which could be visited in automated tours, complete with animatronics and light and effects. Like most Christian attractions in the U.S., Holy Land Experience was a crossover of infotainment and ministry. Over the course of its two-decade lifespan, changing management continuously transformed the park's message and appearance. In *Roadside Religion* (2005, 55), Beal observed a strongly supersessionist agenda in the park, with Christianity represented as the truer form of Judaism. In 2016, although ownership and management had changed over the years, supersessionism was still the core theme, with gift shops filled with Jewish paraphernalia such as menorahs and tallits (Molter 2018; see also Callahan 2010).

As with other worldviews, the presence of religion as an underlying issue in the theme park is not mutually exclusive to a regionalist or nationalist agenda. For example, just outside Ho Chi Minh City, in Vietnam, is the hugely popular Su´ôi Tiên ("Fairy Stream") theme park, often called a Buddhist theme park. In reality, its central theme is Vietnamese tradition and popular culture, and Buddhism and folk beliefs are prominent because they are central to both. Snuggling under the roller coaster and between the crocodile lake, the dolphinarium, 4D cinema, paddle boats, water park, cable car, ice world, and so on, are a number of substantial working temples, statues of the Buddha and of Avalokitesvara, and many other sites referencing Vietnamese traditions and beliefs.

This is indeed how religious worldviews most often arrive in theme parks—as an important feature of traditional culture, as parks themed to local or exotic foreign cultures both frequently feature religion as part of that culture. One example is the Thirteenth Century Theme Park, 100 km outside Ulaanbaatar, the capital of Mongolia, which celebrates the great Mongol leader Genghiz Khan. Among the groups of tents scattered across a large area of desert is the Shaman's Camp. Sheltered in a gully, the camp comprises six tents from different Mongolian clan traditions, in which some visitors leave offerings. At the center of the camp is a lightning-struck tree (seen in Mongolian Shamanism as particularly sacred), honored with blue scarves and protected by 365 outward-pointing sharpened stakes. This is periodically the focus of various shamanic rituals. More typical perhaps is Europa-Park, where a Roman Catholic and an Evangelical Protestant chaplain serve the park's five churches and hold weddings, christenings, and other services, welcoming youth groups and church groups in French, Italian, German, and English. Here, Christianity in some of its different forms is represented as being integral to Europe.

Sometimes religious worldviews appear in theme parks as a result of negotiations—or struggles—between the park's commercial or governmental promoters and the religious interests they want to recruit. Thus, the Beijing Ethnic Park's Tibetan Kalacakra temple was constructed by monk artisans under the direction of a *trulku* from Lhasa, and consecrated with a full ritual by lamas and monks from the nearby state-sponsored Buddhist Studies Institute. Charlene Makley (2010) points out that this reflected highly complex power relations not merely between Tibetan Buddhist leaders and the Chinese government and other stakeholders, but also between various Tibetan Buddhist authorities.

One of the oddest features of theme park worldviews is the frequent association of religious parks with right-wing politics. This tends to reflect the views of their often eccentric founders. Compare, for example, Gerald Smith, the founder (in the 1960s) of The Great Passion Play Theme Park at Eureka Springs in Arkansas, with Suchart Kosolkitiwong, who in 1997 founded the Guan Yin Inter-Religious Park at Phetchaburi (Thailand). Smith was a paranoid and anti-intellectual demagogue who hated communists, but above all hated Jews and campaigned for the release of Nazi war criminals. Kosolkitiwong was a similarly rabid anti-communist, and a supporter of the Thai military, who learned early on the value of religion in the anti-communist struggle and later in the struggle against the threat of extra-terrestrials. Such extreme attitudes quickly fade away after their founders' death, but many religious parks still ally themselves with right-wing attitudes, albeit comparatively moderate ones.

2.6 Environmentalism

Theme parks and nature have a complicated relationship. Since theme parks share a history with botanical parks and zoos (see ANTECEDENTS), the display of "picturesque" nature for pleasure is ubiquitous in theming. However, as in most of their antecedents, in theme parks, nature (as opposed to culture) tends to be constructed as wild and adventurous and thereby situated according to the (neo)colonial gaze in the non-Western, "less civilized" areas of the world, like Africa (see, e.g. Phantasialand or Disney's Animal Kingdom). As utopic representations of a "perfect" world, theme parks simultaneously tend to evade problematic issues like environmental destruction.

And yet, with sustainability becoming increasingly important to corporate identity, theme parks also tout their supposedly green thinking: "The Walt Disney Company is committed to protecting the planet and delivering a positive environmental legacy for future generations as we operate and grow our business," the Walt Disney Company asserts on its official homepage, alongside a photograph of a Mickey Mouse-shaped solar facility (Walt Disney Company 2020; see ECONOMIC STRATEGY). Nevertheless, the fact that a fast food-serving capitalist happy-place raises awareness for sustainable behavior might seem somewhat paradoxical:

> As of the 1990s, together with the consolidation of green thinking among broad sectors of society and the appearance of alternative kinds of tourism reflecting the greater awareness and concern by tourists conscious of the problems of the environment, theme park operators not only incorporated environmental quality criteria into their facilities but parks were even developed that adapt the planet's environmental conditions into the theme. (Anton Clavé 2007, 253)

The solution to the conundrum is narratives that transform consumption into salvation. Rutherford (2011, 65–71) explains how Disney's Animal Kingdom patrons are inscribed with neocolonial roles of Western adventurers, ecotourists, or scientists who save nature through capitalist investment (or, at least, their sheer presence). This includes reality-based places like Africa (Animal Kingdom's "Harambe"), where visitors chase illegal poachers on a safari ride, and fictional places like "Pandora," where patrons assume the role of ethnographers.

The Mexican "eco-archeological nature theme park" (Torres and Momsen 2005, 323) Xcaret mirrors this saviorist ecotourist gaze: "You are a witness to this land. A place of wonders, where culture becomes inspired by nature," announces the off-screen voice in a promotional video (Xcaret 2019). The same video shows white Western tourists who marvel at nature and enjoy activities such as zip-lining and snorkeling and "natives" in colorful costumes, who perform dances. Xcaret is themed as a primeval rainforest oasis, aimed at the tourists who visit the Cancún hotel riviera.

Claiming to promote Mexican heritage in both nature and culture, Xcaret thus offers the general theme park version of Mexico, consisting of the jungle and ancient culture, as its own theme park—which may well be considered authentic, since it fulfills Western expectations (Slater 2003, 14; see AUTHENTICITY). And unlike the "real" wilderness and native culture, where tourists might be considered intruders, at Xcaret, they assume the role of ecotourists or saviors who sustain the environment by consuming it. The colorful imaginations of ancient Maya people that populate Xcaret serve to intensify this self-image of the park visitors: "An eco-park that celebrates a 'nature-loving' civilization, its secondary icon is thus an ancient people whose environmental sensitivities happen to coincide almost exactly with those of twenty-first-century tourists" (Slater 2003, 13). Grupo Xcaret, the corporation that runs the park and a number of other entertainment facilities, has developed its own "Xustainability" model built on the three pillars of "People, Prosperity, and Planet" (Xcaret 2020). In this model, education and environmental sustainability are intertwined with economic growth: together, they serve to boost the Mexican economy (see also Anton Clavé 2007, 187–189).

As the examples show, environmentalism usually intersects with all of the other ideologies and worldviews dealt with above. Moreover, the topic resonates with the theme park imperative of being family-friendly. Accordingly, religious parks also follow an environmentalist agenda: at Ark Encounter, for example, the park creators even get to define nature by displaying the animals that presumably populated the earth in Noah's days, according to Creationist beliefs. Ultimately, family-friendly environmentalism is a worldview that is found in every theme park, more or less subtly. Sometimes strongly "environmentalist" attitudes are carried by traditional stories, as in Efteling's "Fairytale Forest," which includes an attraction based on a story by Hans Christian Andersen, the Chinese Nightingale, in which the dying emperor of China is brought peace by his tame nightingale. But one day the emperor of Japan sends him a bejeweled animatronic nightingale, which sings even better, and the real bird flies away. Soon, though, the false bird stops working, and the emperor is assailed by nightmares and pain. Only when the real nightingale returns to sing outside his window, does peace return.

3 Disseminating Worldviews

Among the many activities shared by all visitor attractions is education. Theme parks also have a strong interest in disseminating their worldviews outside the park, and have thus developed many different strategies to achieve this. A serious approach to disseminating information beyond the theme park experience is taken, for example, by Answers in Genesis in their Creationist Creation Museum and Ark Encounter theme park. In addition to a gift shop, both venues feature a bookstore, where titles on every imaginable aspect of Young Earth Creationism are sold—most of which are printed by Answers in Genesis's own publisher. Among these are school textbooks and homeschooling curricula on comparative religion, chemistry, math, and biology. The curricula are also sold via AiG's online shop, but with the Creation Museum and Ark Encounter offering special admissions for student groups and homeschooling parents, the immersive experience is an extra incentive for potential customers. Unlike in other Christian parks like Holy Land Experience, which sell Christian (and Jewish) items for religious practice and worship, the Creationist gift shops resemble those in natural history museums with their science kits, dinosaur excavation sets, and elaborately designed posters with world maps and historical timelines from Genesis to present (see, e.g. Bielo 2018; Trollinger and Trollinger 2016).

In combining entertainment with education, theme parks are taking their cue from museums, which in turn are focusing more and more on theming. Modern museums descend from funfairs as much as they do from scholarly collections. In the 1970s, "Disneyfication" was a pejorative term used by curators who feared that a popular approach threatened scholarship and was a sign of "dumbing-down." Attitudes have changed, however, and over the past generation, it has become increasingly harder to say where the line between a theme park and a museum lies. Museums can be more light-hearted, and theme parks more serious. What in English is called an open-air museum, is a *tema paku* in Japanese. Design companies work happily in museums and theme parks alike, using many of the same techniques, and trade magazines like Attractions Management or Blooloop report on both (see METHODS).

Many, and perhaps most theme parks, also offer some sort of service to local schools and colleges. At the very least, a "Park Education Service" offers a reduced-price group entry to educational groups, preview opportunities for teachers, and somewhere for groups to eat lunch. Many theme parks offer a lot more, however, including age-appropriate educational visits run by qualified park staff, as well as printed and online resources for teachers and students (see PARATEXTS AND RECEPTION). Phantasialand, for example, offers adventure rallies for students from grade 3 to 10, that "teach knowledge in the fields of mathematics, physics, biology, geography and history" (Phantasialand 2020). Moving through the themed areas, students have to answer questions that are mostly on ride physics, but also on iconic symbols for the themed areas, like naming the "Big Five" for "Deep in Africa" and identifying a Buddha statue in "China Town." The rallies are neither long nor complicated, nicely illustrated with the park's dragon mascots, and clearly designed not to interfere with the fun of a theme park visit. One could argue that the questions reinforce the attributes of myth/nature/past to the areas of "Mexico," "China Town," and "Deep in Africa," but the focus on ride physics suggests an awareness of the pitfalls of the cultural themes. Yet surprisingly, the fantasy-themed area "Wuze Town" is not included in the rally, which may imply a lower degree of "authenticity" in that area compared to those modeled on real places—a questionable symbolism.

For many decades, theme parks have also been very important ways of transmitting history at a popular level. Indeed, all theme parks are "historical," as the themes are either located in real or imaginary time levels or—even when they relate to completely imaginary parallel worlds—nonetheless always adopt visual languages which are derived from historical styles and monuments: there is no theme park without time (see TIME). Even if historians have tended to dismiss the impact and the relevance of this form of knowledge transmission (at least until the 1990s), often representing the parks' depictions as purely wrong and dangerous (Carlà 2016; Carlà-Uhink 2020, 3–20), popular culture is often the primary means by which historical knowledge is gathered, and theme parks are undoubtedly very important vectors of popular culture (Francaviglia 1995, 34).

Theme parks act upon history as much as upon any other theme—and as with all popular culture, they follow the same basic principle: what is represented must be recognizable, otherwise there will be no immersion (see IMMERSION and THEMING). Images and concepts that are already known from a historical period (or that clearly imitate and recall previously known ones) give a sense of satisfaction, of "knowing" and learning even more, and thus the possibility of "time travelling" (Holtorf 2009) while being immersed in another period. However, being confronted with reconstructions which might be more "accurate" or more "correct" from a scholarly point of view risks creating a sense of alienation that is entirely extraneous to the aims of the park, as it makes the visit uncomfortable and unpleasant. This is thus highly counterproductive, regardless of whether the theme park is a commercial enterprise (and thus attracts fewer and less satisfied visitors) or an "educational" institution (as, for instance, in the case of Mini Indonesia discussed earlier), as the sense of discomfort can lead to a much lesser degree of "absorption" of the transmitted content, and even to forms of rejection. If immersion takes place, however, theme parks can be highly efficient vectors of historical information, with which history books or other media cannot compete.

This is true on two levels: the mere size of the audience reached, and the way in which the information is transmitted. Even less successful theme parks are visited by hundreds of thousands of people every year, numbers that even the most famous and globally renowned publications in the field of popular history can reach only very rarely. On the other hand, "time travel" generates a form of knowledge that is not intellectual-argumentative: this kind of knowledge is "mythic" rather than "scientific." The ensuing historical images are interiorized by visitors in a way that makes them, through direct experience, basically indisputable. Every rational argument aimed at deconstructing this form of immersive experience ultimately clashes with the knowledge derived from "having been there" and having personally experienced the environment with all the senses (Carlà-Uhink 2020, 20–25). As noted before, the precondition for such a successful form of time traveling is a high level of recognizability. Yet this also implies the reinforcement of existing stereotypes, the eventual perpetuation of erroneous representations and interpretations, an extremely selective choice of the elements to be shown, and general removal of all aspects which might appear disturbing, uncanny, or not family-friendly, such as war, slavery, poverty, famine, etc. (Carlà and Freitag 2015). This is a point from which it is necessary to move on in order to be able to integrate the theme park and its representations into educational offers and history teaching.

Terra Mítica has responded to this challenge by integrating the role of theme parks in popularizing history into a special offer for schools. The park developed a series of booklets for school teachers and pupils, explaining the civilizations of the different themed areas (with the exception of "Ibéria"), showing the buildings and archaeological sites which had been copied and reconstructed in the park. These booklets also introduce some basic elements of the culture, daily life, history, and art of the Egyptians, Greeks, and Romans. Yet this is still a selection that leaves out the most "uncomfortable" sides of history, and which simply provides a somewhat patchy legitimation for a visit to the theme park, as it does not truly integrate the theme park, its immersive environment, and its way of communicating into a well-conceived didactic program (Carlà-Uhink 2020, 84–85).

4 Conclusion

What sets theme parks apart from other media is the level of immersion by which they affect their visitors' memory and perception of the world. For this reason, theme park ideology is a promising field of study and exploration. When analyzed through the lens of worldviews and ideologies, theme parks show us how the geography of the world as we perceive it is socially constructed, and as simulacra, they reinforce our perceptions (and stereotypes) of history and geography. Geographical knowledge and imagination organize the way people comprehend their local, regional, and global environments. For instance, the opposition between urban and rural space, or between natural and cultural landscapes, is not an "objective" structure organizing the reality of the world, but rather a dichotomy deeply rooted in Western culture, a lens through which Westerners perceive, understand, and practice the world they live in.

Such worldviews discretely appear in theme parks where designers try to incorporate a village, a street, a forest, a mountain, a river, an island, etc.: each of these geographical objects results from a social construction and thus refers to a worldview. Forests are different in the Latin and the German worlds—not because of the climate, but because they mean something different, as is clearly illustrated by their huge importance in Germanic mythologies and their quasi-absence in Greek and Latin mythologies. In that sense, including even a simple tree in a theme park—whether basic, or spectacular as in Disney's Animal Kingdom—expresses a worldview. Each and every plant, landscape, building, music, food, etc. in a theme park has been designed or chosen more or less consciously to reflect a worldview, in that they refer to a specific place or kind of place.

Theme parks not only implement worldviews, but as a medium, they have also created their own frame for viewing the world. As a metaphor (see PARATEXTS AND RECEPTION), theme parks stand for a family-friendly capitalist consumer culture, which is not only spread throughout the world within the medium, but also transferred to other realms of social life, such as community building or politics of commemoration (as in the "Reichspark" example). As immersive mass media, theme parks are powerful tools for manifesting worldviews and ideologies through shared experience. New national, regional, or religious theme parks may suggest a trend toward increasing use of the medium to this end, but historically, the medium is already rooted in ideology.

References

Anton Clavé, Salvador. 2007. *The Global Theme Park Industry.* Wallingford: CABI.
Arbeiter, J. Patrick, and Sebastian Teitge, dir. 2017. Unternehmen Reichspark. *Neo Magazin Royale,* October 6. Cologne: Bildundtonfabrik.
Ariès, Paul. 2002. *Disneyland: Le royaume désenchanté.* Villeurbane: Golias.
Beal, Timothy K. 2005. *Roadside Religion: In Search for the Sacred, the Strange, and the Substance of Faith.* Boston: Beacon.
Bielo, James S. 2018. *Ark Encounter: The Making of a Creationist Theme Park.* New York: New York University Press.
Bryman, Alan. 2004. *The Disneyization of Society.* London: Sage.
Callahan, Sarah B. Dykins. 2010. *Where Christ Dies Daily: Performances of Faith at Orlando's Holy Land Experience.* University of South Florida: Scholar Commons.
Carlà, Filippo. 2016. The Uses of History in Themed Spaces. In *A Reader in Themed and Immersive Spaces*, ed. Scott A. Lukas, 19–29. Pittsburgh: ETC.
Carlà, Filippo, and Florian Freitag. 2015. Ancient Greek Culture and Myth in the Terra Mítica Theme Park. *Classical Receptions Journal* 7 (2): 242–259.
Carlà-Uhink, Filippo. 2020. *Representations of Classical Greece in Theme Parks.* London: Bloomsbury.
Carlà-Uhink, Filippo, and Florian Freitag. 2022. Theme Park Imitations: The Case of Happy World (Happy Valley Beijing). *Cultural History* 11 (2): 181–198.
Eagleton, Terry, ed. 2013. [1994]. *Ideology.* New York: Routledge.
Feige, Michael. 2017. Mini Israel and the Subversive Present. In *Time and Temporality in Theme Parks*, ed. Filippo Carlà-Uhink, Florian Freitag, Sabrina Mittermeier, and Ariane Schwarz, 155–168. Hanover: Wehrhahn.
Fjellman, Stephen M. 1992. *Vinyl Leaves: Walt Disney World and America.* Boulder: Westview.
Francaviglia, Richard V. 1995. History after Disney: The Significance of "Imagineered" Historical Places. *The Public Historian* 17 (4): 69–74.
Freitag, Florian. 2016. Autotheming: Themed and Immersive Spaces in Self-Dialogue. In *A Reader in Themed and Immersive Spaces*, ed. Scott A. Lukas, 141–149. Pittsburgh: ETC.
Gottdiener, Mark. 1997. *The Theming of America: Dreams, Visions, and Commercial Spaces.* Boulder: Westview.
Griffin, Sean P. 2000. *Tinker Belles and Evil Queens: The Walt Disney Company from the Inside Out.* New York: New York University Press.
Holtorf, Cornelius. 2009. On the Possibility of Time Travel. *Lund Archaeological Review* 15: 31–41.
Hom, Stephanie Malia. 2013. Simulated Imperialism. *Traditional Dwellings and Settlements Review* 25 (1): 25–44.
Lennon, John, and Malcolm Foley. 2000. *Dark Tourism.* London: Continuum.
Lukas, Scott A. 2016. Dark Theming Reconsidered. In *A Reader in Themed an Immersive Spaces*, ed. Scott A. Lukas, 225–235. Pittsburgh: ETC.
Makai, Péter. 2019. Three Ways of Transmediating a Theme Park: Spatializing Storyworlds in Epic Mickey, the Monkey Island Series and Theme Park Management Simulators. In *Transmediations: Communication Across Media Borders*, ed. Niklas Salmose and Lars Elleström, 164–185. London: Routledge.
Makley, Charlene. 2010. Minzu, Market and the Mandala: National Exhibitionism and Tibetan Buddhist Revival in Post-Mao China. In *Faiths on Display: Religion, Tourism, and the Chinese State*, ed. Tim Oakes and Donald Sutton, 127–156. Lanham: Rowman & Littlefield.
Martin, Jean-Clément, and Charles Suaud. 1992. Le Puy du Fou. *Actes De La Recherche En Sciences Sociales* 93: 21–37.
Mittermeier, Sabrina. 2017. Utopia, Nostalgia, and Our Struggle with the Present: Time Travelling through Discovery Bay. In *Time and Temporality in Theme Parks*, ed. Filippo Carlà-Uhink, Florian Freitag, Sabrina Mittermeier, and Ariane Schwarz, 171–187. Hanover: Wehrhahn.

Mittermeier, Sabrina. 2021. *A Cultural History of the Disneyland Theme Parks: Middle Class Kingdoms*. Chicago: Intellect.

Molter, Céline. 2018. The Jerusalem Experience: Comparing Theme Park Versions of the Holy Land. In *Doing Cultural History: Insights, Innovations, Impulses*, ed. Judith Mengler and Kristina Müller-Bongard, 97–108. Bielefeld: Transcript.

Montalvo, Raúl, and Eileen Daspro. 2018. MultiMexicans in the Entertainment Industry: KidZania and Cinépolis. In *Mexican Multinationals: Building Multinationals in Emerging Markets*, ed. Alvaro Cuervo-Cazurra and Miguel A. Montoya, 494–521. Cambridge: Cambridge University Press.

Paine, Crispin. 2019. *Gods and Rollercoasters: Religion in Theme Parks Worldwide*. London: Bloomsbury.

Phantasialand. 2020. Klassenfahrten: Interaktiv lernen im Phantasialand. https://www.phantasialand.de/de/themenpark/gruppen-und-schulklassen/klassenfahrten/. Accessed 28 Aug 2022.

Rutherford, Stephanie. 2011. *Governing the Wild: Ecotours of Power*. Minneapolis: University of Minnesota Press.

Said, Edward. 1978. *Orientalism*. New York: Pantheon.

Sanchez, Sandra Inés. 2012. La construcción de un parque temático religioso como ciudad análoga. *Avá* 20. http://www.scielo.org.ar/scielo.php?script=sci_arttext&pid=S1851-16942012000100008&lng=en&nrm=iso. Accessed 18 March 2022.

Slater, Candace, ed. 2003. *In Search of the Rain Forest*. Durham and London: Duke University Press.

Steinkrüger, Jan-Erik. 2013. *Thematisierte Welten: Über Darstellungspraxen in Zoologischen Gärten und Vergnügungsparks*. Bielefeld: Transcript.

Stone, Philip. 2006. A Dark Tourism Spectrum: Towards a Typology of Death and Macabre Related Tourist Sites, Attractions and Exhibitions. *TOURISM: An Interdisciplinary International Journal* 54.2: 145–160.

Torres, Rebecca M., and Janet D. Momsen. 2005. Gringolandia: The Construction of a New Tourist Space in Mexico. *Annals of the Association of American Geographers* 95 (2): 314–335.

Trollinger, Susan L., and William Vance Trollinger, Jr. 2016. *Righting America at the Creation Museum*. Baltimore: Johns Hopkins University Press.

Tuan, Yi-Fu. 1976. Geopiety: A Theme in Man's Attachment to Nature and to Place. In *Geographies of the Mind*, ed. David Lowenthal and Martyn Bowden, 11–39. New York: Oxford University Press.

Turner, Victor. 1969. Liminality and Communitas. In *The Ritual Process: Structure and Anti-Structure*, 94–130. Chicago: Aldine.

Wallace, Mike. 1985. Mickey Mouse History: Portraying the Past at Disney World. *Radical History Review* 32: 33–57.

Walt Disney Company. 2020. Environmental Sustainability. https://thewaltdisneycompany.com/environmental-sustainability/. Accessed 28 Aug 2022.

Walt Disney World. N.D. Harambe Market. https://www.disneyworld.eu/dining/animal-kingdom/harambe-market/. Accessed 28 Aug 2022.

Xcaret. 2019. MEXICO'S MAJESTIC PARADISE: Xcaret Mexico! Cancún Eco Park. https://www.youtube.com/watch?v=-6Y5F_Feepg. Accessed 28 Aug 2022.

Xcaret. 2020. Corporate Xustainability. https://www.grupoxcaret.com/en/sostenibilidad-corporativa/. Accessed 28 Aug 2022.

Zavadski, Katie. 2014. Russia is Building a Pro-Russia Theme Park in Crimea. *Intelligencer*, October 10. https://nymag.com/intelligencer/2014/10/russia-building-pro-russia-theme-park-in-crimea.html. Accessed 28 Aug 2022.

Printed in the United States
by Baker & Taylor Publisher Services